RECENT ADVANCES IN DEVELOPMENTAL BIOLOGY

Recent Advances in Developmental Biology

Vol. 2
Principles of Embryology

A.P. DIWAN
N.K. DHAKAD

ANMOL PUBLICATIONS PVT LTD
New Delhi-110 002

ANMOL PUBLICATIONS PVT LTD
4374/4B, Ansari Road
Daryaganj, New Delhi-110 002

Recent Advances in Developmental Biology

First Edition 1995
ISBN 81-7488-039-9 _2

PRINTED IN INDIA

Published by J.L. Kumar for Anmol Publications Pvt. Ltd. and Printed at Mehra Offset Press, Delhi.

Preface

The fundamentals of Developmental Biology are presented within the framework of scientific discovery. Researches in Embryology have made almost incredible strides in the past few decades. Consequently, the existing concepts of the developmental biology have expanded. There has been a revolution indeed in this direction.

The text integrates the descriptive, experimental and biochemical approches into a conceptual framework for the analysis of development. All important points are illustrated diagrammatically. The title is not intended to be comprehensive, nor could it be at this length, but concentrates as putting across the basic principles of the subject as briefly and lucidly as possible. It does this with the aid of careful selected examples, some recent and other classic of the field and with numerous illustrations. The aim is to enthuse the reader with this active and exciting area of research and to lay a solid foundation on which further study of its various facts may be based.

Authors

Contents

1

Introduction

Embryology deals with the study of development of animals. But the term *development* is used in various senses in biology, but in its broadest sense it deals with the complex changes which an individual organism undergoes in its life cycle from fertilization to death. Development normally involves both quantitative changes i.e., *growth,* and qualitative changes leading to increased specialization in the various cells, tissues and organs of the body, which we refer to as *differentiation.* The processes leading to and determining the form and structcure of the organisms are included in the general term *morphogenesis.*

The development of many animals is divided by the incident of birth or hatching into *prenatal* or *postnatal periods.* For a long time attention was focussed on the events taking place before birth, when the most striking advances occur in these animals as a whole. Only gradually was it realized that important changes, beyond were growth, continue to occur even to the adult state. This broader concept of embryology brings into its range all the developmental events from sexual reproduction, it is sometimes designated as *outogeny.*

The Value of Embryology

A general conception of how man, like other animals, develops from a single cell should share in the cultural background of every educated mind. Everybody gives one an understanding of some of the ways of life itself. The processes by which a fish, or an alligator, or a chick grows from a single fertilized egg cell to its fully eleborated adult structure are fundamentally the same as those involved in our own development. And these growth processes hold for us something definite and tangible in answer to that ever-recurring question, 'Whence come we, and how?" From the philosophical side, embryology is a key that helps unlock such secretes as heredity, the determination of sex and organic evolution.

Muller in 1804 for the first time propounded the general idea of recapitulation on the basis of his studies of the development of invertebrates. In 1868, *Haeckel* formulated its principles much more fully and gave it the name *biogenetic law.* In essence the law is that an animal in its individual development passes through a series of constructive stages like those in the evolutionary development of the race to which it belongs.

From embryology we can learn in one short, uninterrupted story how each individual grows into an adult. We can see this process going on in the laboratory under our very eyes. Embryology is not an easy subject. It requires a high type of visual imaginations. The student of embryology must bear in mind that he is dealing with living objects, three dimensional and continually changing in volume, shape and constitution. Much of his attention must be given to the cells of the embryo, as they multiply, migrate, take on different appearance, and carry on different functions. He must not lose sight of the embryo as a whole.

Historical Background of Embryology

Before the 17th century embryological knowledge was based on the writings of *Aristotle* and *Galen*. Embryology as a branch of biology was initiated by the famous Greek philosopher *Aristotle* (384—322 B.C.). He was the first embryologist to describe the development and reproduction of many kinds of organisms in his book entitled *Degeneratione Animalium*. He believed firmly that the complex adult organism develops from a simple formless beginning. Thus, he laid the foundation for the basic principles of *epigenesis*, a theory postulated after 2000 years. For this *Aristotle* is honoured as the *father of embryology.*

Aristotle has written that the male contributes the *semen* and the female contributes the *catamenia*. The semen is secreted by the testis and it is hot and active. The catamenia is carried through very fine vessels into the uterus. It is equivalent to the menstrual blood. He had no knowledge of ovary. He thought that the female reproductive system is formed of uterus only and not of anything else.

The catamenia represents the material foundation of the embryo. The development is activated and guided by the semen. The semen contributes nothing material to the embryo. Thus according to *Aristotle* no ovum is required for development.

The human ovaries were first described in 300 A.D. by *Herophilus*. He said that it is equivalent to the testis of the male and he named it *testiculi foeminis* and this name was retained for centuries. He denid that the uterus produces the female generative materials.

Galen (200 A.D.) followed the views of *Herophilus*. He claimed the ovary as the structural and functional homologues of the testis. It contributes the foundation material for embryogeny.

Galen's time was followed by more than a thousand years of stagnation and retrogression in many fields of intellectual development. From Galen's day until the 16th century embryological thought made no substantial advance.

William Harvey (1578-1657) believed that all animals come from egg. *Malpighi* (1628-1694) explained that the egg contains in it the miniature adult animal. He formulated a hypothesis called *preformation theory*.

THEORIES OF EMBROLOGY

Preformation Theories

During seventeen century a Dutch scientist *Swammerdam* (1637-80) theorized that sex-cells contained preformed embryos or miniatures of the adult. He and many other biologists, namely *Haller* (1708-77): *Bonnet* (1720-93); *Spallanzani* (1729-99) and *Malpighi* (1673), thought that the ovum contained a transparent, highly folded, small and unobservable miniature of the adult, which was in some way stimulated to growth by the seminal fluid. Those who believed that the miniature organism was present in the egg were known as *ovists*. Some other microscopiss, such as, *Leeuwenhoek* (1632-1723), *Hartsoeker* (1656-1725), etc., believed that preformed miniature organism presented in the head of the sperm and not in egg and that was called *homunculus* or *animalcule* (animalcule means spermatozoon). These scientists were called *animalculists* or *spermists*. Spermists, thus, theorized the sperm as the seed and eggs as the soil in which the seed was planted.

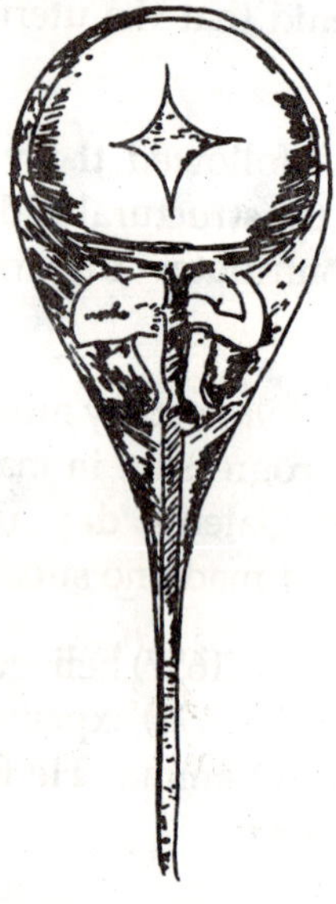

Fig. 1.1. Hartsoeker's imaginative illustration of the human sperm with a homunculus.

Some ovists namely *Swammerdam* and *Bonnet* advocated and extreme form of preformation theory called *encasement* or *'emboitement' theory*. This theory hold that successive generations of individual organisms pre-existed one inside the other in the germ cells of the mother. It was estimated that, as many as 200 million years of human beings were present, already delineated in the ovaries of Eve.

Such theories of preformation presisted well in the eighteenth century, by which time the German investigator *Caspar Friedrich Wolff* (1759) offered experimental evidence that no preformed embryo existed in the egg of the chicken. He suggested that during embryonic development the organs formed successively in an epigenetic manner.

Theory of Epigenesis

We may now turn to the other great theory of development: *epigenesis*. This theory assumes that embryonic development and differentiation originate in a homogeneous mass of living matter in which there is no preformed material, neither tissues nor organs. This theory is very old. It was advanced in a simpler and more primitive form by *Aristotle* and was kept alive during the Middle Ages up to the time of *William Harvey*, who was an ardent advocate of epigenesis along Aristotelian lines. *Harvey* and his contemporaries and successors supplied some additional concepts to this ancient theory, until *Casper* F. *Wolff* in 1759 put it on a firm and acceptable basis. *Harvey*, in addition to his well-known and important physiological contributions to biology, did a great deal of work in the embryology of mammals, of the chick, and of other forms. He wrote a fine comparative embryology, contrasting embryos of such widely different groups as vertebrates and invertebrates. To him we owe the familiar aphorism *"Omne vivum ex ovo"*, which at his time was a revolutionary biologic generalization. Without the advantage of a microscope, he surmised correctly that the living homogeneous part of the chick ovum was centered in the blastodisc, which is the small whitish spot on the top of the yolk. It is therefore hard to understand that an outstanding anatomist like *Malpighi*, about sixty years later, could make the error, in his observations, of identifying a preformed embryo in the blastodisc.

Although the epigenetic theory was more or less known from the time of *aristotle* it is usually ascribed to *C.F. Wolff* (1733-1794) who published the results of his observations on the problem in 1759 in a comprehensive volume entitled *Theoria Generationis*. He saw in the chick egg that the organs form, one after another, from a homogeneous living substance. There was no semblance of a preformed miniature chick in that substance, and in describing the appearance of tissues he speaks of small vesicles, blisters and ampullae, thus coming very close to the discovery of the cell theory, which was to be announced eighty years later. Of course, in justice to other investigators, as for example *Malpighi, Leeuwenhoek, Grew, Brown, Dutrochet* and several others, it should be noted that they also came close to the recognition of cells.

Wolff was vigorously opposed by *Haller* and *Bonnet* and both being eminent naturalists and philosophers their objections were considered with respect by their contemporaries. It was not until 1821, when the famous anatomist *Johann Friedrich Meckel* (1781-1833) pointed out the accuracy of Wolff's observations and his logical conclusions, that the theory of epigenesis was accepted and looked upon with favour. The acceptance of this theory spread rapidly after the discory of protoplasm in animals in 1835 by *Felix Dujardin* (1801-1860) and in plants in 1846 by *Hugo von Mohl* (1805-1872).

Although to-day we accept the theory of epigenesis, we do not accept it in the form offered by *Wolff*. We know that the genetic constitution of an organism is rigorously fixed by heredity, and that it is laid down in the fertilized ovum from which the organism develops. On the basis of its hereditary constitution—its genes—the organism will develop along a specific pattern, and the ever present external environment exerts its modifying influence upon that pattern. So long as the environment is normal for the developing ovum or embryo for a particular species, a normal individual will result. However even the so-called "normal" embryo is highly variable in its finer details, for there are untold numbers of delicate changes in the environment which escape crude methods of observation, but which exert a continual modifying action in the expression of the genes. Moreover, when the environment is intentionally altered by the experimenter, this modification may have a far reaching effect on the structure of the organism. We shall return to this problem toward the end of this chapter.

A great impetus was given to embryological investigation by *Karl Ernst von Baer* (1792-1876). His work touched on so many phases of the science that it is impossible to enumberate and evaluate them here in detail. Of outstanding merit must be considered his germ layer theory, the discovery of the mammalian ovum, his anticipation of the biogenetic law and his correlation of contemporaneous embryological data into one integrated account.

After the formation of the cell theory, embryological research entered an entirely new avenue. It was discovered that the ovum was a single cell and that fertilization consisted of the union of the ovum and spermatozoon in the formation of the zygote. Embryological development was now recognized as a continuous process of all division with progressive cell differentiation.

Baer's Law

K.E. von Baer (1792-1876) contributed much to the field of embryology. His investigatins are tremendous. He published the problems of embryology in an ordinary manner. Hence, he has been accepted as the *founder of comparative embryology*. After a thorough stydy of the development of various animals, *von Baer* arrived at four important conclusions, which are collectively known as the laws of *von Baer*. They are as follows:

1. 'The more general features of any large group of animals appear in the embryo earlier than the more special features.'
2. 'After the more general features, those that are less general arise, until the most special features appear.'

3. 'The embryos of different species are more alike than the adults and depart progressively from each other during ontogeny.'
4. 'The embryo of a higher species may resemble the embryo of lower species, but the embryo of the higher species never resembles the adult of the lower species.'

For example, in the development of chordates, the more general features like the brain, notochord, segmental muscles etc., develop first. Then the features distinguishing the classes of vertebrates like the limbs in quadrupeds, feathers in birds, hairs in mammals etc., develop. Finally, the characters, distinguishing the families, genera and species develop.

The Biogenetic Law or the Recapitulation Theory

This theory was postulated by *Ernst Haeckel* (1868) and *Muller* (1864). It is aa modified theory of Baer's law. It is based on the contributions of evolutionary theory. This theory states that higher animals in their development passes through stages which are similar to the adult stages of lower animals which were their ancestors. That is the *ontogeny* briefly *repeats* or *recapitulates the phylogeny*. In other words, the ancestral characters reappear in the developmental stages of an individual. Thus the biogenetic law states that the *embryonic stages of a higher animal resemble the adult stages of its ancestor*.

For example, in the ontogeny of frog there appears a tadpole larva. This larva in every respect resembles an adult fish. According to the *biogenetic law,* the fish is the ancestor of amphibians. That is why, in the development of amphibian, there is a fish-like tadpole stage. Though there are direct evidences for biogenetic law, there are some objections also.

Weismann's Germplasm Theory

The contentions of His fell right in line with the theories and speculations of *August Weismann* (1834-1914), whose work was along genetical and evolutionary lines rather than in the newly founded science of experimental embryology. *Weismann* held the attention and the respect of biologists all over the world through his sweeping generalizations, and it is pertinent here to mention his germ-plasm theory, though, at the present time it has lost some of the force which it had at the turn of the last century.

In his studies in insect and other eggs of the invertebrate animals, *Weismann* discovered that the primordial germ cells which are destined to give rise to the gonads of the organism are segregated very early in cleavage, and that they contain *all* the hereditary *"determinants"* necessary for the development of another individual. In certain forms, as for example in *Ascaris,* the single primordial germ cell is set aside in the second cleavage of the ovum, and while this cell continues to divide, it does so at a retarded tempo. The other cells keep on cleaving at their usual rate and form eventually the body or *soma* of the animal, while the retarded germ-cells become enclosed in the body to develop the *gonads*. The somatic cells do not have *all* the *"determinants"*; they are different in that each has a specific assignment in the development of the organs and organ systems of the body (*soma*). It is the function of these somatic cells to carry the germplasm and nourish and protect it. The *soma* or *somatoplasm* has to die sometime and revert to the inorganic world. The *germplasm,* however, is protected by and parasitic on the somatoplasm and is immortal so long as, during the life of the individual, one or a number of cells from the germplasm (the gonad) with the total number of determinants would be liberated and become activated by fertilization or by parthenogenesis and develop again into somatoplasm (body) and germplasm (gonad) in the next generation, and so *ad infinitum.* Life is without end. Any crature existing to-day carries the spark of life that was inherent in the first *Protista* or the first sub-cellular life, whatever its form may have been.

The Experimental Embryology

During the middle and later part of the nineteenth century embryologists approached the problems with a practical mind, and that led to the development of *experimental embryology*. It began with the remarkable invention of microtome in 1870 by *Wilhelm His* (1831-1904). With this newly invented instrument, embryologists have been able to section the embryos into thin slices, arrange them in series, mount them on a slide, stain and examine them under a microscope conveniently.

Wilhelm His, the first experimental embryologist attacked the Biogenetic law of *Haeckel* Vehemently. He pointed out that the causative agent for the development and differentiation of cells, tissues and organs could never be located by mere speculation along phylogenetic lines. He was of the opinion that the formation and development of organs depend on what kind of functions these organisms are called upon to perform and not upon their phylogenetic history. He advanced the bold idea that cleavage, blastulation, gastrulation and all succeeding stages in the development of the embryo could be explained on a purely mechanic basis, and that any stage in the development is the immediate cause for the formation of the next stage. Thus *His* thought that there is a *chain of reactions* in the development. There is no beginning and no end to this chain, since it began with the fertilization of ovum which had to come into existence and these gametes (egg and sperm) again had to come from gonads that were differentiated early in the parent organism.

Mosaic Theory

Wilhlem Roux in 1888 performed a classic experiment which may be viewed as marking the begining of the science of experimental embryology. He took a frog's egg at the two-cell stage of cleavage and touched one of the two cells with a hot needle thus destroying the nucleus of that cell. He observed that the uninjured cell continued dividing and developed into what he interpreted to be a one half blastula, a one half gastrula and ultimately a one-half embryo. His work seemd to confirm the hypothesis that the early cell divisions of the egg are differential, a process of the orderly dispersal of the materials of the nucleus, and at blastula stage each cell, so he supposed, possesses just its own potentialities for development. Thus, the pigmented cytoplasm of animal pole of frog's unfertilized egg chiefly develops into the *head* region of the animal, while the yolky cytoplasm of vegetal pole into the *posterior region*. Before fertilization, the egg is *radially symmetrical* but after fertilizatiion it becomes *bilateral symmetrical*. As cleavage continues the blastomeres gradually become smaller and smaller and more areas for specific tissues and organs are delineated on the blastula, so that, finally it comes to resemble a mosaic with regard to the potencies of its blastomeres. The mosaic theory of *Roux* thus, holds that *the ontogenetic development consists of differentiation of cells from a more or less undifferentiated and homogeneous egg with crude and simple pre-localization of materials to greater differentiation and specialization of cells as development went on.*

Regulative Theory

In 1891, the German Scientist, *Hans Driesh* performed experiment on sea urchin eggs, simialar to *Roux*. Blastomeres of the two cell stage of the sea urchin each of which normally gives rise to one half of the embryo and larva, behave differently when separated by him, from each other, and each gives rise to a whole larva, though of somewhat reduced size. Blastomeres isolated at the four cell stage showed the same results. Together they form one whole; separately each becomes a whole. Visualizing these peculiar results, he reasoned that the early cleavages of the egg are equal (equational), *"a quantitative division of homogeneous material."* Hence, at an early stage the blastomeres have equal potencies. Their fate, he said, is determined by their position in the whole. Thus, he regarded a cleaving egg, as a harmonious, equipotential system, in which every part has potentially the properties of the whole.

Development as observed by *Driesch* in sea urchin eggs was to be called *regulative development* and the eggs which were capable of performing this were called *regulative eggs*.

The Gradient Theory

Roux's experiments created much curiousity among the embryologists in the practical field. *Boveri* in 1901 demonstrated that the vegetal pole of the sea urchin egg exerts a dominant influence and this influence gradually decreases towards the animal pole. This was the fundamental idea for the postulation of a new hypothesis called the *gradient theory*. However, *Boveri* was not able to give a plausible proof for his hypothesis.

Later in 1940, *M.C. Child* strongly supported the gradient theory with his experimental evidence. He said that the animal pole is different from the vegetal pole in its rate of metabolism. The rate of metabolism is higher at the animal pole and this gradually decreases towards the vegetal pole where the rate of metabolism is lesser. *Child's* hypothesis is generally called *metablic axial gradient theory*.

Recently the gradient theory was highly modified by embryologists like *Horstadius* (1955), *Runnstrom* (1967) and *Josefeson* (1969). *Horstadius* and *Runnstrom* demonstrated the existence of a double gradient system in the sea urchin egg. They are the animal gradient and the vegetal gradient. The *animal gradient* is highest at the animal pole and fades towards the vegetal pole. The *vegetal gradient* is the highest at the vegetal pole and fades towards the animal pole. It has been shown that the animal gradient is due to the presence of *tryptophan* and the vegetal gradient is due to the presence of *nucleotide*.

Spemann's Theory of Organizers

From the time of the first experiment concerning development, it is evident that there are interactions between the parts of the developing embryo. These interactions assure that the parts will develop in their proper positions as well regulate their relative sizes. This is true even when the parts have been experimentally disarranged.

A classical example of interaction, or at least of action and reaction, was the discovery by *Spemann* (1901 and 1912) of the development of the lens nof the vertebrate eye. The retina and lens come from separate embryonic sources. The retina from the sides of the fore-brain, which in turn comes from the neural plate while the lens from head epidermis. When the neural plate folds to form the brain, the epidermis comes into contact with the bulges of brain tissue which are destined to form the optic cups (future retinas). *Spemann* and others found that the epidermis of the future lens region possesses a competence to form a lens; but if no contact is made between the epidermis and the bulges of brain tissue, a normal lens does not usually form. The optic cup is said to induce or evoke the epidermis to form a lens.

In 1918, *Spemann* reported that small pieces of presumptive neural plate and skin ectoderm exchanged between early salamander gastrulate, each one of them developed according to their new location, whereas a piece of dorsal lip of the blatopore developed independently of its new surroundings into an embryo-like body. He suggested that the dorsal lip might represent a "centre of differentiation" from which spreads an influence determining the development of the gastrular ectoderm. He further concluded that the organizing effect of the dorsal lip might be related to an effect of archenteron roof upon the overlying ectoderm comparable with that of the optical vesicle upon the overlying "lens" ectoderm.

Later, in 1924 *Spemann* and *Hilde Mangold* published a classical paper providing definitive proof of the organizing action of transplanted dorsal lip in the production of secondary embryos, establishing firmly the concept of induction as a basic mechanism in the embryonic development. Later on, other workers found such an organizing capacity in the chorda-mesoderm cells of many other animals. These cells are capable of inducing the development of notochord, neural plate and mesoderm with all its differentiations. To this organizer, the name *chorda-mesodermal field* has been given. It is the first or *primary field* to exist in the embryonic germ. Gradually more and more areas, the *secondary organizers* or *subordinate fields* become organized and the whole process of differentiation become established and fixed. Thus, "when one embryonic tissue transmits a stimulus that influences another tissue to produce a structure that otherwise would not come into being, then the former tissue is called *inductor* or *organizer* and its morphogenetic effect is called *induction.*"

Branches of Embryology

Embryology has many sub-branches in it. They are :

1.Descriptive Embryology

This term is applied to the method of study concerned with the direct observation and description of embryological development. Embryology in ancient times started as a branch of study based on the direct observation and description by scientists like *Aristotle* (340 BC). *Fabricius* (1537-1619) *Harvey* (1578-1657) and so on.

2.Comparative Embryology

In experimental embryology, the embryological development of different animals are studied and compared. Comparative embryology throws much light on the understanding of evolution and phylogenetic significance. It also gives some ideas on the developmental stages of certain animals in whose case the study of development is difficult.

3.Experimental Embryology

In experimental embryology experiments are used for studying the developmental stages. It helps to understand the fundamental developmental mechanisms. In experimental embryology the various parts of developing embryo are removed, transplanted, parts exchanged or the environmental conditions altered. This helps to understand induction, gradient system, etc. *Roux* (1850-1924) is the pioneer in the field of experimental embryology. Experimental embryology is also called *casual embryology of analytical embryology.*

4.Chemical Embryology

Here the developmental stages are studied by biochemical,

biophysical and physiological techniques. It is also called physiological or biochemical embryology *Needham* (1931) is the pioneer in this field.

5. Teratology

It is a branch of embryology concerned with the study of malformations.

6. Developmental Biology

It includes not only embryonic development but also postnatal processes such as normal and neoplastic growth, metamorphosis, regeneration and tissue repair.

THE NORMAL SEQUENCE OF EVENTS IN EMBRYOLOGY

Gametogenesis

The embryogenesis is started from the time of differentiation and organization of haploid male and female sex cells, *viz.*, *sperm* and *ova* respectively, from diploid somatic cells of each parent during a process called *gametogenesis*. The gametogenesis includes *spermatogenesis* and *oogenesis*. The spermatogenesis is a process which occurs in male gonads (testes) and produces small-sized, motile, haploid sex cells, the *sperms* or *spermatozoa*. A *spermatozoon* is devoid of excess cytoplasm and has an unspecialized nucleus, a generally differentiated middle piece and tail, and a highly specialized distal acrosome, the latter being the derivative of Golgi-complex, containing enzymes for the penetration of sperm into egg cell. The oogenesis occurs in female gonads (ovaries) and produces a large, non-motile, nutrient-filled, haploid sex cell, the egg or *ovum* and three haploid, small sized *polar bodies* or *polocytes*. In both spermatogenesis and oogenesis, meiosis by discarding half of the chromosomes, singles out the sets of genes which are to operate in the development of a particular individual.

Fertilization

The process of fertilization involves a number of

independent biological and physiological events—(i) the nucleus and cytoplasm of spermatozoon fuse with the nucleus and cytoplasm of the egg; (ii) the cortical reaction in the egg cytoplasm to elevate *a fertilization membrane* outside the plasma membrane; (iii) activation of egg to start its rapid cleavage by mitosis. The process of fertilization also involves adaptation on the part of parents ensuring that they meet during breeding season and discharge their sex-cells simultaneously in cases of *external fertilization* or copulate in cases of *internal fertilization* to bring spermatozoa into proximity with the eggs.

Cleavage

During third phase of embryogenesis, the *cellulation, segmentation* or *cleavage,* no growth of egg-cytoplasm (*ooplasm*) takes place, but, rate of synthesis of some macromolecules such as DNA and proteins is increased at the expense of reserve food substances of egg (*viz.,* glycogen, fats and yolk), and the fertilized egg undergoes repreated and successive mitotic cell divisions and produces a compact heap of cells or *blastomeres,* called *morula.* The blastomeres of morula soon get arranged in a hollow spherical body, a *blastula,* with a layer of blastomeres, called *blastaderm,* surrounding a fluid-filled central space or cavity, called *blastocoel.* The conversion of a fertilized egg into a multicellular blastula is called *blastulation.*

The fate of cleavage cells. What determines the fates of the blastomeres? Generally speaking, there are two possible answers (1) the fate of a blastomere is decided by the "stuffs" which it contains; or (2) the fate of a blastomere is determined by its location with respect to other blastomeres.

Experiments on the eggs of annelids and mollusks seem to indicate that the first answer applies. If a blastomere, or group of blastomeres, is isolated, it becomes just what it would be expected to become by reason of the material within it. Such eggs, made up of self-differentiating parts, are called mosaic eggs. The egg of the toothshell, *Dentalium,* may be cited as an example. E.B. Wilson (1904) found that it possesses, near its vegetal pole, a stuff which ultimately becomes incorporated into the mesoderm of the larva. At the first cleavage this material bulges cut as a first "polar lobe" and is passed on to only one of the first two blastomeres, the blastomere designated CD. The lobe is then withdrawn. At the second division a second polar lobe is formed and passed on to one of the four resulting blastomeres, the blastomere labeled D. Now, if at either stage the lobe is cut away, an imperfect larva is formed which lacks mesoderm. The entire posterior portion of the embryo is missing, and the anterior part is abnormal. Observations such as these by Wilson, Conklin, and others laid a firm foundation for the mosaic theory of development.

In the eggs of echinoderms and vertebrates, on the other hand, the fate of a blastomere is decided, not only by its stuffs, but also by the position which it occupies with respect to other blastomeres and their stuffs. Sven Horstadius has found it possible, by careful hand-dissection with finely drawn glass needles, to separate the individual blastomeres or groups of blastomeres of sea urchin eggs and then to recombine them in various combinations. Now at the 16-cell stage the sea urchin egg consists of eight mesomeres surrounding the animal pole, four macromeres below the equator, and four micromeres at the vegetal pole. In normal development, the eight mesomeres become ectoderm, and produce a small apical tuft of cilia, a ciliary band, and a stomodaeum (the oral cavity lined by ectoderm). When, however, the mesomeres are cut away and reared as an isolated group, they become a ciliated ectodermal vesicle with a greatly exaggerated apical tuft of cilia but with no ciliary band or stomodaeum. But if micromeres are grafted to the group of mesomeres, the combination may become an essentially normal larva. In this case, the mesomeres contribute to the gut endoderm, a thing they normally never do.

The cells of the vegetal hemisphere (macromeres and micromeres), cut off by themselves, give rise to an unbalanced larva with very little ectoderm but with a huge gut (endoderm). The gut may be unable to cave in, as it normally does, by the process of gastrulation. Instead, it may bulge outward (exogastrulate).

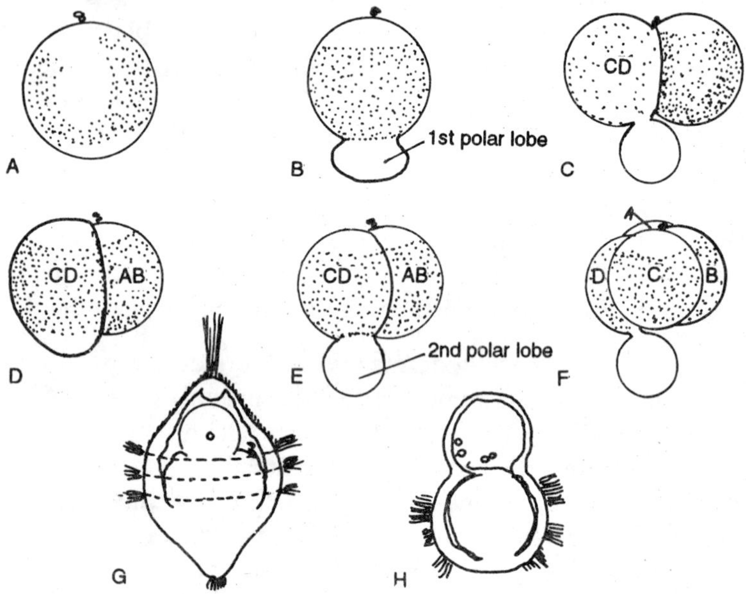

Fig. 1.2. Wilson's experiment which demonstrated the importance of a cytoplasmic "stuff" in the development of the tooth shell Dentalium. A to F. The first two cleavages, showing the first and second polar lobes. G. A normal larva, H. An abnormal larva which developed when a first polar lobe was cut off. The result of removing the second polar lobe is similar.

Now a strange chemical effect has been discovered: If sea urchin eggs are treated with sodium thiocyanate, zinc or mercury salts, or trypsin or various other agents, they become "animalized"; that is, they develop into ectodermal vesicles comparable to those which are produced by isolated animal mesomeres. There is an exaggerated apical tuft of cilia, but no ciliary band, no stomodaeum and no mesenchyme. The cells which normally would have become endoderm and mesoderm become ectoderm. If, on the other hand, sea urchin eggs are exposed to lithium salts, sodium azide, or dinitrophenol, they tend to be "vegetalized." They develop like isolated vegetal halves (macromeres and micromeres). Only a small region at the animal pole becomes ectoderm.

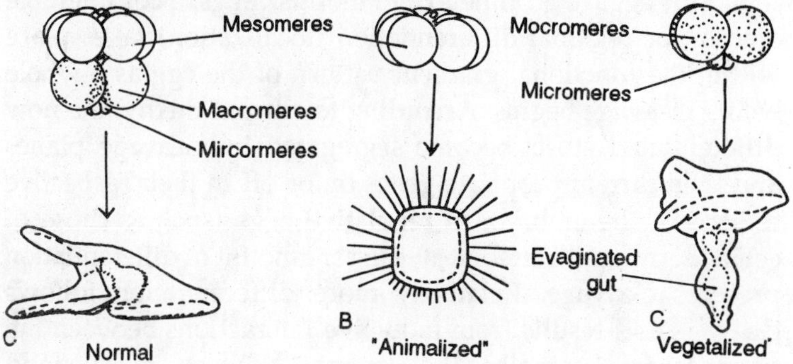

Fig. 1.3. Examples of "animalized" and "vegetalized" embryos of the sea urchin.

What would happen if an isolated animal half (mesomeres only) of an egg were to be treated with a vegetalizing chemical? This experiment was performed by von Ubisch, who found that in favourable cases abnormalities are avoided and a small but fairly normal larva may result.

These and other observations have led to speculation concerning a functional gradient pattern in the sea urchin egg: an animal gradient with its center at the animal pole, and a vegetal gradient centering in the micromeres at the vegetal pole. According to another interpretation, there is only a single bipolar gradient, one which produces unlike effects at its opposite poles. In any case, normal development requires a proper balance between what seem to be two influences. Animalizing agents act by suppressing the vegetal influence. Vegetalizing agents suppress the animal influence. What a cell becomes, therefore, is determined not only by the stuffs its cytoplasm, but also by its position in the functional gradient pattern of the whole.

How does one resolve the apparent conflict between experiments on mosaic eggs and experiments on regulative eggs? The difference in part is one of timing, and in part it is a question as to which morphogenetic factors are emphasized. In mosaic eggs (e.g., annelid and mollusk eggs) a considerable amount of regional differentiation (localization) takes place within the functional gradient pattern of the egg as a whole before cleavage begins. According to current theory, the now differentiated stuffs become segregated by cleavage planes and then turn appropriate genes on or off in their respective regions of the embryo. In regulative eggs (such as those of echinoderms and vertebrates) a lesser amount of differentiation prececes cleavage. Relatively more differentiation follows cleavage and results from inductive interactions between the germ layers. Actually, the contrast between mosaic and regulative eggs has been grossly exaggerated. There is localization and interaction in both.

Gastrulation

The result of cleavage is a *blastula,* typically a ball of cells surrounding a central cavity, or *blastocoel.* Cell divisions by mitosis continue after the blastula is formed, but, generally speaking, the divisions alternate with intervals of cell growth. Hence no further decrease in the size of cells takes place. Instead, mass movements of cells commence by which the single-layered ball of cells, the blastula, becomes transformed into a two-or three-layered ball of cells (Two-layered in coelenterates), the *gastrula*. The process is known as *gastrulation,* and the layers of cells are called *germ layers.*

We shall first describe gastrulation as it occurs in echinoderms and amphibians. One side of the blastula folds inward and partially obliterates the original cavity or blastocoel. A new cavity open to the outside is formed. It is known as the *archenteron* or primitive gut. The opening from the archenteron to the outside is the *blastopore* (primitive pore). The cells which border the blastopore constitute the *blastoporal lips.* The outer germ layer of the gastrula, when gastrulation is finally complete, is the *ectoderm.* The inner germ layer is the *endoderm.* The third germ layer which in vertebrates includes the *notochord,* is the *misoderm.* It occupies a position intermediate between the ectoderm and endoderm. Loose cells which are commonly present between the germ layers are termed *mersenchme.* They are usually thought of as mesoderm, although they may arise from any of the three germ layers.

In echinoderms and amphibians, gastrulation is essentially an inward folding, and is accomplished by three easily defined processes invagination, involution and expansion.

1. *Invagination,* or "caving in," is the process by which cells of the vegetal area pocket inward to form the archenteron. Note that the invaginating layer of cells is concave when looked at from the outside.

2. *involution* or "rolling under," is the process by which the cells which border the vegetal area move toward the blastopore and progressively roll around and beneath the lips of the blastopore, and so come to lie within the egg. Note that during involution the layer of cells which is in process of passing around the lips of the blastopore is convex when seen from the outside.

3. *Expansion* is the process by which the cells of the animal hemisphere extend and converge toward the blastopore. As a result of the co-ordination between these three processes, an echinoderm or amphibian egg during gastrulation does not cease to be a sphere.

In echinoderm eggs, such as those of a starfish or sea urchin, the movements of gastrulation are fairly equal on all sides. The newly formed archenteron lies near the center of the vegetal area. In vertebrates, however, because of the greater mass of yolk in the vegetal hemisphere, the process of gastrulation takes place predominantly on the dorsal side; involution is mainly around the dorsal and dorsolateral lips of the blastopore, and expansion is more of the dorsal than of the

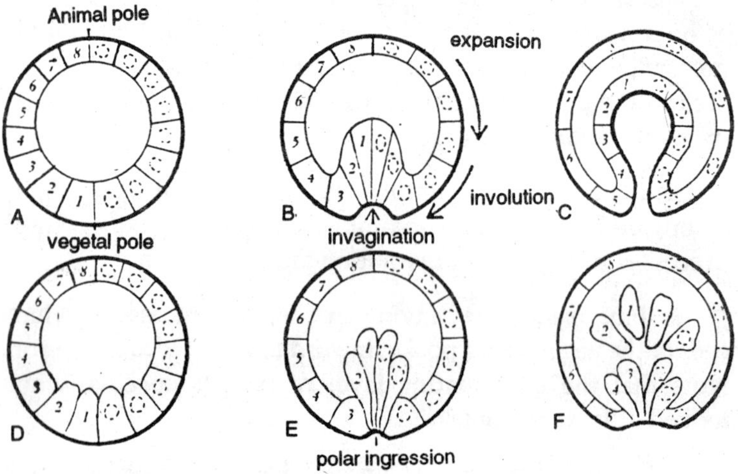

Fig. 1.4. Schema illustrating types of cell movements which takes place during gastrulation and mesoderm formation. The heavy line indicates, the original cortex. *A* to C. In invagination and involution the cells retain connection with the cortex. *D* to *F*. In polar ingression, attachment to the cortex is broken.

ventral surface. As a result, the archenteron is dorsal in position. It is almost as though the future endoderm of the vegetal area swings inward upon the ventral lip of the blastopore as though on a hinge, and so comes to lie within the ventral half of the gastrula.

There are, however, other methods of gastrulation in the animal kingdom than the method just described. In coelenterates, the stage which corresponds to the gastrula is usually a free-swimming larva known as a *planula*. It consists of an outer layer of ectoderm and an inner solid core of endoderm. A cleft later appears in the endoderm (cf. the archenteron) which forms a mouth by breaking through the ectoderm to the outside (cf. blastopore).

In annelids, mollusks, and certain other animals (mainly the animals which undergo spiral cleavage), there is an inward migration of cells at the vegetal pole. This process is termed *polar ingression.*

In the yolk-rich eggs of reptiles and birds, the endodermal cells appear to split away from the inner surface of the outer layer by a process known as *delamination.* The same process takes place in the eggs of mammals which behave in gastrulation as though yolk were still present. There is some question, however, as to just what the cells which delaminate represent.

Now it is cuite possible that these various processes are modifications of a single type of gastrulation, although it is not easy to understand how this could be so.

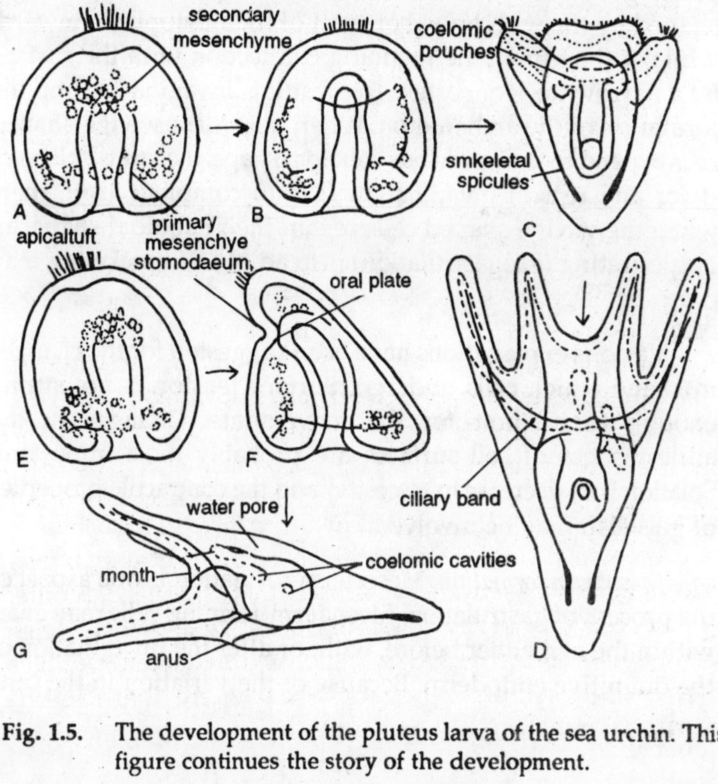

Fig. 1.5. The development of the pluteus larva of the sea urchin. This figure continues the story of the development.

The mechanisms of gastrulation. Because of the variety of processes of gastrulation, the following account is based largely on the sea urchin. During gastrulation a ball of cells folds and differentiates into the first tissues of the embryo. How is this accomplished? It seems to be a movement of groups of cells. Actually it is, at the start, the result of individual cells changing their shape. This in turn may well be the result of changes in the adhesive properties of the cell surfaces. It is not an osmotic phenomenon, that is, a caving in of the blastocoel due to the withdrawal of fluid from the cavity. This is proved by the fact that it will take place even when the blastocoel is opened. In some cases, small groups of cells will perform their characteristic movements even when separated from the rest of the egg.

The outer layer of the blastula (cortex) plays an important part in gastrulation. The cells of the vegetal pole become wedge-shaped by the withdrawal of most of their cytoplasm inward, while at the same time retaining connection with the surface by a narrow neck of cortex. The result is invagination and the formation of the archenteron. When next these wedge-shaped cells expand to their previous cuboidal shape, the gut is enlarged. If for any reason invagination fails to complete itself, then, when the wedge-shaped cells regain their cuboidal shape, an outpocketing (exogastrulation) instead of an inpocketing may occur.

Various explanations have been suggested for the changes in shape which cells undergo. Surface tension is not strong enough to account for the movements. Changes in the adhesiveness of cell surfaces are probably most important. Solation (i.e., decrease in viscosity) and the contractile properties of gels also may be involved.

Mesoderm formation. Mesoderm formation is one aspect of the process of gastrulation. Mesoderm-forming cells may enter within the egg either before, with, or after the invagination of the definitive endoderm. Because of the variation in the time and place of mesoderm origin, the inner cells at the two-layer stage of development are often given the non-committal name of *hypoblast,* and the outer cells are called *epiblast.*

In sea urchin development certain loose cells (primary mesenchyme) wander inward from the vegetal pole in advance of the invagination of the future gut endoderm. Unlike the wedge-shaped cells of invaginating endoderm, these do not remain attached to the surface coat and to each other, but instead they move independently. They throw out filamentous pseudopods which attach to the inside of the wall of the blastula and then contract. In starfish development similar cells separate from the tip of the gut after in has formed. In both cases, the inwandering cells take a position in the space between ectoderm and the gut. Here they play a part in weaving the skeletal structure of the larva. It is an impressive fact that these cells, although apparently free to wander wherever they will within the larva, nevertheless go to the proper place, at the proper time, and do the proper thing. In sea urchins they form a ring at the equator and then break into two groups of cells, a group on the left and a group on the right. Each group becomes triangular, and beginning at its center calcareous spicules form.

Later in the development of echinoderms, oouches form from the sides of the gut and give rise to the definitive mesoderm. The cavities of the pouches become the body cavities (coeloms) and the water vascular system. Similar pouches give rise to the mesoderm in some of the lower chordates. In these the roof of the gut between the pouches becomes the notochord.

Mesoderm forms in a different manner in annelids, arthropods, and mollusks. Typically, two cells, known as *mesoblast,* one on each side of the vegetal pole, enter the blastocoel, proliferate, and form two longitudinal bands of mesoderm. As the bands grow forward they segment and produce a series of clefts which become the coelomic cavities.

Formation of Mouth and Anus

There are two ways in which mouths develop in the animal kingdom. In echinoderms and chordates, the mouth is a new formation. It is produced when the anterior tip of the gut grows forward and comes into contact with a shallow pocket of ectoderm (stomodaeum) on the ventral surface. The *oral plate,* which is thus formed by the fusion of ectoderm and endoderm, breaks through so that the mouth opens into the gut. Those animals which form their mouths in this fashion are known as *deuterostomes*. They are, generally speaking, the same animals which cleave in radial fashion and form their mesoderm from the sides of the gut.

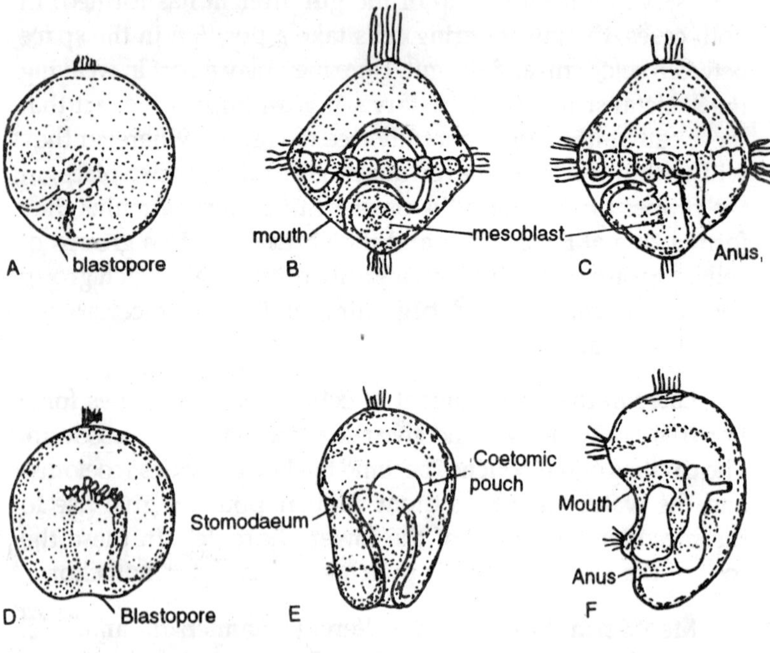

Fig. 1.6. The development of protostomes and deuterostomes (idealized). *A* to *C*. Gastrulation, mesoderm formation, and the trochophore larva of a protostome. *D* to *F*. Gastrulation, mesoderm formation, and an echinoderm-type larva of a deuterostome.

In annelids and mollusks, on the other hand, the blastopore becomes the mouth. The anus is either a new formation, or else both mouth and anus are formed from the blastopore by the coming together of the lateral lips of the blastopore. (In some cases, the mouth closes and later reopens as an apparently new formation.) Animals in which the mouth is formed from the blastopore are termed *protostomes.* They are, generally speaking the ones whose eggs cleave spirally and which form their mesoderm from posterior mesoblasts.

Embryos and Larvae

At every stage of development, an organism must have a source of energy. If the egg is very small and there is little energy stored away as yolk, then the organism must hatch while yet it is very immature, and must begin to feed for itself. Such an immature stage, specialized for feeding (or for finding a new location where it can feed), is known as a larva. Larvae are nature's way of briding the gap between a tiny egg and a complex adult. The life of a larva is always precarious. The chances that it will become a morsel of food for another animal are tremendous. Hence, these animals which produce small eggs and larvae always produce them in great numbers. Often a profound metamorphosis takes place betweeen the larval stage and the young adult stage, during which the organism undergoes a radical change in its structure and mode of life.

One direction of progress in the animal kingdom has been toward providing better methods for the r ourishment and protection of the young. Stages in development, which in more primitive creatures are passed as larvae, are in these more advanced forms passed as embryos. In this case hatching does not take place until the young organism is far enough advanced to be termed a juvenile. Take the crustaceans as an example. The lower members of this class have a larval stage known as a nauplius in which three are there pairs of bristly appendages for swimming and chewing. In the higher crustaceans (lobsters. crabs, etc.) the "nauplius stage" is passed over rapidly while the organism is yet a yolk-rich embryo.

Similarly, the frog hatches as an immature tadpole, while the bird and mammal go through comparable "tadpole stages" rather early in their embryonic development.

The opposite evolutionary trend may have taken place in the case of the higher insects. Instead of larval stages becoming embryonic stages, embryonic stages may have become larval. It is possible to think of caterpillars, maggots and grubs as embryos which have adapted to free life in order to make use of certain rich sources of nutrition. If this interpretation is correct, then the profound metamorphosis which these insects undergo is in truth a continuation of their embryonic development.

Metamorphosis

Sometimes, the animal emerging from the egg possesses special organs which are absent in the adult, but are necessary for the special mode of existence of the young animal called the larva. Now the larva acquires reactivation of morphogenetic process called the *metamorphosis.* During metamorphosis the larva is transformed into an animal similar to the adult and involves more or less drastic morphological and physiological changes in the organization as well in its behaviour, depending on the degree of difference between the larva and the adult. During metamorphosis new organs may develop so that morphogenetic processes become active once again after a more or less prolonged period of larval life. The process of metamorphosis is regulated by hormones.

Regeneration

Many animals processes a considerable plasticity and may be able to repair injuries sustained to them from the environment or caused experimentally. Lost parts or tissues may be regenerated i.e., the developmental processes may sometimes be repeated in an adult or adolescent animal.

Blastogenesis

During blastogenesis or asexual reproduction, the

morphogenetic processes become active once again after sometime. It involves the development of new parts or organs in animals that had already achieved the adult stage or new individuals develop from a bud or from the rudiment of parent's body consisting of somatic cells.

Protostomes and Deuterostomes

We have already noted that many-celled animals above the coelenterates develop in two diverse manners. One assemblage, which includes the annelids, arthropods, mollusks, and several phyla of lesser rank is known as the protostomes. Three eggs cleave in spiral fashion, they form their mesoderm as strands which bud off from mesoblast cells at the posterior end; and their blastopores become a mouth (or mouth and anus). When they produce skeletons, these are mostly cuticles or exoskeletons of ectodermal origin.

The other assemblage stage deuterostomes, to which belong the echinocerms, chordates, and a few minor phyla. Their eggs cleave in a radial or even an irregular manner; they form mesoderm (theoretically at least) from paired pouches at the side of the gut; their mouths are not derived from the blastopore but in stead are new formations; and when they produce skeletons, these are for the most part internal skeletons of mesodermal origin.

Trochophores and nauplii. The typical larvae of protostomes are the *trochophores* of annelids and mollusks and the *nauplii* of crustaceans. Trochophores are essentially spheres with a sensitive fuft of cilia at one pole (apical organ) and a single or double girdle of cilia (prototron) around the equator. A mouth is present on the ventral side, just below the ciliated girdle, and an anus occupies a position near the pole opposite the apica tuft. Generally the whole trochophore rotates like a wheel as it swims—hence the name. The trochophores of mollusks develop a muscular region (foot) on the ventral surface between the mouth and the anus and a shell on the dorsal surface between the girdle and the anus. Annelids, on the other hand, although they may start with an a most identical structure. constrict off a series of body segments anterior to the anus, and *thus* they elongate into segmented worms.

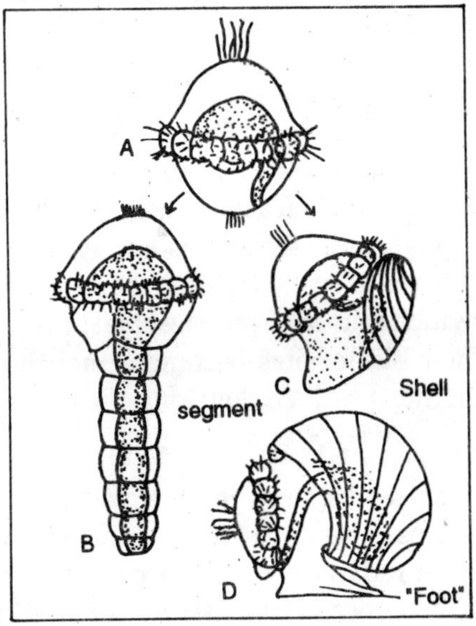

Fig. 1.7. Further development of a trochophore, a type of larva found in both annelids and mollusks. *A* and *B*. In annelids the trochophore strobilates (segments transversely) and becomes a segmented worm. *A, C* and *D*. In mollusks a similar trochophore develops a "foot" and a mantle which usually secretes a shell.

The crustaceans pass through a stage comparable to a trochohore while they are yet embryos. When the more primitive crustaceans hatch, they are nauplii larvae, which may be compared to annelid larvae that have formed three body segments, and in which three pairs of lateral appendages have already been developed. The nauplius, however, has no cilia. Instead, it swims with a jerky motion produced by the strokes of its appendages. Facts such as these seem to link the annelids with the molusks, and suggest—but do not prove—that the arthropods are descended from specialized annelid-like ancestors. The similarities could, of course, be a result of convergence in evolution.

The dipleurula. The hypothetical larva of ancestral deuterostomes, namely echinoderms and chordates, has been visualized as something like the larvae of present-day ehinoderms. A generalized larva of this sort has been called a *dipleurula.* It has been pictured as a sort of common denominator of the bipinnaria of a starfish, the plutei of sea urchins and brittle stars, and the auricularia of sea cucumbers. Like these, the blastopore becomes the anus, and the mouth would be a new formation of the ventral side. Hypothetically, it possessed a band of cilia; but instead of girdling the larva, the band formed a border between the dorsal and ventral surfaces. The dipleurula presumably glided smoothly through the water, sweeping plankton into its mouth by means of currents set up by the ciliary band.

The echinoderms seem to have evolved from such an ancestral type by attaching themselves to some solid object by their anterior end, then undergoing a torsion of their bodies (possibly originally so as to bring their mouth and anus closer together), and finally developing outgrowths from their body for "arm feeding." Particles which came into contact with the arms were entrapped in mucus and then driven by ciilia along grooves of the arm into the mouth. The most ancient echinoderms, which are known only by their fossils, were quite certainly of this sort. Indeed the most primitive group of echinoderms living to-day, namely, the crinoids or sea lilies, are for the most part, sedentary arm feeders. Even present-day free-living echinoderms—starfish, sea urchins, and sea cucumbers—show the marks of their sedentary ancestry by their centrally located mouths and general radial symmetry. Radial symmetry is an adaptation which is evolved when an animal uses all its sides alike.

Some of the startish in their development actually recapitulate the history of their race—or at least this is a possible interpretation. First they are free-swimming, bilaterally

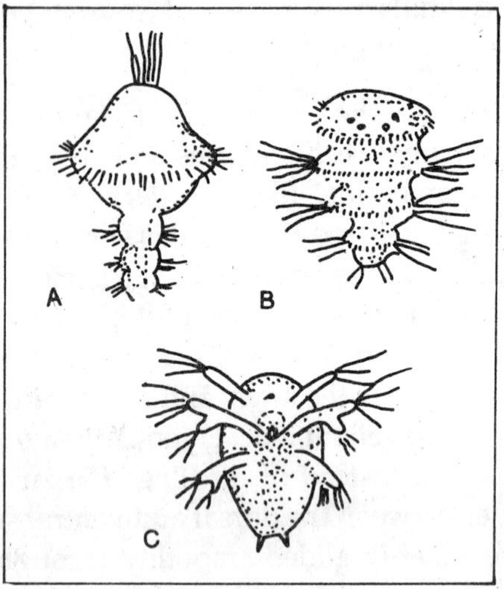

Fig. 1.8. The nauplius larva of a crustacean compared to an annelid larva of three segments. A. Larva of a primitive annelid. B. Larva of the clamworm *Nereis.* C. Nauplius larva of one of the lower crustaceans.

symmetrical larvae known as. These then settle down, attach to the substrate by their anterior ends, and become sedentary larvae known as *brachiolaria*. A most remarkable metamorphosis now takes place. The anterior region of the larva regresses. The left side turns forward the substrate and surrounds the mouth. The right side turns away from the substrate. The midregion of the left coelom gives rise to the water-vascular system. Starfish arms grow out which have no relation whatever to the original bilateral symmetry of the larva. Finally, the stalk by which the brachiolaria was attached to teh substrate regresses, and the starfish becomes free to approach food or retreat from danger by moving in any direction.

The ancestral chordate. We may suppose that the first step toward the chordate type was the development of a dipleurula-like creature with gill clefts at the sides of the gut (pharynx) just posterior to the mouth. Some have speculated that the gill clefts indicate a sedentary stage in the ancestry. It is true that one group of primitive chordates, the tunicates, are sedentary and have a remarkable development of gill clefts. But their larvae are typically chordate and form gill clefts before they settle down. In any event. It is from forms like the free-swimming larvae, not from sedentary adult ancestors (if such ever existed), that the vertebrates are descended. (Reproduction by larval stages is known as neoteny or paedogenesis). From the first, the gill clefts were important for plankton feeding, that is, for filtering out particles of food from the water about them. Their role in respiration came later.

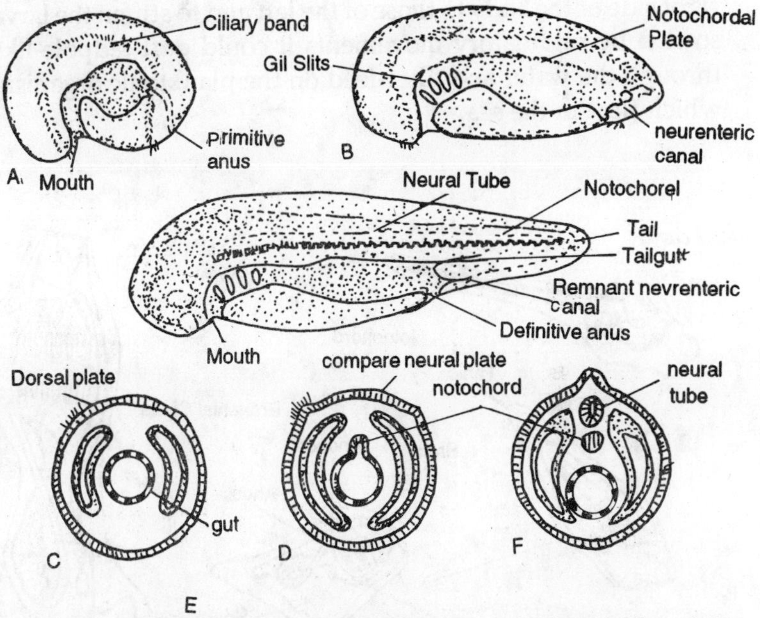

Fig. 1.9. Diagrams which illustrate Garstang's hypothesis of the origin of the chordate body plan.

The primitive chordate larva has been pictured by Garstang as a modified dipleurula. The ciliary band which surrounds the dorsal area is envisioned as having become the neural folds, the dorsal area as the neural plate. According to this hypothesis, the ciliated bands rose up and met dorsally so that the neural plate became transformed into the neural tube, just as it is transformed in the development of most vertebrates. The posterior ends of the neural folds enclosed the blastopore. As they came together, they supposedly formed a canal (neurenteric canal) which connected the cavity of the neural tube with that of the gut. This also occurs in the development of vertebrates. Both mouth and anus in this case were new formations. According to Gar stang's theory, the notochord, so characteristic of chordates, was a flexible, longitudinal rod formed from the top of the gut, just as it is formed in vertebrate embryos today. It served to appose the muscle fibers of the right side of the body to those of the left and to stiffen the larva so that by undulatory movements it could dart tadpole-like through the water and filter-feed on the planktonic organism which flcat in the sea.

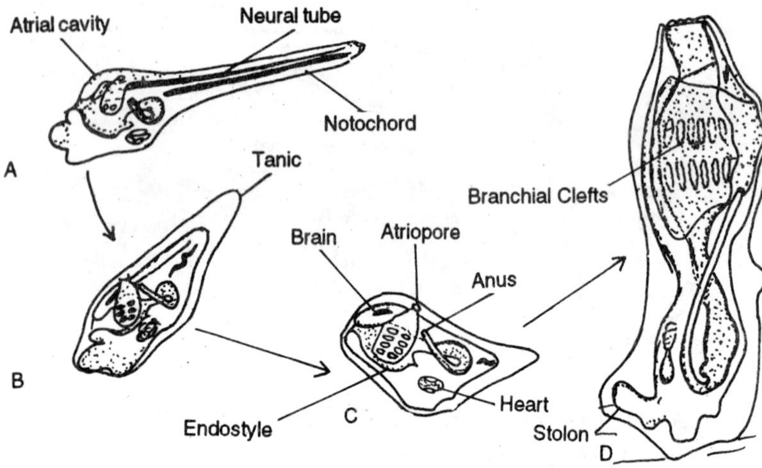

Fig. 1.10. Metamorphosis of a tunicate. *A*, the "tadpole larva" of a tunicate showing its chordate characteristics. *B* to *D*. The larva attaches to a surface by its "chin," develops a tunic around itself and a large gill basket with numerous branchial clefts. It then proceeds to lose its tail, notochord, and neural tube, except for a tiny.

Such evolutionary speculations do not assume that a larval or embryonic stage represents an adult ancestor. These is really no way of knowing what an ancestor looked like. But it is good sense to suppose that if two present-day animals have similar larvae, then they may have had a common ancestor with a similar larva. And may we not go further and speculate that the common ancestor was not specialized in the direction of either of these descendants?

Gastrulation

Following Blastulation a part of the blastoderm disappears from the surface of the blastula by *morphogenetic movements* of cells and is enclosed by the remainder of the blastoderm. The part of the blatoderm that remains on the survace becomes *ectoderm;* the part migrating into the interior becomes *endoderm* and *mesoderm*. In this way a simple spherical body becomes converted into two or three layered embryo known as the *gastrula*. The process involved in gastrula formation is called the *gastrulation,* and the cellular layers of gastrula are known as the *primary germinal layers*. The germinal layers are complex rudiments from which various organs of the animal's body are derived.

The fully formed gastrula has a cavity called the *archenteron* which is lined by endoderm. The opening leading from this cavity to the exterior is called the *blastopore*. During later development, the archenteron or part of it eventually gives rise to the cavity of the alimentary canal; The fate of blastopore differs in the three main groups of metazoa:

(i) In *Coelenterata* it becomes the mouth. (ii) In *Protostomia* (iincluding Annelida, Molluusca, Arthropoda and allied groups), it becomes subdivided into two openings, one of which becomes the mouth and other the anus. (iii) In *Deuterostomia* (including Echinodermata and Chordata), only the anus is formed.

2

The Cell

The cell is a unit of biological activity delimited by a semipermeable membrane and capable of self-reproduction in a suitable non living medium. The body of all living organism (bacteria, blue green algae, plants and animals) except viruses has cellular organization and may contain one or many cells. The organisms with only one cell in their body are called unicellular organisms (e.g., bacteria, blue green algae, some algae, protozoan, etc.). The organisms having many cells in their body are called multicellular organisms (e.g., most plants and animals). Any cellular organism may contain only one type of cell Prokaryotic cells, or Eukaryotic cells.

It was *Robert Hooke* who first of all in 1665 observed under a microscope certain honey comb like structures in cork, and he applied the term 'cell' (L., *cella;* Small room) for the same. Previously it was believed that their is only one component of the cell which is cell wall. But in 1831, one more peculiar structure was observed by *Robert Brown.* He gave the name 'nucleus.'Later on in 1835 the word 'Sarcode' was proposed for the jelly like material present inside the cell by *Dujardin.* In 1940 *Purkinje* replaced the word sarcode by Protoplasm which is now used universally.

Prokaryotic Cells

The prokaryotic (Gr., *pro;* primitive *Karyon;* nucleus) cells are the most primitive cells from morphological point of view. They occur in bacteria and blue green algae. Prokaryotes are small, single cell organism, usually less than a micrometer (abbreviated µm; 1000 µm = 1 millimeter, abbreviated mm) are generallyy not lenger than 3 µm. Their structural organizations is simple than that of the eukaryotes. They do not have a distinct nucleus, and they are lacking in many membranous structures found in the more complex cell types. The hereditary information is contained in a single circular structure that lacks many of the molecular components of the eukaryotic hereditary structure called the chromosome. In addition, the prokaryotes seem to lack any kind of special apparatus for cell division that is why they divide amitotically or by fission.

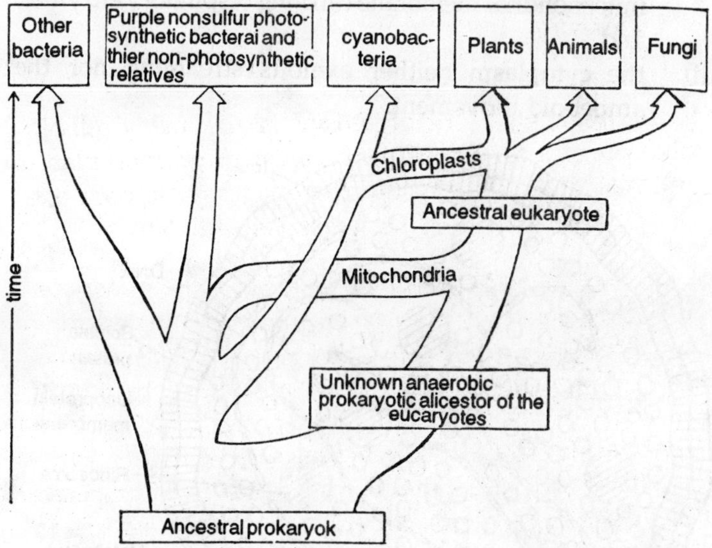

Fig. 2.1. The postulated origin of the eukaryotic cell from ancestral prokaryotic.

The prokaryotes are characterized by:

(i) the absence of a membrane around the nuclear material,

(ii) the absence of clearly defined membrane-limited organelles like mitochondria, chloroplasts, Golgi body and lysosomes,

(iii) the genetic material is located on a single chromosome which consists of a circular double strand of DNA,

(iv) the basic protein—histones, which are one of the most important constituents of chromosomes of eukaryotic cells, are absent in prokaryotic chromosomes,

(v) the absence of nucleolus and mitotic apparatus,

(vi) the cell wall is non-cellulosic, being formed of carbohydrates and amino acids,

(vii) the plasma membrane which lies below the cell wall is produced into the cytoplasm and acts as the mitochondrial membrane carrying respiratory enzymes, and

(viii) the cytoplasm neither exhibits streaming nor the amoeboid movement.

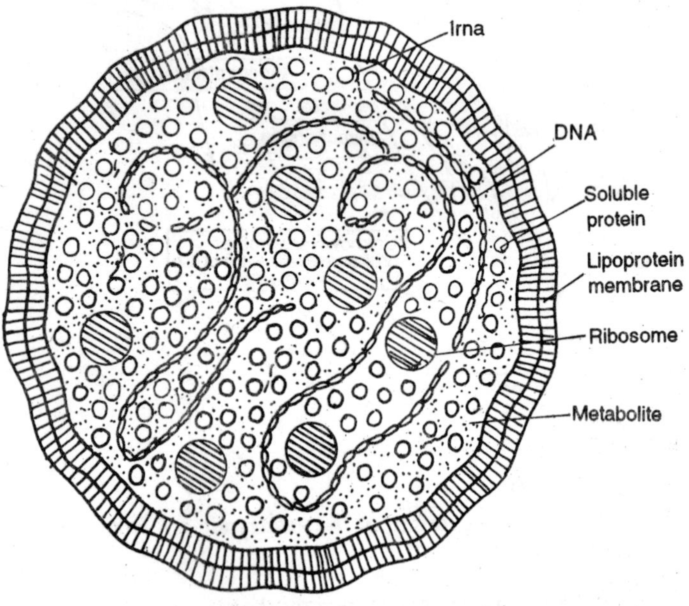

Fig. 2.2. A prokaryotic cell.

The prokaryotes include bacteria, blue green algael, spirochaete, *Mycoplasma* or pleuropncumonia like organism (PPLO).

Bacterial Cell

The bacteria are microscopic, cellular, achlorophillous, asexually reproducing prokaryotes which occur in fresh and salt water, in soil and in plants and animal. They lead either a saprophytic or parasitic mode of existence. The saprophytic species of bacteria are of great economic significance for man. Some parasitic species of bacteria are pathogenic (disease producing) to plants, animals and man.

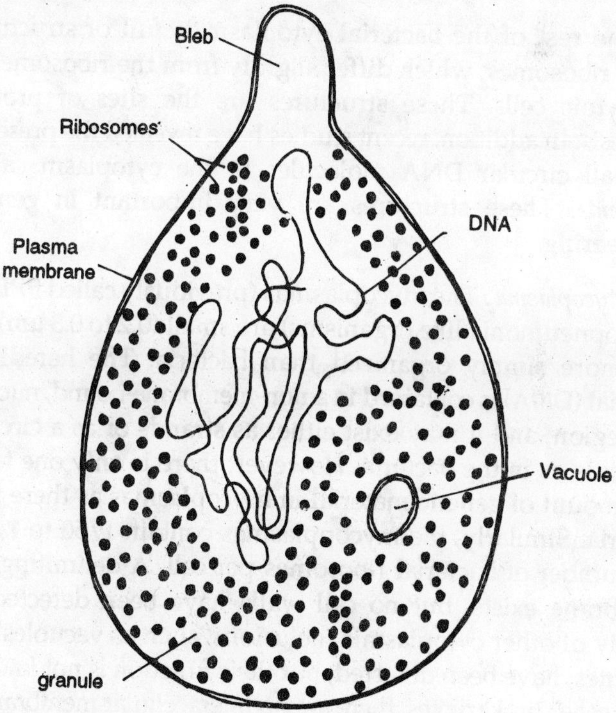

Fig. 2.3. Diagram of a *Mycoplasma.*

Bacteria are very small having an average diameter of 1.25μ.

The smallest bacterium is *Dialister pneumosintes* (0.15 to .3m) and largest bacterium is *Spirillum volutans* (13-15m) in length.

One of the best studied bacteria is *Escherichia coli*. It is about 2 μm long and 1 μm wide and contains several thousand different kinds of molecules. Structurally, *E. coli* is rod like and has a typical cell wall and underlying cell membrane. The membrane contains the special regions called *mesosomes,* which apparently have a role in cellular respiration and cell division. There may be one or several "nucleoid" areas, which contain a single, circular molecule of hereditary information (DNA).

The rest of the bacterial cytoplasm is full of structures called ribosomes, which differ slightly from the ribosomes of eukarytoic cells. These structures are the sites of protein synthesis. In addition, recent studies have revealed the presence of small circular DNA molecules in the cytoplasm called *pplasmids*. These structures are very important in genetic engineering.

Mycoplasma : The mycoplasmas (previously called PPLOs, pleuropneumonia like organisms) are small (0.2 to 0.3 μm) are and more simply organized than bacteria. The hereditary material (DNA) is contained in a non-membrane-bound, nucleus like region, and it may exist either as strands or as a circular molecule as in the bacteria. However, there is only one tenth the amount of genetic material in mycoplasmas as there is in bacteria. Similarly, the mycopplasmas contain 1/50 to 1/100 the number of bacterial ribosomes per cell. A delimiting cell membrane exists, but no cell walls have been detected. A variety of other cytoplasmic inclusions, such as vacuoles and granules, have been detected, but their function is not known. As in other prokaryotes, there are no interacellular membranous structures.

Mycoplasmas have been known to cause certain respiratory diseases in animals and man. They also occur as routine containants in eukaryote tissue culture and currently are of much concern to researchers who use these culture procedures in their studies.

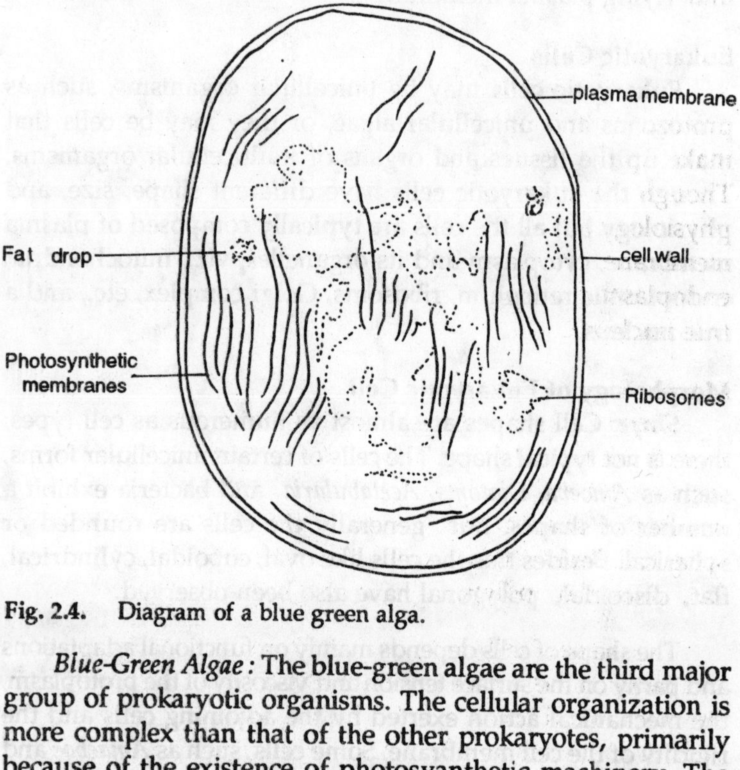

Fig. 2.4. Diagram of a blue green alga.

Blue-Green Algae : The blue-green algae are the third major group of prokaryotic organisms. The cellular organization is more complex than that of the other prokaryotes, primarily because of the existence of photosyanthetic machinery. The blue-green algae contain chlorophyll as well as other more unsual pigments called *phycobilins*. The chlorophylls are bound to membrans in flattened rows called lamellae, which give the algal cytoplasm a rather organized and structured appearance. The phycobilins appear to be located in granular structures called *cyanosomes,* which apparently are attached to the outer lamellar membrane surface. Like the other prokaryotes, the blue-green algae do not have a membrane-bound nucleus or other membrane-delimited organelles. The genetic material exists as free strands of DNA, and the ribosomes are free in the cytoplasm as in the bacteria and mycoplasmas. There appears to the other cytoplasmic granules of a protein and phosphate nature, and the entire cell is surrounded by a cell wall with an underlying plasma membrane.

Eukaryotic Cells

Eukaryotic cells may be unicellular organisms, such as protozoans and unicellular algae, or they may be cells that make up the tissues and organs of multicellular organisms. Though the eukaryotic cells have different shape, size, and physiology but all the cells are typically composed of plasma membrane, cytoplasm and its organelles, viz., mitochondria, endoplasmic reticulum, ribosoms, Golgi complex, etc., and a true nucleus.

Morphology of Eukaryotic Cell

Shape: Cell shapes are almost as numerous as cell types; there is not typical shape. The cells of certain unicellular forms, such as *Amoeba, Diatoms, Acetabularia* and bacteria exhibit a number of shapes. But generally the cells are rounded or spherical. Besides this the cells like oval, cuboidal, cylindrical, flat, discoidal, polygonal have also been observed.

The shape of cells depends mainly on functional adaptations and partly on the surface tension and viscosity of the protoplasm, the mechanical action exerted by the adjoining cells and the rigidity of the cell membrane. Some cells, such as *Amaebae* and leucocytes, can change their body shape very frequently.

Number: Some organisms like protozoans are single called and others are multicellular. The body of human being is composed of about 26 trillions of cells. In human blood the number of crythrocytes is about five million per cubic ml of blood. About 10 billion neurons constitute the nervous system in human being.

Size: The size of different cells ranges within broad limits. Some plant and animal cells are visible to the naked eye, such as eggs of certain birds have a diameter of several centimeters.

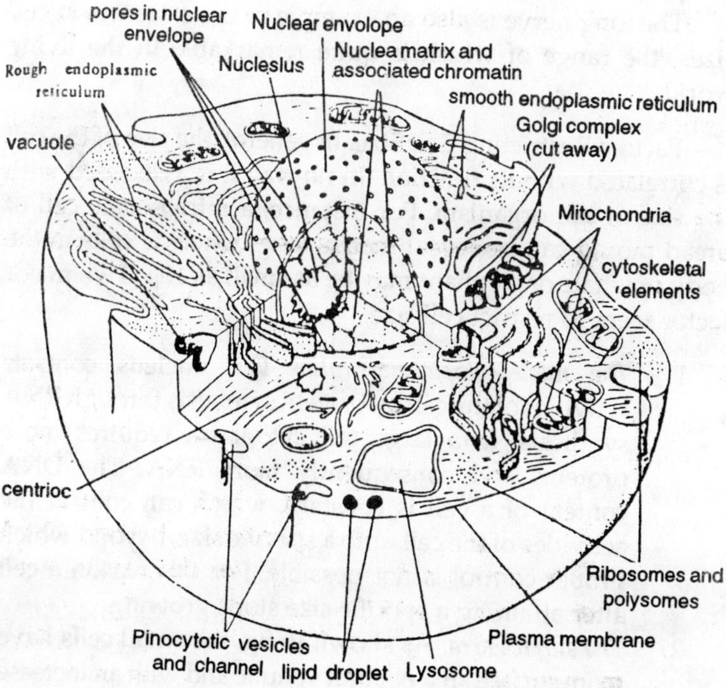

Fig. 2.5. A three dimensional view of a typical animal cell.

But the great majority of cells are visible only under microscope. The smallest living cells are found among bacteria and viruses where the size ranges 0.1μ to 1μ (1μ = 0.001 mm). Organisms like Pleuropeneumonia, so-called "elementary bodies" have been observed with diameters of 100 mμ. These appear to be a resting from of the bacteria which may grow into bodies of 250 mμ in diameter during the active metabolism of the organism. It would seem that 200—250 mμ in diameter is the lower limit for the size of inactive, living cells, this lower limit may be set by the minimum number and size of the component necessary for independent cellular existence.

Usually the vast majority of cells lie in the range of .5 to 20 μ in diameter. The length of unicellular diatoms is about 100μ. And the size of *Amoeba proteus* is 1 mm (100 μ) in length. The size of human R.B.C., is 7-8μ in diameter.

The long nerve is also an example of the extremes in cell sizes, the range of which is quite remarkable in the living world.

Factors Restricting Cell Size. In general, the size of the cell is correlated with its function. In no way it is associated with the size of the organism. For instance, a microscopic cell of bread mould can be seveal metres long, whereas cells in the body of whale or elephant may be as small as 10μ. Two major factor seem to restrict cell size.

1. *The nucleocytoplasmic ratio:* The nucleus controls activities of the cell by protein synthesis through RNA synthesis. In a large cell cytoplasm requires more proteins and consequently more RNA. The DNA content of a cell is constant, which can control the activities of the cell upto a specific size, beyond which proper control is not possible. For this reason a cell after attaining a specific size stops growing.
2. *The surface area:* As shown in fig, the small cells have more surface area per unit volume and with an increase in size, this ratio between surface area and volume gradually decreases. However, a large cell naturally requires more nutrients and oxygen for its metabolism, which diffuse into the cell through its surface. It means, the surface area of a cell affects its ability to exchange materials with its environment and puts a check on cell volume. For this reason, the cells that the metabolically more active are small in size.

Similarly, large cells face problem in the diffusion of excretory products from cell cytoplasm to the exterior.

Structure of Fixed Cells

Examination of the living cell is limited to light microscopy and is based mainly on the differences in refrative index of the different cell components. Sometimes the use of stains that act on the living organism (vital staining) facilitates observation of the living cell. However, more important in the morphologic study of the cell are methods of fixation, by which cell death results in such a way that physiologic structure and chemical composition are preserved as much as possible.

The complexity of the structural organization in a cell of the higher plants and animals is most impressive. Although there are great differences between the primitive forms of life, and the higher plant and animal cells, the similarities between primitive and advanced cells are also notable. Eukaryotic cells are characterized by a true nucleus with a nuclear membrane or envelope which divides the cell into two main compartments: nucleus and cytoplasm. The cytoplasm in turn is limited by the plasma membrane. In a plan, cell the plasma membrane is covered and protected on the outside by a thicker cell wall through which there are tunnels, the *plamodesmta,* by which the cell intercommunicates with neighbouring cells via fine cell process. In animal cells, parts of the plasma membrane are covered by a thin layer of material, which is generally described as the extraneous coat of the plasma membrane. The so-called *basement membranes* correspond to this extraneous coat.

In every stained cell usually one nucleus is present. It is central organele of cell and is the main site of hereditary material. The nucleus is surrounded by a nuclear membrane but in bacteria and blue green algae it is without nuclear membrane. Within the nucleus a single nuclelus and network of chormatin is present. These two materials remain suspended in the nucleoplasm.

A small cytoplasmic organoid 'centriole' is present in the majority of animal cells and in some cells of the lower plants. The position is generally fixed for each type of cell. Generally the cell center or centriole occupies an axial position behind the nucleus. In some cells, it has the tendency to occupy the geometrical center i.e., in leucocytes it occupies this position within the nucleus which is horseshoe shaped. The morphology of centriole changes with the functional state of the cell.

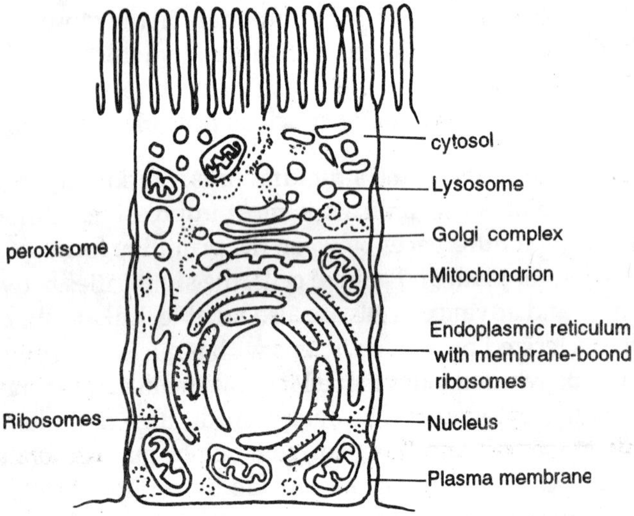

Fig. 2.6. A typical animal cell.

In the cytoplasm of the cell, there are several organelles out of them, only mitochondira and Golgi bodies are visible under light microscope. Mitochondria or chondriosomes are scattered in the cytoplasm of all kinds of cells. Their shape in the fixed cells depends upon the fixation and handling of the tissues before preparation they may appear as long rods, short rods or as small granules. Golgi material appears in its very prominent form in nervous and secretory cells in animals. Plastids are cytoplamic organelles of plant cells may or may not contain pigments. In green plants chlorphyll is present in these organelles and thus termed as chloroplasts.

Electron Microscopic Structure of a Typical Cell

Under ordinary light microscope only few cell organelles

like mitochandria, Golgi complex, chloroplasts and nucleus are visible. However, under electron microscope, several other cytoplasusmic organelles such as endoplasmic reticulum, ribosomes, lysosomes, nuclear membrane, plasma membrane, etc., are seen.

Some important cellular components are being described here under the following heads:

Plasma Membrane : The outermost boundary of the animal cell is called the 'plasma membrane.' It is invisible under a light microscope but under electron microscope, it appears to be composed of two dense layers which are called the outer dense layer and inner dense layer. Both these layers are about 20A° in thickness, and are separated from each other by a less dense area of about 35A^0 in thickness. By this way the total thickness of plasma membrane is about 75A°. Before the advance of electron microscope people used to think of the plasma membrane as been stretched tightly over the cell. But now we know that a single plasma membrane can have as many as 3000 microvilli. Another preculiarity of the structure of the membrane is that, it remains connected with the endoplasmic reticulum.

Plasma membrane is lipo-protein in composition.

Functions:

(1) Gives as identity to the cell.
(2) Protects the cell organelles.
(3) Helps in digestion through phagocytosis and pinocytosis.
(4) Being permeable to water, water can pass through it.
(5) Helps in the transport of ions.
(6) Helps in the formation of cytoplasmic organelles.

Mitochondria : the term mitochondria was for the first time used by *Benda* (1902) to designate the filamentous and granules of cell cytoplasm. Most of the knowledge related to mitochondrial fine structure is due to the studies of *Palade* (1952, 55, 56) and *Sjostrand* (1955). Due to these studies, now mitochondria are defined as the double walled structures of cell organelles within which the membranous structures called *Cristae* are present. The outer chamber of mitochondira is made up of two dense layers and a less dense layer having the total thickness of 140 A° to 180 A° while the outer and inner dense layers very in their measurements from 35 A° to 60 A° in thickness depending upon the physiological state of the cell. The less dense layer is 40-70 A° in thickness.

The matrix of the mitochondira may have some fine opaque granules of various sizes. *David E. Green* (1964) observed in a very highly magnified electron microscope that the inside of the inner membrane and the outside of outer membrane are covered with very small particles. They measure about 90-100A° in diameter. The particles on the outer membrane differ from those on the inner membrane. The former generally appear as simple spheres packed closely together which give to the surface of the mitochondrion, a rough or pimpled appearance. The particle on the inside of the inner membrane has the more complex configuration. In some cells, but not all, this particle is seen to have a base, a stalk and a spherical head. Mitochondria represent the power generating centres of the cell. They can bring about oxidation of the intermediates of carbohydrates, fat and amino acid metabolism in the presence of O_2 releasing energy. The released energy is trapped in the form of adenosine triphosphate (ATP).

Golgi bodies : Golgi bodies have probably been first observed earlier in 1967 by *La, Vallete St. George,* but any how credit goes to *Camillo Golgi* (1898) who described it as internal reticulate apparatus' and after it, was named Golgi apparatus. It is variously called Golgi element, Golgi complex, Dalton's complex etc. Each Golgi complex is composed of many lamella, tubules, vesicles and vacuoles. The membranes of Golgi complex are of lipoproteins and these are supposed to be orginated from the membranes of endoplasmic reticulum. The function of the

Golgi Complex is storage of proteins and enzymes which are synthisized by ribosomes and transported by endoplasmic reticulum to them. Further the Golgi complex has most important scretory function. It secretes many secretory granules and lysosomes. In plant cells the Golgi complex is known as dictyosome and secretes necessary materials for cell wall formation during cell division.

Lysosomes : The lysosomes (*Gr. lyso;* digestive + *some;* body) are also found scatered in the cytoplasm. By gently centrifuging lysosomes can be separated from other cell orgganelles. They are usually spherical but may be irregular in shape, covered with a membrane of lipoprotein. Inside structure is always variable. Lysosomes contain bydrolysing enzymes, such as acid phosphatase, acid ribonuclease, acid deoxyribonuclease and cathespins. But they do not have oxidiative enzymes. This property distinguishes them from the mitochondira. Lysosomes are polymorphic in nature and are commonly known as suicide bags and are responsible for postmartem changes in cells.

Endoplasmic Reticulum: The term "Microsome" was introduced to the modern cytology of *Claude* (1943) to represent one of the submicroscopic cellular components isolated by centrifugation. These are now known to be the broken parts of the endoplasmic reticulum after the studies of *Kollman* (1953). *Porter* (1945) was the 1st man to describe their electron microscopic structure in cultured cells. He described them as lace like reticulum. There are membrane bounded sacs in the form of double membrane or cisternae. The term cisternae was introduced by *Sjostrand* (1953) to describe long or elongated rod like parts of the endoplasmic retculum measuring 50-200 A°. *Wiess* (1953) described the vesicles of the endoplasmic reticulum which usually has a diameter of 25-500 A° while *Bradfield* (1913) came across of another type of endoplasmic reticulum termed them as tubules, they measure from 50 to 100 A° in diameter.

On the basis of presence or absence of ribosomes or RNP

particles the endoplasmic reticulum (E.R.) is distinguished in two varieties. The *rough walled endoplasmic reticulum, and smooth walled endoplasmic reticulum.* They are also known as granular or α-β type of endoplasmic reticulum.

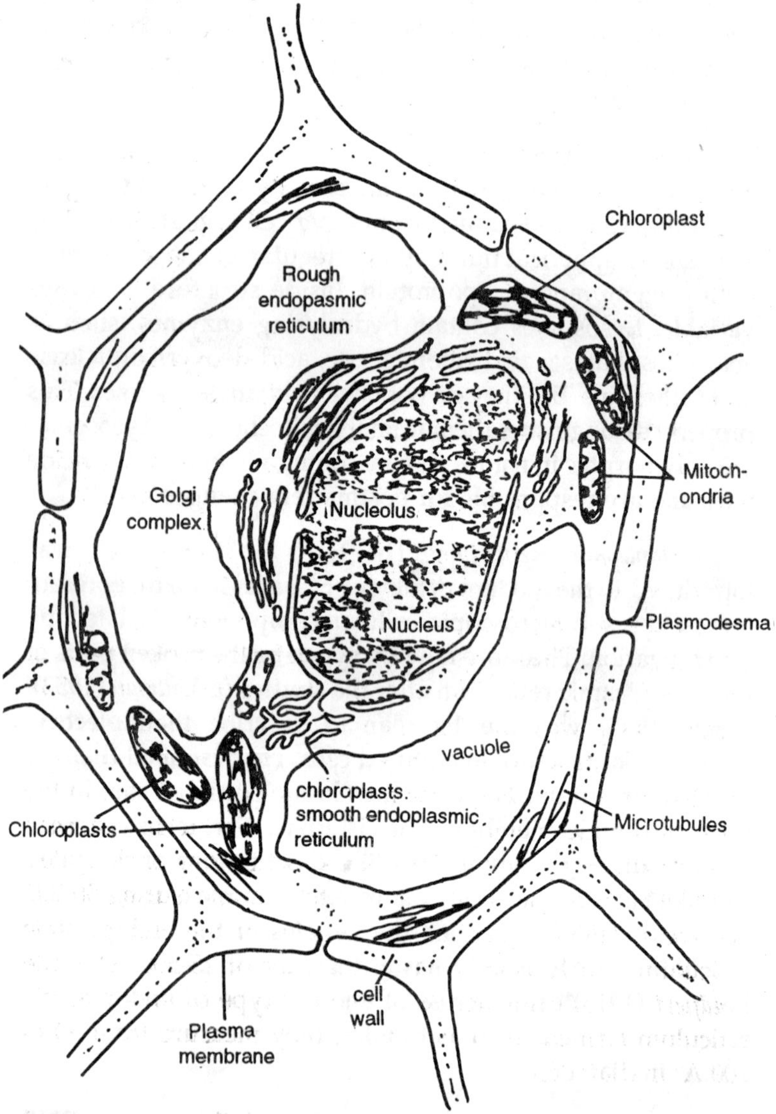

Fig. 2.7. Typical plant cell.

Rough Walled Endoplasmic Reticulum: Rough walled endoplasmic reticulum (R.E.R.) is that variety which bears the ribosomes at the external surface of the cisternae. The distribution of the ribosomes can be circular, spital, or rosette type. These particles may be induced to leave the surface of the cisteme without damaging the structure of the cistemae or the capacity to associate again with the particles. Usually the rough walled endoplasmic reticulum richly distributed in those cells which are engaged in the synthesis of the protein.

Smooth Walled Endoplasmic Reticulum: The other large division of the system owes its identity in parts to the absence of the particles and therefore commonly referred to as the smooth surfaced agranular form. Like the R.E.R., the S.E.R. shows a characteristic morphology which is tubular rather than cisternae. The tubules have the diameter nearly of 50-100 mu. As studies by *Fwcett* (1960) S.E.R., richly distributed in those cells which are engaged in the synthesis of steroids.

Continuity between the S.E.R., of a particular cell type prossess a characteristic pattern of structure. Thus it is typical for rat liver cells to show groups of eight or ten slender profiles in parallel array.

Plastids : Plastids are present only in plant cells (not in animal cells). These are double-membrane structure relatively larger than the mitochondria. Some of the plastids are colourless and store starch (leucoplasts). Others contain various pigments and impart different colours (red, brown, yellow, etc.) to different parts of the plant. These are collectively called chromoplasts. Those containing green pigment—cholorophyll, are called cholorplasts. These contain several enzymes in their matrix necessary for photosynthesis. The chloroplasts capture light energy of the sun and convert it into the chemical energy—ATP, which is used in the biosynthesis of glucose from CO_2 and H_2O.

Rhiosomes : Many minute, spherical structures known as ribosomes are remain attached with the membrane of endoplamic reticulum and forms granular or rough type of E.R. The ribosomes are originated in the nucleolus and consist of mainly the ribonucleic acids (RNA) and proteins. Each ribosome is composed of two structural units, smaller subunit known as 40 S subunit and a large subunit known as 60 S subunit. The ribosomes remain attached with the membranes of endoplasmic reticulum by the 60 S subunits. The 40 S subunits occur on the larger sub unit and form a cap-like structure. The ribosomes consists of 3 types of RNA's known as ribosomal RNA's or rRNA, viz., 5S 185 and 28 S rRNA's. The 28S and 5 S rRNA's occur in large (605) subunit, while 18 S r RNA occurs in the smaller ribosomal subunit. The ribosomes also occur freely in the cytoplasm. They are the sites of protein synthesis.

Peroxisomes : These are also bounded by a single membrane. These contain enzymes for the breakdown of hydrogen peroxide to form water and oxygen. These are protective organelles of the cell.

Cilia and Flagella : The cells of many unicellular organisms and ciliated epithelium of multicellular organisms consists of some hair-likecy-toplasmic projections outside the surface of the cell. These are known as cilia or flagella. They help in locomotion of the cell. The cilia and flagella consist of nine outer fibrils around the two large central fibrils. Each outer fibrils consists of two microtubules. The two microtubules of each peripheral pair or *doublet* remain morphologically distinct. The central fibrils are separate but remain enveloped in a common *central sheath.* Cilia and flagella are envolved in locomotion, cleansing etc.

Vacuoles : These structures are fluid-filled "bubbles" found in the cytoplasm. Vacuoles are bounded by a membrane, called *tonoplast,* which is structurally similar to, if not identical with, a cell-membrane. Inside the vacuoles, food materials or wastes are found.

Vacuoles are consipicuous in plant-cells but they are found less in animal cells. Vacuoles are absent in actively dividing cells. In a young plant-cell, there are many vacuoles but, when a cell matures, they unite to form a single, large, central vacuole.

Guilliermlond (1941) was of the opinion that vacuoles are colloidal with a strong imbibing capacity. Vacuoles arise *denovo* i.e., spontaneously formed from pre-existing vacuoles.

In many protozoan, vacuoles are formed at the bottom of the gullet. In these vacuoles, food-materials, such as bacteria, are collected. Some vacuoles are formed by cell-drinking. This type of vacuoles are called *pinosomes.* Food-vacuoles formed by evagination of cell-membrane surrounding food-particles with some fluid, are called *phagosomes.* In animals cells, the vacuoles are bounded by a lipoprotein membrane that helps in storage and transport of materials and in the maintenance of internal pressure of the cell.

Vacuoles are not a living part of the protoplasm but are a storing place of reserve materials and waste products.

Crystals and oil droplets : In certain plant cells the food or waste material is found deposited in the form of crystals. The oil droplets are round in both plant and animal cells, which appear as tiny glistening white spheres. These serve as a reserve supply of energy-rich fuel.

Nucleus : It is the most vital part of cell which is invariably present in all the cells of higher animals and plants. Normally one nucleus is present in a cell but more than one may be present in certain cells.

Structurally it is enveloped by a porous *nuclear membrane* that separates it form the surrounding cytoplasm. It has the characteristic permeability. During cell division, it breaks up and soon after the mitosis it is reconstituted. The inside of ucleus is filled with *nuclear sap or karyolymph.* In nuclear sap remain embedded other nuclear constituents.

The *chromatin* is the Feulgen positive material in the interphase nucleus which during division becomes the definite chromosomes. In short it represents the chromosome substance. The sex-chromatin are the specific heterochromatic bodies, found generally near the periphery of the nucleus. In nucleus usually one sex chromatin body is found but more than one can also be observed. *Nucleolus* is relatively large, generally spherical ball like body. It has a high concentration of RNA and thus is an important structure involved in protein synthesis. Its number in a nucleus vary in different cells and depend on either the species or the sets of chromosomes.

The nucleus co-ordinates the many cell activities. DNA Molecules are the source of coded instructions, which pass to RNA molecule which then act as messenger, moving out of the nucleus to one of the various structures of the cytoplasm. It also controls the cell reproduction. Here too, DNA is the source of coded instructions.

The Cytosol, Hyaloplasm of Ground Substance: The material that remains suspended under conditions sufficient to sediment even the ribosomes is referred to by biochemists as the *soluble fraction* or *cytosol*. This fraction usually contains some material lost from the organelles in the course of homogenization and centrifugation. However, it also contains other molecules, including some enzymes, that are not known to be part of the intracellular structures identifiable in the microscope. Such molecules are usually thought to derive from the hyaloplasm (also called ground substance, cell matrix, or cell sap), which was originally defined by microscopists are the apparently structureless medium that occupies the space between the visible organelles. As will be evident in subsequent chapters, with the improved techniques less and less of the cell appears structureless. Most investigators continue to believe that there does exist in the cytoplasm a "soluble phase"—a solution containing inorganic ions and small molecules, materials in transit from one cell region to another and dissolved or suspended enzymes that function as individual molecules or perhaps as parts of small groups of molecules rather than as components of large, microscopically recognizable organelles. But there is a lively debate going on over proposals that many components once though to be suspended in this solution are actually bound, perhaps loosely, in structured arrangements that do not survive the disruption involved in cell fractionation.

TABLE 2.1

The Relative Volumes Occupied by the Major Intracellular Compartments in a Typical Liver Cell (Hepatocyte)

Intracellular Compartment	*Percent of Total Cell volume*	*Approximate Number of Cell*
Gytosol	54	1
Mitochondria	22	1700
Rough endoplasmic reticulum cistemae	9	1
Smooth endoplasmic reticulum cistemae plus gogi cistemae	6	
Nucleus	6	1
Peroxisomes	1	400
Lysosomes	1	300

TABLE 2.2

Comparison of Prokarytoic and Eukaryotic Organisms

	Prokaryotes	*Eukaryotes*
Organisms	becteria and cyanobacteria	protists, fungi, plants and animals
Cell Size	generally 1 to 10 μm in linear dimension	generally 10 to 100 μm in linear dimension
Metabolism	anaerobic or aerobic	aerobic
Organelles	few or none	nucleus, mitochondira. chloroplasts, endoplasmic reticulum, etc.
DNA	circular DNA in cytoplasm	very long DNA containing many non-coding regions; organized into chromosomes and bounded by nuclear envelope

RNA and protein	RNA and protein synthesized in same compartment	RNA synthesized and processed in nucleus; proteins synthesized in cytoplasm
Cytoplasm	no cytoskeleton, cytoplasmic streming, endocytosis, or exocytosis	cytoskeleton composed of protein filaments; cytoplasmic steaming; endocytosis and exocytosis
Cell division	by-binary fission	by mitosis and (or meiosis)
Cellular organization	mainly unicellular	mainly multicellular, with differentiation of cells

TABLE 2.3

Comparison of Plant and Animal Cells

Plant cell	*Animal cell*
1. Generally larger than the animal cell.	1. Usually smaller than the plant cell.
2. Plant cells are surrounded by a rigid cellulose wall.	2. Cellulose wall is not found in animal cell.
3. Majority of the mature plant cells have a large central sap vacuole.	3. Vacuoles in animal cells are small
4. Plastids, especially choloroplasts are found usually in cells of green plants.	4. Choloroplast is not found in animal cell except in some Protozoa.
5. In the cell division, the division of cytoplasm of plant cells take place by formation of a partition called cell plate.	5. Animal cells divide by a constriction during cell division.

TABLE 2.4

Functions of Cellular Organells

Cell Organelles	*Functions*
Cell or Plasma membrane	Serves as a differentially premeable membrane through which extra-cellular substances may be selectively sampled and cell products may be liberated. Gives external form. Provides mechanical support and thus protects the cell.

Cell Wall (in plants)	Thick cellulose wall around the cell membrane gives strength and rigidity of the cell.
Cytoplasm	Provides greatly expanded surface Endoplasmic reticulum area for various biochemical reactions which normally take place at or across membrane surfaces.
Ribosomes	Sites of protein synthesis and known as "Protein factories on the cell"
Mitochondria	Oxidation of food substances, changes potential energy into a form of energy to be used by cell. Energy Production (Krebs' cycle, electron transport chain, beta oxidation of fatty acids etc.). Power house of cell.
Centrioles (usually note seen in plant cells)	From poles for the cell division. Capable of replication usually.
Golgi body or apparatus	Produce cellular secretions known as dictyosomes in plants.
Lysosome (in animals)	Produce intracellular digestive enzymes which help in disposal of bacteria and other foreign bodies. If ruptured, may cause cell destruction.
Plastids (in plants)	Serve for storage of starch, pigments and other cellular products. Photosynthesis takes place in chloroplasts.
Vacuoles	Store excess water, waste products, soluble pigments etc.
Hyaloplasm	Contains engymes for glycolysis and structural materials, viz., sugars, amino acids, water, vitamins, nucleotides etc.
Cilla and flagella	Serve to move the animals or to move the material.
Nucleus	Provides selective continuity
Nuclear membrane	between nuclear and cyto plasmic materials.

Nucleoplasm	Contains materials, for building DNA and messenger molecules which serve as intermediates between nucleus and cytoplasm.
Nucleolus	May synthesize ribosomes Disappears during cellular replication. Function uncertain.
Chromosomes	Bearers of heriditary instructions. Regulation of cellular processes (evident only during nuclear division). Control the nucleus and metablism.

3

Chromosomes

Chromosomes were first seen by *Hofmeister* (1848) in the pollen mother cells of *Tradescantia* in the form of darkly stained bodies. The term chromosome (Gr : *chrom*= colour; *soma*= body) was used by *Waldeyer* (1888) to designage their great affinity to basic dyes. Their functional significance was described by *W.S. Sutton* (1900) when he traced parallelism between segregation of chromosomes during meiosis and transmission of hereditary factors during gametogenesis. General reviews on the morphology of chromosomes have been published by *Heitz* (1935), *Kuwada* (1939), *Gitler* (1940) and *Kaufmann* (1948).

Chromosomes are the most significant components of the cell, particular they are apparent during mitosis and meiosis. Their presence was demonstrated long before they were named "chromosomes" by *Waldeyer* in 1888. A chromosome can be considered as a nuclear componant having special organisation, individual and function. It is capable of self-reproduction while maintaining its morphologic and physiologic properties through successive cell divisions.

MORPHOLOGY

The morphology of chromosome can be best studied at the metaphase or anaphase of mitosis when they are present as definite organelles, being most condensed or coiled.

Number

The number of chromosomes in a given species is usually

constant containing diploid number (2n) of chromosomes in their somatic cells and haploid (Gametic or reduced) number (n) of chromosomes in their sex cell (sperms and ova). The number of chromosomes is variable from one to several hundreds among different species. For example, in pinworms the number of haploid chromosome is 1, in *Ascaris megalocephala* it is 2, while in certain protzonans (*Aggregta*), there are more than 300 chromosomes; in *Paramecium* 30 to 40, in radiolarians as many as 1600, in *Hydravulgaris* 32, *Musca domestica* 12, *Ranaesculenta* 26, *Columbalivia* 80, *Orycotolagus cuniculus* 44, *Gorilla gorilla* 48 and *Homo sapiens* (man) 46.

The chromosome numbers are also helpful for taxonomy. In the angiosperms the most frequent haploid number is 12 and members of this group have a range from 3 to 16. Similarly, in fungi, haploid number ranges from 3 to 8. In primates this haploid number is from 16 to 30. This haploid set of chromosomes present in the nucleus of gametes is generally called *genome* while in a diploid cell there as will be two genomes. The diploid cells are the somatic in body cells. The diploid cells get the diploid set of chromosomes by the union of the haploid male and female gametes in the sexual reproduction.

Size

Chromosomes range, on an average from 0.5 to about 30μ in length and from 0.2 to 3* in diameter. The relative number of chromosomes generally differ in the nucleus but at a time all chromosomes of a cell may be of the same size. Plant cell normally posses larger chromosomes then animal cells. *Trillium* has chromosomes which may reach upto the length of 32μ at metaphase. Monocotyledones plants usually have larger chromosomes, than the dicotyledon which contain greater number of chromosomes. Among the animals, grass hoppers, crickets, mantids, newts and slamanders have large chromosomes.

Variation in size of the chromosome can be induced by a number of environmental agents:

(1) Cells dividing at low temperature have shorter, more compact chromosomes than those dividing at high temperature.

(2) Colchicine is an alkaloid which interferes with spindle formation and cell division. It tends to shorten the chromosomes.

(3) Rapid and repeated division tend to result in smaller chromosomes. It appears that rate of cell division proceeds more rapidly than the formation of chromatin material as usual.

(4) In plants, the amount of phosphate in the nutritory medium has marked effect on the size of the chromosomes; high concentration gives larger chromosomes than those plants deficient of phosphates. Since the phosphate is an integral part of nucleic acid molecule, it would seem that the amount of nucleic acid in the chromosome can be varied to give alterations in size.

Shape

The shape of the chromosomes is changeable from phase to phase in the continuous process of the cell growth and cell division. In the resting phase or interphase stage of the cell, the chromosomes occur in the form of thin, coiled, elastic and contractile, thread-like stainable structures, the *chrommatin threads.* In the metaphase and the anaphase, the chromosomes become thick and filamentous. Each chromosome contains a clear zone, known as *centromere* or *kinetocore,* along their length. The centromere divides the chromosomes into two parts, each part is called *chromosomes arm.* The position of centromere vary from chromosome to chromosome and it provides different shapes to the later which are following:

1. *Telocentric:* The rod-like chromosomes which have the

centromere on the proxmal end are known as the telocentric chromosomes.

2. *Acrocentric:* The acrocentric chromosomes are also J-like in shape but these have the centromere at one end and thus giving a very short arm and an exceptionally long arm. The locusts (Acrididae) have the acrocentric chromosomes.

3. *Submetacentric:* The submetacentric chromosomes are L-shaped. In these, the centromere occurs near the centre or at medium portion of the chromosome and thus forming two unequal arms.

4. *Metacentric:* The metacentric chromosomes are V-shaped and in these chromosomes the centromere occurs in the centre and forming two equal arms. The amphibians have metacentric chromosomes.

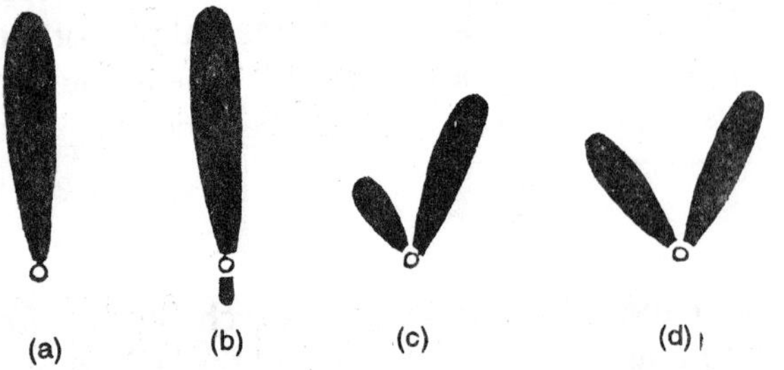

Fig. 3.1. Four morphologic types of chromosomes (a) Telocentric (b) Acrocentric (c) Submetacentric (d) Metacentric.

Structure of the Chromosome

In earlier light microscopic descriptions the chromosome, or chromatid, was though to consist of a coiled thread called the chromonema lying in a matrix. The chromosome was supposed to be covered by a membranous pellicle. Electron microscopic studies later showed that there is no definite membranous pellicle surrounding the chromosome. Other structures present in the chromosome include the chromatids, centromere, secondary constrictions, nucleolar organizers, telomeres and satellites as given under the following heads:

TABLE 3.1

Showing the Diploid (2n) Numbers of Chromosomes in Some Important Animals

Common name	*Zoological name*	*Chromosomes*
Man	*Homo sapiens*	48 or 46 (?)
Rhesus Monkey	*Maccaca mulatta*	48
Horse	*Equus caballus*	66 or 64 (?)
Wild horse	*Equus przewalshii*	66
Pig	*Sus scrofa*	40
Ox	*Boa taurus*	60
Sheep	*Ovis aries*	54
Cat	*Felis maniculata*	38
Dog	*Canis familiaris*	78
Goat	*Capra hircus*	60
Ass	*Equus asinus*	66
Rat	*Rattus rattus*	42
Domestic mouse	*Mus musculatus*	40
Guinea pig	*Cavia cobaya*	64
Rabbit	*Oryctolagus cuniculus*	44
Opossum	*Didelyphys vigrginiana*	22
Frog (Indian)	*Rana tigrina*	24
Frog(British)	*Rana esculanta*	26
Hyla(Tree frog)	*Hyla sp.*	24
Toad	*Bufo vulgaris*	22
Platyfish	*Platypoecilus maculatus*	48
Gold fish	*Carassius auratus*	94
Pigeon	*Columba livia*	79,80
Turkey	*Meleagris gallopavo*	81,82
Chicken	*Gallus domesticus*	77,78
Duck	*Anas indicus*	80
Aulocantha	*Aulocantha*	1600

(Protozoan)		
Hydra	*Hydra vulgaris*	32
Ascaris	*Ascaris megalocephala*	2
House fly	*Musca domestica*	12
Honey bee	*Apis mellifica*	32,16
Cricket	*Gryllus*	23,24
Mosquito	*Culex sp.*	6
Lac insect	*Bombyx mori*	32
Hermit crab	*Eupagurus*	254

TABLE 3.2

Showing the Diploid Numbers of Chromosomes in Some Plants

Common name	*Botanical name*	*Chromosomes*
Yellow pine	*Pinus ponderosa*	24
Cabbage	*Brassica oleracea*	18
Radish	*Raphanus sativus*	18
Cocumber	*Cucumin sativus*	14
Cotton	*Gossipium hirsutum*	52
Plum	*Prunus domestica*	48
Cherry	*Prunus carasus*	32
Apple	*Malus sylvestris*	34,51
Pear	*Pyrus communis*	34,51,68
Pea (Garden)	*Pisum sativum*	14
Sweet pea	*Lathyrus odoratus*	14
Bean	*Phascolus vulgaris*	22
Scarlet	*Quercus coccinea*	24
Orange	*Citrus sinensis*	18,27,36
Sun flower	*Heliannthus annus*	34
Tobacco	*Nicotiana tobaccum*	48
Potato	*Solanum tuberosum*	18
Banana	*Musca paradisiaca*	22,44,55,77,78

Cron	*Zea mays*	20
Onion	*Allium cepa*	16
Berley	*Hordeum vulgare*	14
Wheat (Summer)	*Triticum dicoccum*	28
Rye	*Secale cereale*	14
Rice	*Oryza sativa*	24

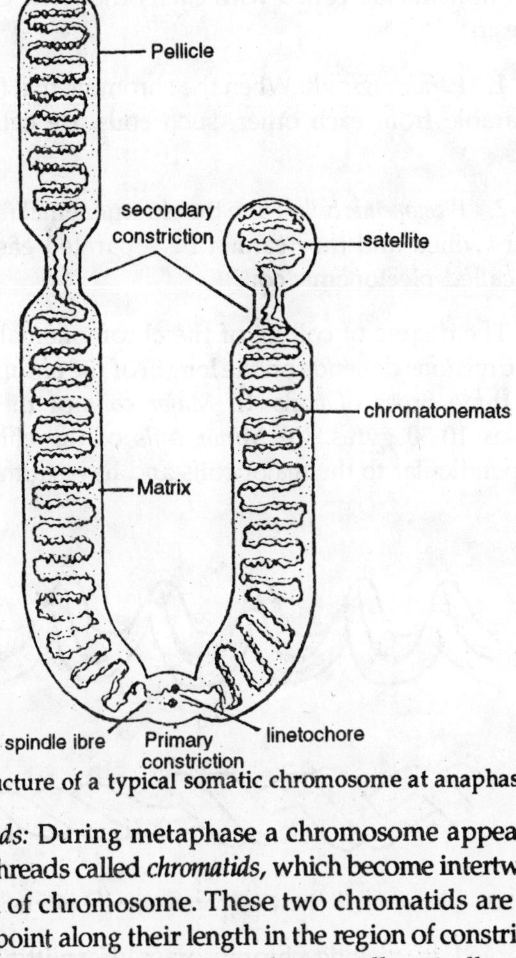

Fig. 3.2: Structure of a typical somatic chromosome at anaphase.

Chromatids: During metaphase a chromosome appears to possess two threads called *chromatids,* which become intertwined in the matrix of chromosome. These two chromatids are held together at a point along their length in the region of constriction of the chromosome. These chromatids are really spirally coiled *chromonemata* (sing., chromonema) at metaphase. The coiled filament was first of all observed by *Baranetzky*, in 1880, in the pollen mother cells of *Tradescantia* and was called chromonema by *Vejdovsky* in 1912. The chromonema may be composed of 2,4 or more fibrils depending upon the species. This number of fibrils in the chromonema may depend on the different phases since at one phase it may contain one fibril and at other phase it may contain three or four fibrils. These fibrils of the chromonema are coiled with each other. The coils are of two types:

1. *Paranemic coils:* When the chromonemal fibrils are easily separable from each other, such coils are called paranemic coils.

2. *Plectonemic coils:* Here the chromonemal fibirls are closely inter-twined and they cannot be separated easily. Such coils are called plectonemic coils.

The degree of coiling of the chromonemal fibrils during cell divisions depends on the length of the chromosome. There are three types of coils: (i) *Major coils* of the chromonema posses 10-30 gyres. (ii) *Minor coils* of the chromonema are perpendicular to the major coils and have numerous gyres as

(a)

(b)

Fig. 3.3. Diagram to show (a) paranemic (b) plectonemic coils.

observed in meiotic chromosomes. If splitting has not yet occurred at this stage, there will be a single chromonema, if it has already taken place there would be two chromonemata. (iii) *Standard* or *somatic coils* are found in the chromonema of mitosis where chromonemata possess helical structures, resembling the major coils of the meiotic chromosome.

Chromomeres: The chromonema of thin chromosomes of mitotic and meitoc prophases have been found to contain alternating thick and thin regions and thus giving the appearance of a necklace in which several beads occur on a string. The thick or bead-like structures of the chromonema are known as the *chromomeres* and the thin region in between the chromosomes is termed as the *inter-chromomeres.* The position of the chromosomes in the chromonema is found to be constant for a given chromosome.

The cytologists have given various interpretations about the chromomeres. Some consider chromomeres as condensed nucleo-protein material, while other postualted that the chromomeres are regions of the super-imposed coils. The latter view has been confirmed by the election microscopic observations. For long time most geneticists considered these chromosomes as *genes,* i.e., the units of heredity.

Centromere: It is an indispensable part of the chromosome and forms the *primary constrictions* of metaphase. Without centromeres chromosomes are unable to orient themselves properly on the metaphase plate; come, later on, at anaphase and ultimatley, are eleminated. As centromeres occupy a constant position, therefore, the centromeres are responsible for the shape of the chromosomes. Thus, the shape of the chromosomes is determined by the primary constriction, situated at the meeting point of the arms of the chromosomes. Within the primary constriction, there is a clear zone, having a small granule or spherule. This clear region is known as centromere (*Gr. meros,* part) or *kinetochore* or *kinetomere.* Its function is in terms of movement. They are responsible for the formation of chromosomal fibers in the spindle. The structure of the centromere is ovoid, non-stainable, with a large diameter, as in maize or may be like small granules or spherules, as in *Tadescantia*. In the centromere, there may be one or more small granules or spherules, called *chromomeres* and the spindle—bibres. Usually, each chromosome has only one centromere. In such cases, chromosome is called *monocentric*. There may be two i.e., *dicentric* or more *polycentric* or with a *diffused centromere,* as found in *Ascaris megalocephala* and Hemiptera.

After recent studies, it has been known that the centromere is made up of three zones present in duplicate. The middle zone maintain the relation of the chromosomes to the spindle. The diagram given below slows two sister-chromatids forming each metaphase-chromosome held by a region having a special divis

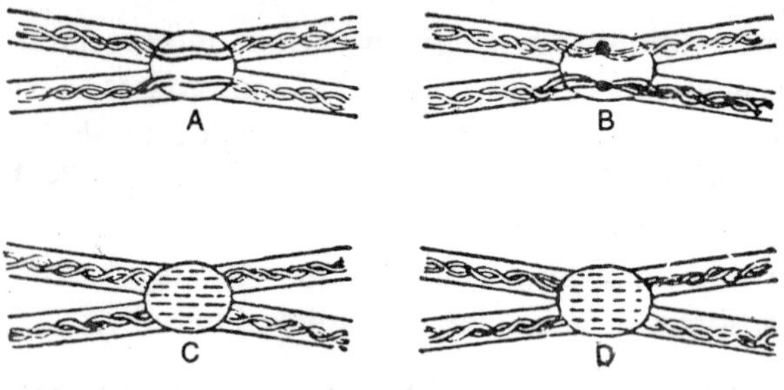

Fig. 3.4. Diagram, illustrating the organization of the centromere.ion cycle.

The centromere is regarded to become functionally divided to the longitudial axis of chromosome at the beginning of the anaphase. Its movements towards poles are governed by its attachment to the spindle. Sometime, divisions occur also at right angles to the longitudinal axis forming two segment-centromeres, to which are attached the two chromatids of each arm. This structure, made up of two arms, is known a *isochromosome;* the name was suggested by *Darlingtion* in 1939. Mc *Clintock* (1932) has reported that such breakage is also possible by x-rays. In such cases, each fragment-part of the centromere is functional. It is also known that the centromere is a compound structure, whose parts are coordinated in division and movement. *Schrader* (1936) and *Darlington* (1939) have suggested that the centromere can be considered to be homologous with the centrioles on structural as well as on theoretical basis.

Secondary Constrictions. In addition to the primary constriction or centromere the arms of the chromosome may show one or more secondary constrictions (called *secondary constrictions II)*. These are different from nucleolar organizers (called *secondary constrictions I)*, although some cytologists also refer to the nucleolar organiser as the secondary constriction. The location of the secondary constriction II is constant for a particular chromosome, and is, therefore, useful for identification of chromosomes. It has been suggested that the secondary constrictions represent sites of breakage and subsequent fusion. In man secondary constrictions II are found on the long arms of chromosomes 1,10,13,16 and Y.

Nucleolar Organiser (secondary constriction I): Normally in each diploid set of chromosomes, two homologous chromosomes have additional 'constrictions' called nucleolar organisers. These are so-called because they are necessary for the formation of the nucleolus. The nucleolus is formed in the post-mitotic reconstruction phase. Under the light microcope the nucleolar organiser appears as a 'constriction' near one end of chromosome. The part of the chromosome beyond the nucleolar organiser is very short and appears like a sphere *(satellite)*. In man chromosomes 13,14,15,21,22 and Y have nucleolar organisers and satellites. Chromosomes bearing satellites are called *SAT-chromosomes*. The prefix Sat-stands for 'Sine Acid Thymonucleinico' (without thymonucleic acid or DNA), since the chromosome on staining shows relative deficiency of DNA in the nucleolar organiser region. There are at least two SAT-chromosomes in each diploid nucleus.

Telomere: The ends of a chromosome act differently from the interstitial portions. If an end or telomere is broken off either spontaneously or by induction, it is usually lost from the nucleus in subsequent cell division because it lacks centromere. The broken end of the remaining free chromosome is unstable and may unite with another broken end of the chromosome in the vicinity. However, the broken end will not unit with normal end. In meiotic prophase the telomeres are sometimes attracted to the centriole and seen to migrate to the nuclear membrane near the centriole. This behaviour results in what has been described as *Bouquet stage*.

Matrix of Chromosome: As supposed by some cytologists that chromonemata are embedded in the chromatic matrix which is bounded by a pellicle. However, recent observation from electron microscopic studies revealed the absence of pellicle. Matrix is now defined as the main mass of the chromosome, which is characteristically Feulgen-positive. It can be removed by enzymatic means leaving behind a Feulgen-negative residual chromosome.

Heteropycnosis

It has generally been observed during various stages of mitosis that certain chromosomes or parts of the chromosomes are not as such, but are more condensed than rest of the karyotype. This refers to the *heteropycnosis*. This phenomenon results in clumping of the chromosomes during cell division. In animals pycnotic sets, on elimination, culminate in a male which was cytologically haploid. Heterpycnosis may be positive, followed by over dondensation or negative, showing under or no condensation. It has been also observed that a particular part of chromosome or entire chromosome may not exhibit the condensation or heteropycnosis in all the phases. Since heteropycnosis is peculiarity of the heterochromatin, it helps in demarcating it from the euchromatin. It is much more prevalent in sex chromosome, though others too exhibit it.

Euchromatin and Heterochromatin

Although, during interphase, the chromatin of the chromosomes spreads out in the form of the fine threads of linin but at certain regions, the cromatin remains condensed in the form of darkly stained chromatin mass. These regions are known as the heterochromatin regions or heterochromatin.

Heterochromatin is of two types:

1. Facultative heterochromatin, and
2. Constitutive heterochromatin.

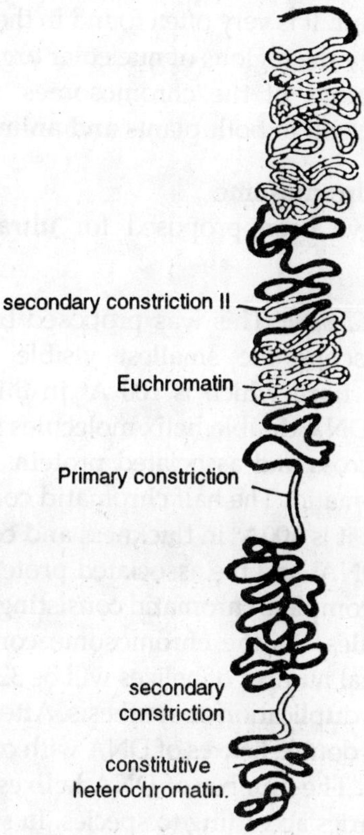

Fig. 3.5. Chromosome with heterochromative and enchromatin.

1. *Facultative heterochromatin:* This represents a temporary state of inactivation of chromatin in which one chromosome of the pair becomes partially or totally heterochromatic. For

example, in mammals one of the two X-chromosomes in female somatic cells becomes heterochromatic and forms the sex-chromatin of *Barr body* (*Barr* and *Bertram*, 1944). In male somatic cell, there is only one X chromosome and remains euchromatic (no barr body).

2. *Constitutive heterochromatin:* This type of heterochromatin presents a more permanent feature and is found in both the chromosomes of a pair. It is very often found in the centromeric regions, telomeres, in the regions of nucleolar organizers or as bands in other regions of the chromosomes. It is closely associated with nucleoli in both plants and animals.

Ultrastructure of Chromosome

Two views have been proposed for ultrastructure of chromosomes:

(a) Multistranded view: This was proposed by Ris (1966). By election microscope the smallest visible unit of the chromosome is the fibril which is 100 A° in thickness. This fibril contains two DNA double helix molecules separated by a space of 25 A° across and associated protein. Next largest unit is the halt chromatid. The half chromatid consists of four 100A° fibrils so that it is 400A° in thickness and contains eight double helixes of DNA and the associated protein. Two half-chromatids from a complete chromatid consisting of 16 double DNA helix molecules. As the chromosome consists of two chromatids, thus total number of helixes will be 32 and dimeter 1600A° thick before duplication or synthesis. After duplication chromosome has 64 double helixes of DNA with corresponding diameter of 3200A°. The number of DNA helixes in each unit about fibril level varies according to species. In summary, the chromosome is composed of numerous microfibrils, the smallest of which is a single nucleoprotein molecule.

(b) Folded-Fibril Model: DuPraw (1965) presented this model for the fine structure of chromosome. According to this model, a chromosome consists of single long chain of DNA and protein

forming what is called as fibril. The fibril is folded many times and irregularly entwinend to form the chromatid. This measure 250-300 A° in thickness.

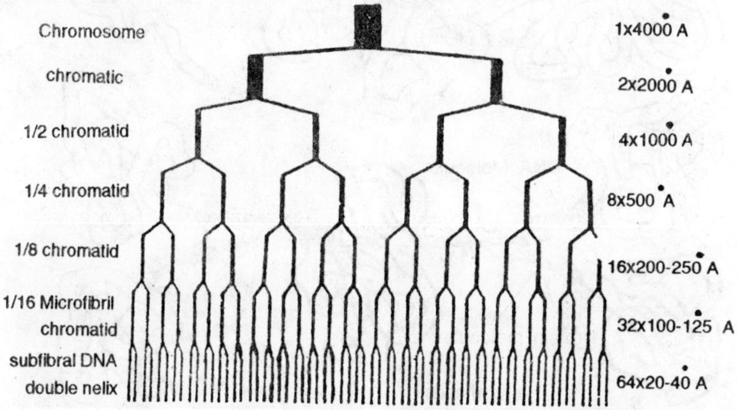

Fig. 3.6. Scheme of chromosome according to multiple strand hypothesis.

The Nucleosome sub-unit of chromatin: The chromatin is formed of repeating units, called *nucleosomes*. The term was given by *Oudet et al.,* (1975). The nucleosome is made of DNA and histone proteins. The proteins forms a *core particle* which is an octomer two molecules of each of the four histone proteins viz. H2a, H2b, H3 and H4. The surface of core particular is surrounded by 1.75 turns of DNA (200 base pairs). The DNA linking the core particle is called linker DNA. An another histone protein, H1, is bounded to the *linker DNA. (Kornberg* and *Thomas,* 1974). The core particle measures 40A° hight and 80A° in diameter. The entire nucleosome measures 55A° in hight and 110A° in diameter.

Salivary Gland Chromosomes

Balbiani in 1881 was the first to observe salivary gland chromosomes in the salivary glands of the *Chrionomous* larva. This kind of giant chromosomes are strictly confined to certain type of somatic tissues in the insects belonging to the order

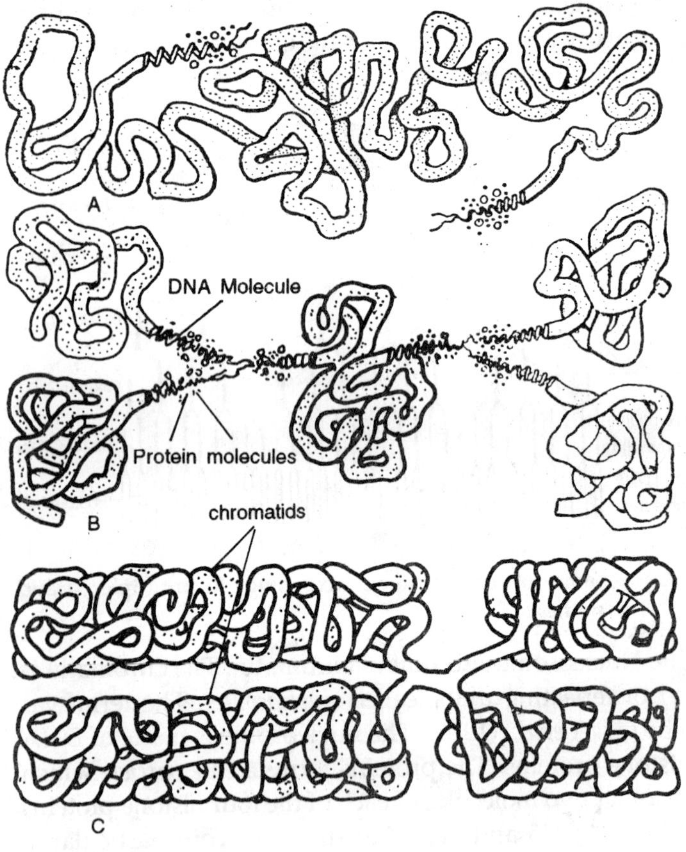

Fig. 3.7. DuPraw's folded fibre model of chromatin in interphase (A&B) and in Metaphase (C).

Diptera. Usually they attain their largest size in the spherical nuclei of the larval salivary gland, but similar nuclei frequently exist in other tissues such as the lining cells of the gut and its derivatives, the Malpighian tubules as well as in muscles and fat cells, etc. Thus the term "polytene" chromosome of *Koller* is more preferable than the general term of salivary gland chromosomes.

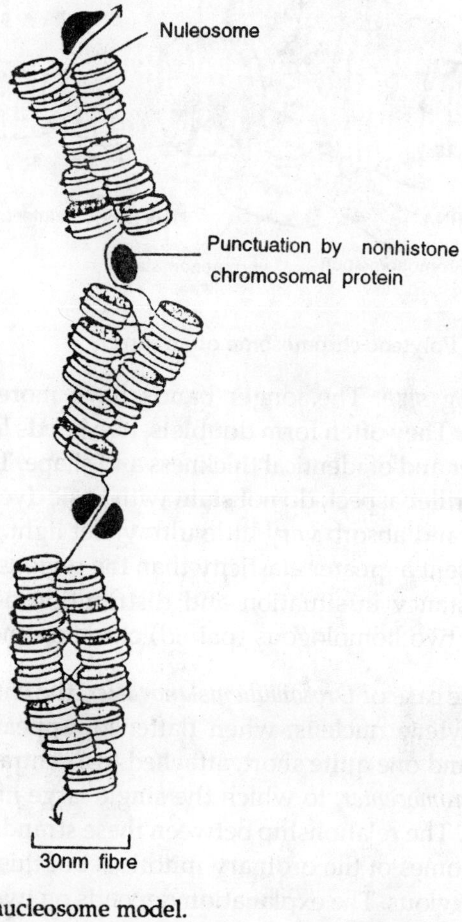

Fig. 3.8. The nucleosome model.

Structure

The structure of the salivary gland chromosome is of great cytogenetic interest. Along the entire length of the chromosome there is a series of *dark* bands alternating into other clear zones called *interbands.* The dark bands stain intensly and are Feulgen positive. Furthermore, they absorb ultraviolet light at 600A°. These bands may be considered as disks, since they occupy the whole diameter of the chromosome. They are

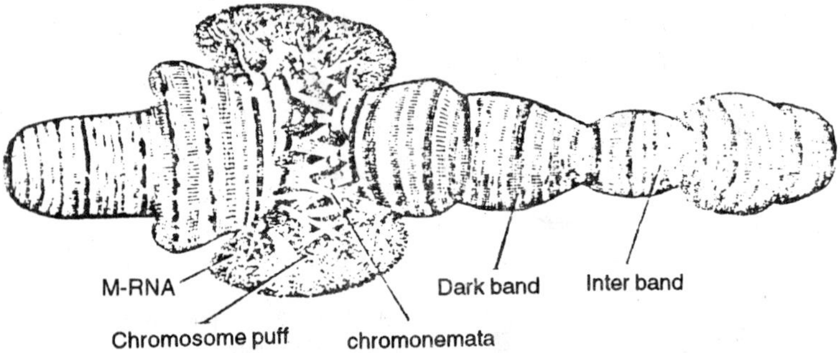

Fig. 3.9. Polytene chromosome of an insect.

of varying size. The longer bands have more complicated structure. They often form doublets, two bands located next to each other and of identical thickness and shape. The interbands are of fibriller aspect, do not stain with basic dyes, are Feulgen negative and absorb very little ultraviolet light. Furthermore, they present a greater elasticity than the regions of the bands. The constancy in situation and distribution of the disks or bands in two homologous (paired) chromosomes is notable.

In the case of *Drosophila melanogaster*, the chromosomes of each polytene nucleus, when flattened, appear as fine long strands and one quite short, attached to a central mass known as the *chromocenter;* to which the single large *nucleolus* is also attached. The relationship between these strands and the light chromosomes of the ordinary mitotic set of this species is not at first obvious. The explanation depends on two facts: (1) The two members of each pair of chromosomes are closely fused through out their length; (2) The centromeres of all the chromosomes together with the heterochromatic segments adjacent to them are all joined together to form the chromocenter.

Thus, of the six strands the short one represent the two fused IVth chromosomes, and longer one represents the X chromosome, while the remaining four are the limbs of the 'V'-shaped second, and third chromosomes. In salivary gland nuclei from female larvae, the strand representing the 'X' is double, like the others, while in nuclei from male individual it is single. The V being quite small and almost completely included in chromocenter.

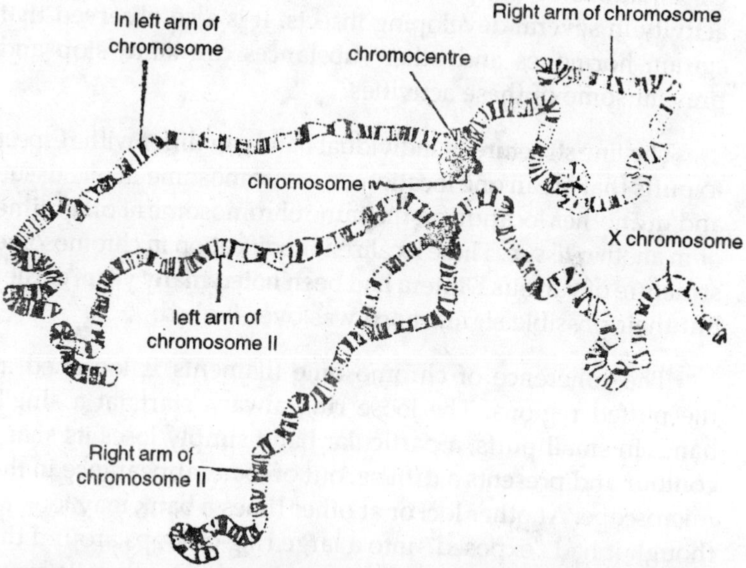

Fig. 3.10. Showing a Giant chromosome from salivary gland of the larva of the fruit fly.

Chromocenter occurs in all species of *Drosophila*, and its size depending on whether the proximal heterochromatic segments are extensive or not. In some other groups of Diptera, of the families 'Sciandae' and 'Chromonidae' a chromocenter is absent.

Puffs and Balbiani Ring and Gene activity: The most important morphological peculiarity of salivary gland chromosome is the presence of bands and interbands. *Brewer, Pawan, Beermann, Mc Chelke* and others have found that at certain stages of larval development some specific hands of the polytene chromosome show an enlargement. These enlarged bands are considered as ultimate units of heredity—the *genes* at work.

These active genes take the form of puffs scattered hear and there along the salivary gland chromosomes. *Beermann* and *Clever* (1964) have found that the puffs produce RNA and the RNA made in one puff differs from the RNA of other puff. Observations of the puffs have shown the patterns of gene activity in several developing insects. It is also observed that certain hormones and other substances can start, stop and prevent some of these activities.

The fine structure of individual band can differ with respect to puffs that are in one location on a chromosome in one tissue and in another location on the same chromosome at other time or in another tissue. These localized modification in chromosome structure of various Diptera had been noted many years earlier but their possible significance was over-looked.

The coherence of chromosome filaments is loosened at the puffed regions. The loose ring always starts at a single band. In small puffs, a particular band simply loses its sharp contour and presents a diffuse, out of focus appearance in the microscope. At other loci or at other times a bank may look as though it had "exposed" into a large ring of loops around the chromosomes. Such nut like structures are called *Balbiani-rings*, after *E.G. Balbiani,* who first described them in 1881; Puffing is thought to be due to the unfolding or uncoiling of individual chromosomes in a band. On observing that specific tissue and stages of development are characterized by definite puff pattern, *Beermann* (1952) postulated a particular sequence of puffs represents a corresponding pattern of gane activity. Of differential gene activation does in fact occur, one could predict that genes in a specific type of cell will regularly puff whereas the same gene in other tissue will not puff.

A gene of the same nature has been described in a group of four cells of salivary glands of *Chironomous. Chironomous pallidivittatus* produces a granular secretion. The closely related species *Chironomous tentatus* gives off a clear, non-granular secretion from the same cells. In hybrids of these two species

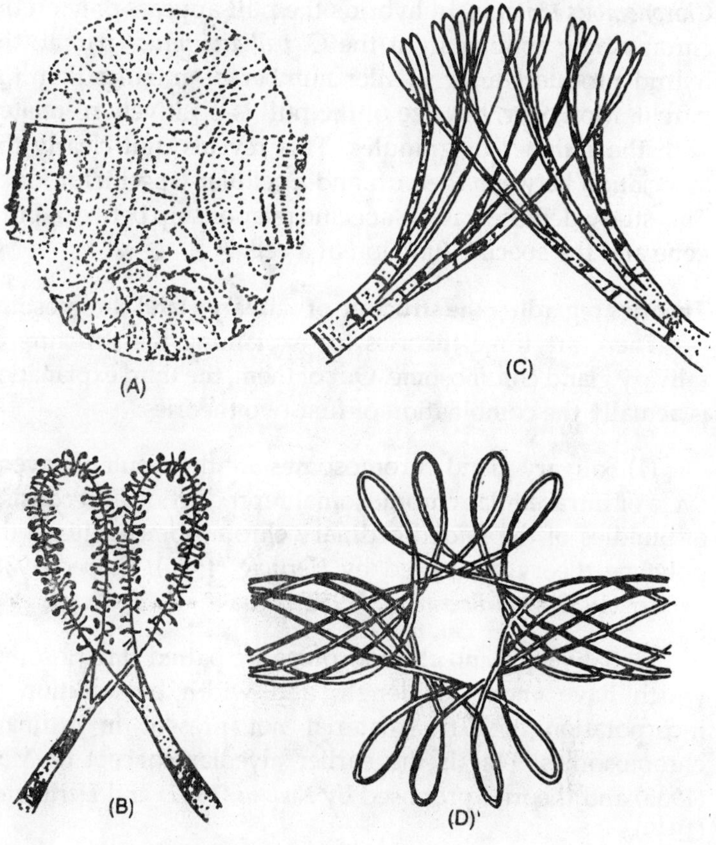

Fig. 3.11. Diagrammatic representation of salivary gland chromosome showing puffs and their formation. (A) showing the large puffs (B) showing the individual loop (C) and (D) fibres and he formation of puff.

this nature follows simple Mendelian laws of heredity. *Beermann* and *Clever* (1964) were able to localize the difference in a group of fewer than 10 bands in one of the chromosomes of *Chironomous* and that chromosome is designated as IV chromosome. The granule producing cells of *C. pallidivittatus* have a puff associated with this group of bands, a puff that is entirely absent at the corresponding loci of chromosome IV in

Chironomous tentatus. In hybrids the puff appears only on the chromosome coming from the *C. pallidivittatus* parent; the hybrid produces a far smaller number of granules than the parent. Moreover, the size of the puff is positively co-related with the number of granules. This reveals quit clearly the association between the puff and a cellular (specific)product. This study demonstrates a specific relationship between puff-gene and the specific function of a cell.

Theories regarding the structure of salivary gland chromosome

There are three theories for explaining the structure of salivary gland chromosome. Out of them, the third explanation is actually the combination of first two theories:

(1) Salivary gland chromosomes are the result of several cycle of intracellular chromosomal reproduction and consists of bundles of the folded ordinary chromosomes. This is the polytene theory sponsored by *Hertwig* (1935), *Cooper* (1938) *Painter* (1939) and *Beermann* (1952).

(2) Salivary gland chromosomes are paired chromosomes which have enormous length and width by addition or incorporation of extra material not present in ordinary chromosomes. This is the earlier alveolar concept of *Metz* (1935) and theories proposed by *Kodani* (1942) and *Darlington* (1949).

(3) Salivary gland chromosomes consist of bundles of chromonema, their size is due at least in part to the accumulation of extra material in the center of chromosomes, and or to an actual growth in length of the chromonema (*Koltzof,* 1934; *Painter,* 1934; *Calvin et.al.,* 1940; *Ris* and *Course,* 1954; *White,* 1945).

Polytene Theory: Painter (1941) believed that the increase in diameter is due to an enlargement and probably by a continuing duplication of the individual chromomeres but without a variable separation of individual chromonemata. Thus, in the course of development each original chromomere becomes resolved by separation through stretching into a number of smaller chromomeres. The enlargement, duplication and aggregation of homologous chromomeres produce the appearance of transverse chromatic bands. The chromosome, therefore, become multistranded as polytene but the individual chromonemata, which according to *Painter* may be as may as 1024, while *Beermann* (1952) estimated the degree of polytene to be as high as 16000 times.

Support to Polytene theory from the Volume of DNA. One important point of evidence may be added in favour of the polytene theory. This concerns the DNA contents of salivary gland nuclei and is complementary to Hertwig's (1935) demonstration of nuclear size classes. *Kurnick* and *Herskowitz* (1952) have estimated from photometric measurements of individual nuclei stained with methyl green that the large salivary gland nuclei in *Drosophila melanogaster* have undergone, about 420 folds increase in DNA content over that the diploid nuclei. Cosrecting few possible errors in the determination of the lowest values, *Kurnick* and *Herskowitz* assume that 8 cycles of internal duplication have resulted in chromosomes consisting of about 1000 strands. *Swift* and *Rasch* (1954) have obtained much more numerous data by photometric Feulgen day determination as nuclei of *Drosophila* salivary gland cells and closely associated cell type. They found a geometrical series of 10 DNA classes, ranging from presumably diploid cell to 1024 ploid salivary gland cells. The DNA values fall into generally non-over lapping groups, indicating that they resulted from a series of successive synchronized duplication.

Although a good number of workers believe in the polytene theory of salivary gland chromosome but several interesting problems remain unresolved. One of these is concerned with the question to what extent different chromosomes in one nucleus or even different regions within a chromosome, can become polytene to different degrees. The "sex ratio" phenonmenon analysed by *Sturtevart* and *Dobzhansky* (1936) represented a well substantiated case of this type. Other class which suggest this possibility have been reported by *Schultz* (1941) and *White* (1948) who discovered that certain chromosomes appear to be out of steps with the rest of the set when judged by their sizes and stainability. It will be eventually determined by means of photometric DNA measurements whether these cases are due to unequal polytene or merely to abnormal states of contraction of certain chromosomes. This later possibility rises during their development. The mechanism of different regions of the same chromosome is still completely obscure. The cause of the break down of certain polytene chromosomes into their constituent fibrils are equally unknown.

Lampbrush Chromosomes

Throughout the vertebrate group of animals the somatic chromosomes present the usual structure, but within developing oocyte of these vertebrates which posses a yolky egg, and during the diplotene stage of meiosis, the same chromosomes undergo a remarkable change, especially characterized by an enormous increase in length and the appearance of radiating hairs or side loops which appear to organize themselves from the chromomeres. This gives a brush-like appearance during meiotic prophase. This type of chromosome was first described by *Flemming* (1882) and *Rukert* (1892) gave the famous name, 'Lampbrush." *Ris* (1951) found similar chromosomes in sharks, birds, amphibians etc. Sometimes these chromosomes attain maximum size up to 800 to 1,000 μ per chromosome.

The lamphbrush chromosome possesses a central chromosomal axis from which project a series of lateral loops. The loops appear to project out from the dense region. According to *Duryee,* each chromosome resembles a single plastic cylinder in which, at specific loci, are embedded chromatin granules. In a bivalent chromosome about 150-200 paired granules are present. These granules are of two sizes, viz., *smaller chromioles* and *larger chromioles,* latter ones are ellipsoidal in appearance as though squeezed inside the matrix.

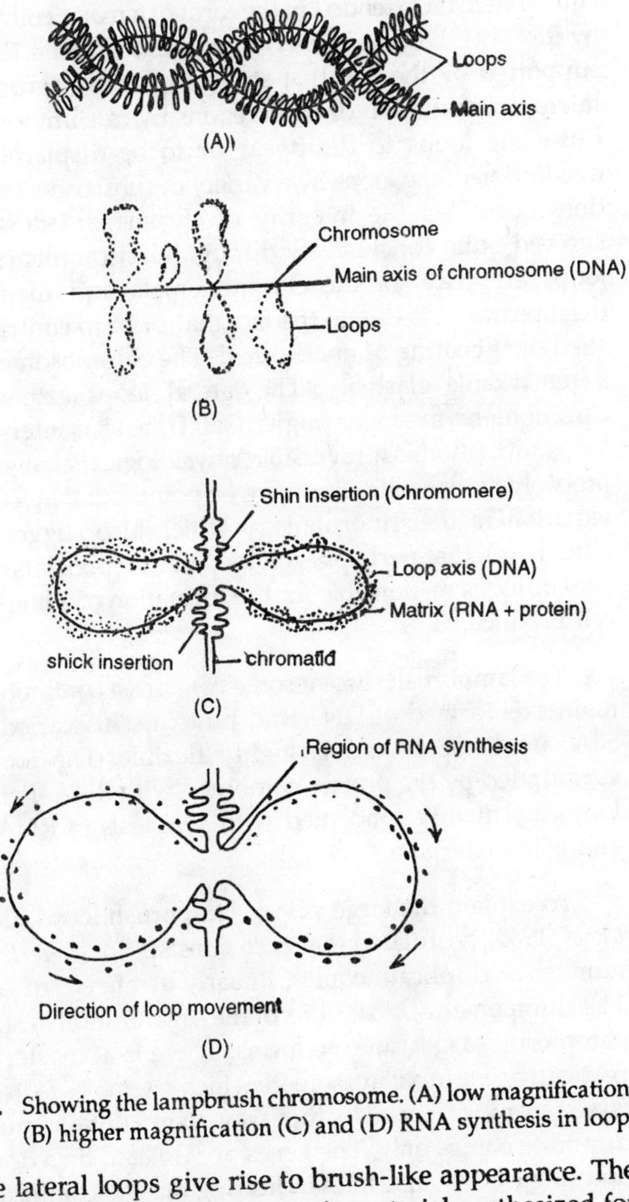

Fig. 3.12. Showing the lampbrush chromosome. (A) low magnification, (B) higher magnification (C) and (D) RNA synthesis in loop.

The lateral loops give rise to brush-like appearance. The loops are considered to be chromatin material synthesized for external utilization, and not an integral portion of the chromonemata extended in the form of a major coil as proposed by *Ris* (1945). *Duryee* (1941) hypothesis of lateral synthesis is supported by the fact that stretching of the chromosome by micro-manipulation or contractions by calcium ions does not cause the loops to disappear or to be displaced and that dissolution of the loops by a variety of substances into granules does not effect the integrity of chromonemata. *Gall* (1956) showed quite conclusively through electron microscopy that loops are parts of the chromonemata and their apparent disappearance is due to the fact that prior to contraction they shed their coating of nucleic acid. The chromosome possesses a remarkable elasticity. The lateral loops extending from chromomeres are more fragile. Gall (1958) has interpreted that loops formation is a reversible physiological change which is probably non-genetic. However, he points out that loops exhibit variation in their morphology which also suggests, on the other hand, that perhaps each loop pair, represents a different genetic locus responsible for the formation of some particular cell product.

The lampbrush chromosome contains a continuous central main zxis formed of DNA and protein. This varied from 30-50A° in diameter and is highly flexible. The loop axis is surrounded by the protein combined with RNA. Perhaps, the loops are mainly concerned with synthesis of RNA, protein and yolk material.

To explain the large size of lampbrush loops, *Callan* and *Lloyd* (1960) postulated that each consists not of gene, but of a number of duplicate copies, linearly arranged, of one gane. The chromomeres, basic units of the organization in lampbrush chromosomes exist in two forms. There is a 'master' copy of particular gene in a chromomere which resembles to its identical 'slave' copy of gene. In this way loop contains number of duplicate copies only. The genes are isolated by a double line and single transverse lines indicating the ends of the duplicate

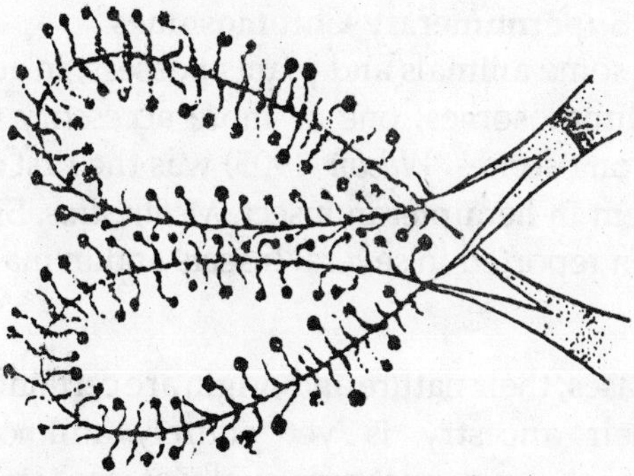

Fig. 3.13. Loops in chromosome puffs.

copies. *Gall* and *Callan* observed that the lateral loops always have one thin and one much thicker end at the point of insertion in the chromomere. It is also believed that loop spun out from the chromomere reunites to it at the thick end, which shows heavy accumulation of RNA. *Callan* further suggested that only 'slave' copies take part in RNA synthesis.

This ensures possibilities to synthesize large amount or ribose nucleic acid.

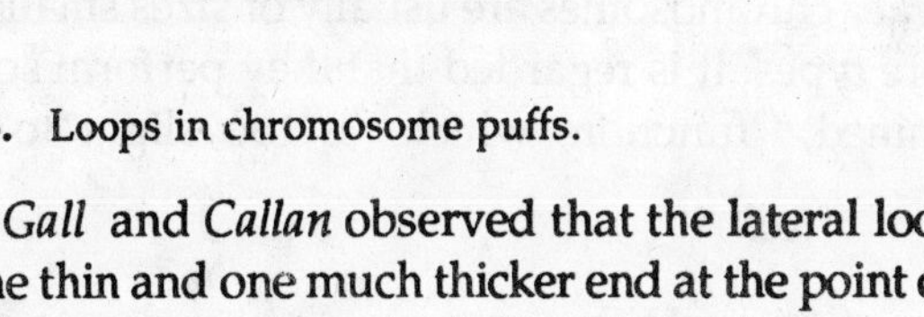

Fig. 3.14. Model of lampbrush chromosome.

Nucleoli formation in lampbrush chromosome shows an unusual pattern. There may be several hundreds of nucleoli floating free in the nucleoplasm. The significance is not well understood, but it is suspected that they must be synthesizing material for growth.

Accessory or Supernumerary Chromosomes

Nuclei of some animals and plants possess, in addition to the normal chromosomes, one or more accessory or supernumerary chromosomes. *Wilson* (1905) was the first cytologist to observe them in hemipteran insect, *Metapodius*. Since then, they have been reported in several insects and in many higher plants, too.

In some cases, their nature and origin are certainly known. However, their ancestry is yet entirely unknown. The supermunerary chromosomes are usually of sizes smaller than those of there types. It is regarded that they perform some, as yet undermined, function, which is too slight to detect genetically.

4

Cell Division

There are two types of organisms-unicellular and multicellular. The growth and development of an individual depends exclusively on the growth and multiplication of the cells. It was *Virchow* who first of all adequately stated the cell division. In animal cell the cell division was studied in the form of segmentation division or leavage by *Prevost* and *Dumos* in 1824. The mechanism of cell division was not precisely investigated until long afterward but *Remak* and *Kolliker* showed that the process involves a division of both the nucleus and the cytoplasm. The term *Karyokenesis* was introduced by *Schleicher* (1878) to designate the changes of nucleus during division, and the term *cytokinesis* was introduced by *Whiterman* (1887) to designate the associated changes taking place in the cytoplasm.

Cell division is necessarily the avoidance of aging, and secondly for the segregation of an individual into semi-independent units which leads to efficiency. Thus, we see that cell division is a widespread phenomenon that is essential not only for the maintenance of life but also for the development of the organism itself.

Cell division can be conveniently described as:

(i) Direct division: Where the nucleus and cell body undergo a simple mass division into two parts. It is also called *amitosis*.

(ii) Indirect division: Here the nucleus undergoes camplicated changes before it is divided into two daughter nuclei.

Amitosis

The amitosis or direct cell division is the means of asexual reproduction in unicellular organisms like bacteria and protozoans and also a method of multiplication or growth in foetal membranes of some vertebrates. In amitosis type of cell division the splitting of nucleus is followed by cytoplasmic constriction.

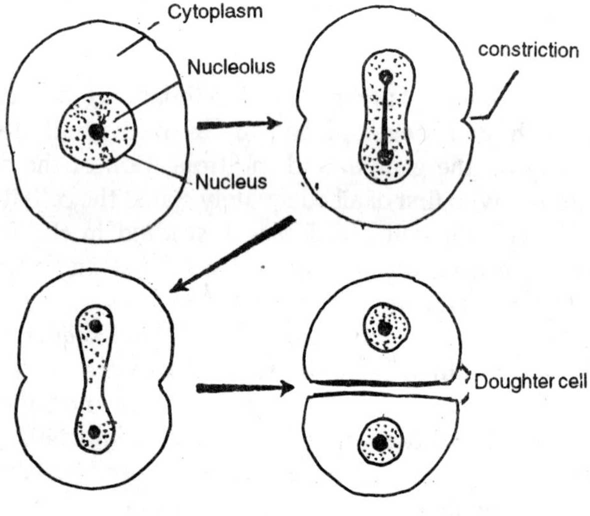

Fig. 4.1. Showing amitosis.

During amitosis the nucleus elongates first and then assumes dumbbell-shaped appearance. The depression or constriction increases in size and ultimately divides the nucleus into two nuclei. The division of nucleus is followed by the construction of cytoplasm which divides the cell into two equal or approximately similar halves. Therefore, without the occurrence of any nuclear event two daughter cells are formed.

Mitosis

in Mitosis, one cell divides into two, both of which are genetically identical to each other and to the parental cell. In other words, both the chromosomes and the genes are the same in all of the cells. This type of all division is necessary if the organisms and/or cell is to persist and survive. There are many facts to the need for cell division, and they very depending on the particular biological function. For example, in liver tissue when some cells die or are damaged, other divide and provide new cell to replenish those that are lost. Other cells in the organism actually grow (increase in size), and it is possible that when they reach a point where there is too much cytoplasm far a given amount of nuclear material, they divide and the whole process begins again. The phenomena of growth also involves an increase in numbers of cells. The increase in size of a tissue or organ is often due to numerical increase in cells rather than an increase in cell size. When subjected to the appropriate environmental and biochemicals signals, these cells may be stimulated to differentiate upto specific cell type. The sum total is that, as a result of all division, a certain degree of plasticity and immortality is provided to the organism. When the plasticity is lost, the organism undergoes the process of aging yet when the process of division is out of control, the organism literally "grows" to death!

The process of mitosis is a continuous one but one the basis of certain changes the same can be divided into several phases or stages.

Characteristic Features of the Mitotic Phases or Stages

In call division, the first visible changes occurs in the nucleus, when the chromonemata condense into *chromosomes* (Coloured bodies). This stage-is-ealled the *Prophase* (Gr. early figure).

Then the nuclear boundary disappears and the chromosomes line up in or near one plane in the cell. This is *Meataphase* (Gr; middle figure).

Following this each chromosome separates into two parts, and these two parts migrate away from each other to the ends of the cells. This is *anaphase* (Gr; up and down figure).

Next the chromosomes at each end of the cell reconstitute a nucleus. This *Telophase* (Gr. final figure).

Then the cytoplasm divides into two daughter calls, bringing about the *Interphase* again.

Interphase: The period of metabolic activity during which call division is not in a process, has been called the 'interphase.' This is frequently referred to as the 'resting phase.' But this term is not appropriate because the call is metabolically most active at this stage that is why *Berril* and *Huskins* (1936) referred it as the *energy phases*. The interphase is the period between the telophase of one division and the prophase of the new cell division. During this phase the cell does every thing except division. It is during interphase that the genes self-duplicate and carry on their function of supervising synthesis.

The chromatin granules in the nucleus are not readily distinguishable in the living cell, but may be brought out by treatment with chemicals which kill, fix and stain them. They appear at first glance to be scattered throughout the nucleus, but careful study has produced evidence that they are arranged in a definite pattern as long coiled strands and appear as a thread like net work in ordinary stained preparation.

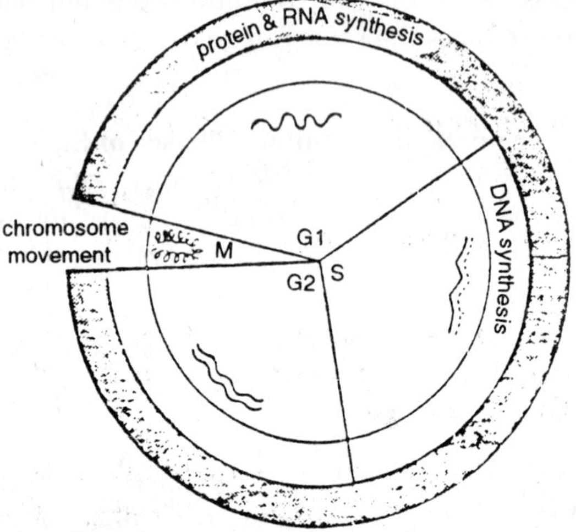

Fig. 4.2. The cycle of events in mitosis.

"Nuclear sap" or, "Karyoplasm" fills the interstics between the chromosomes. One of more rounded bodies, the *nucleoli* are usually present. Between the nucleus and the surrounding cytoplasm, is the *nuclear membrane.* In the cytoplasm adjacent to the nucleus, there is a body, the *central body,* which consists of two granules or, after each granule has replicated, of two pairs of granules, the *centrioles.*

A typical cell cycle, including interphase lasts from 20-24 hrs. Interphase is the longest period in the cell cycle, and may last for several days in cells.

Interphase can be further divided into form sub-phases:

(1) G_1-phase
(2) S-phase
(3) G_2-phase
(4) M-phase

G_1-phase includes the synthesis and organization of the substrate and enzyne necessary for DNA synthesis. Therefore, G_2 is marked by the synthesis of RNA and protein.

G_2-phase is followed by the S phase where the synthesis of DNA occurs.

During G_2 phase, all the metabolic activities are performed M phase is the period of chromosomal division.

The relative lengths of these phases differ in different organisms. A human cell in culture at 37°C, completes the mitotic cycle in about 20 hours and the M-phase lasts for only one hour. Temperature and cell environment plays an important role in determining the rate of cell division. Even the non-meristematic cells can sometimes be made to divide by changing the environmental conditions. Those cells which are not going to divide any more, have the mitotic cycle at the G_1 phase and start differentiating.

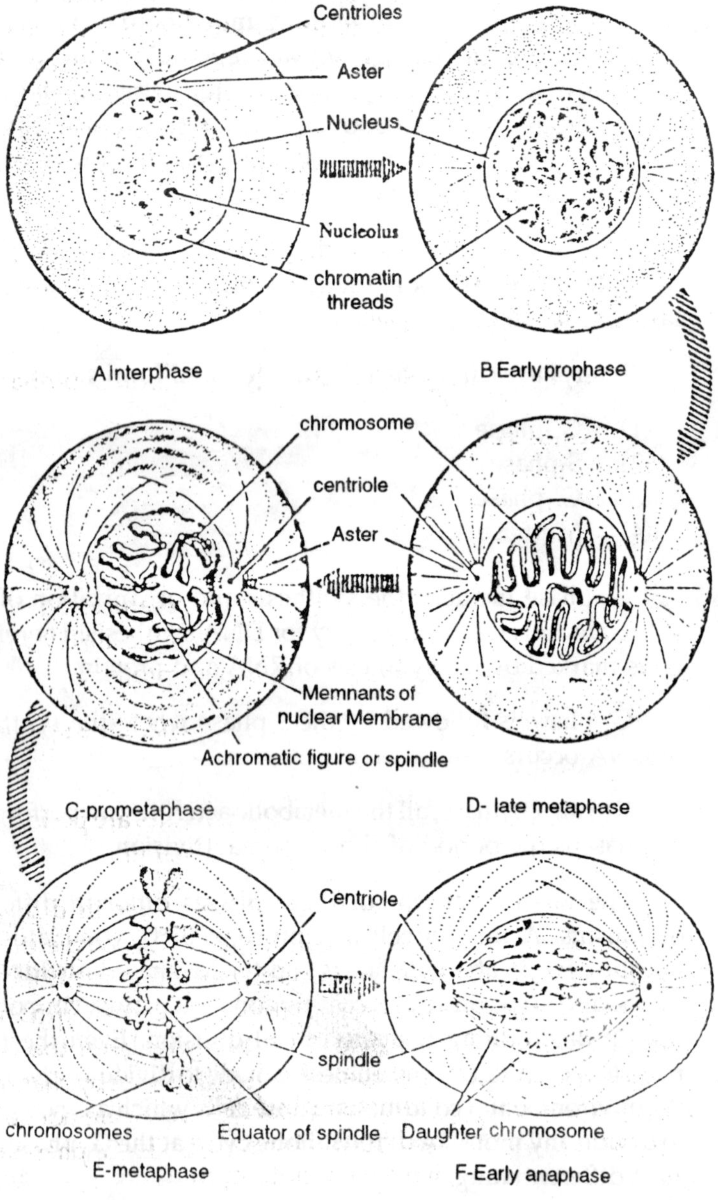

Fig. 4.3. Some early stages of mitosis.

The cell shows following changes:

(1) The cell, as a whole, attains the maximum growth and possesses synthesized proteins for energy for various divisions and processes.

(2) The nuclear membrane is intact and the chromosomes are found in the form of more or less loosely coiled threads, some-what closely appressed to the membrane. In this condition of chromosomes, most of the cytologists regard them to be duplicated, while some workers are of the opinion that they are multipartite.

(3) The two centrioles, which are found at right angles to each other replicate into two each. *Mazia* (1961) has described that, if the replication is checked, then division will not take place.

(4) For the future spindle condensation of protoplasm into a coherent area of jelly-like consistency also takes place. The spindle also starts growing and pushes the centrioles apart.

(5) DNA sythesis occurs during autosynthetic interphase, when the chromosomes are dispersed.

(6) Chromocentre are also conspicuous during the interphase.

Prophase: The prophase is the longest in M-phase and may take about one to several hours. In the neuroblast cells of grasshopper it takes about 102 minutes. Various important changes in this stage are as follows:

(1) The cell tends to become sphenical and increases its refractivity and turgidity by increasing its surface tension by removal of water.

(2) The nucleus imbibes water from cytoplasm, and the chromosomes get themselves organized differently. Each chromosome shows its peculiar structure with marked isolation

of chromatids, each of which undergoes a regular cycle of coiling. At the beginning of division the chromosomes start to contract, get thickened and undergo their coiling. Partially this change appears to be associated with the development, at right angles to old gyres of the coils. *Sparrow* (1941) referred that total end to end contraction is approximately one fifth of the initial length. Physiologically, the coils are formed due to continuous condensation. The coils are of two types: smaller minor coils and larger somatic coils. The mode of twisting during the formation of coils and gyres is also categorized in two distinct varieties :

(i) Plectonemic: Twisted in such a way that it is not easy to isolate them, and

(ii) Paranemic: Coiled chromatids can be separated laterally.

(3) The chromatids are connected with centromere. The chromosomes are separately distributed in the nuclear cavity. The RNA and phospholipid contents gradually increase.

(4) The nuclear boundary becomes disrupted, the spindle apparatus begins to form, and the nucleolus and the centromere generally disappear. The spindle formation takes place in two ways, viz.,

(i) Single centriole divides into two daughter centrioles and on separation, astral rays appear as delicate filaments, called the spindle. The centrioles migrate along with asters until they become located at antipodal positions. This type of spindle is referred as to the *central spindle.*

(ii) The two centrioles are already polarized before the beginning of the division and the spindle formation takes place at metaphase. This type of spindle is known as the *metaphasic spindle.*

The mitosis in which achromatic figure and spindle are formed by centers is called as the *amphiastral mitosis* and whence centers are absent, the mitosis is called as *'anastral.'* The anastral mitosis takes place in plants.

Prometaphase: This stage follows the complete diappearance of the nuclear membrane, the chromosomes tend to aggregate in a central position in the cell, near the equator. The term prometaphase is described by *Coin*(1964), *White* (1963) defined it as the period during which the spindle is being formed and during which the chromosomes give the impression of struggling and jostling one another in an attempt to reach the euqator of developing spindle. At least, in plants, this stage corresponds to the first appearance of an organised spindle. *Wilson* and *Hyppio* (1955) considered this positioning of chromosome to play an important role, both in the development and functional organization. The spindle fibers are tubular, elastic, fiberous and proteinous in nature.

Metaphase: The metaphase chromosomes are sharply defined and descrete bodies and are tightly coiled. In this stage the chromosome number can be easily counted and it is possible to recognise the different chromosomes from their sizes, shape and gross structure. In early metaphase characteristic spindle-shaped figure appears in the clear nuclear region. This consists of fibrous radiations which extend from the central broad-portion called the *equator* and converge at two points on the opposite ends of the nuclear region, called the *poles*. The chromosomes which were till now scattered in a random manner in the central nuclear portion of the cell begin to show peculiar movements and arrange themselves in a single plane midway between the two poles of the spindle, forming an equatorial plate. To be more precise it is the centromeres of the chromosomes that are lined up at the equatorial plate. Usually, the smaller chromosome occur near the center of the *equatorial plate* and the larger ones near the outer ends. It is, however, not necessary that the two homologus should be near each other on the equatorial plate. The position of each chromosome is independent of other on the equatorial plate. The metaphase chromosome is a visibly double structure and is clear seen to be split longitudinally into two exactly identical chromatids.

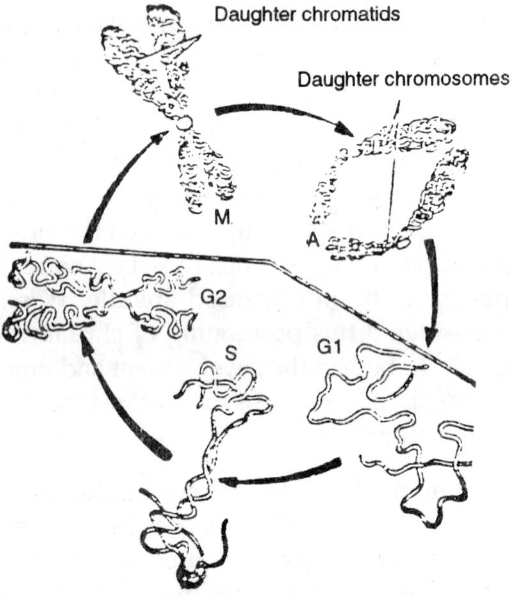

Fig. 4.4. Changes in chromosomes during cell cycle.

The centromere plays an essential role in the development of the spindle and the separation of the daughter chromosomes. The chromosomes are so arranged at the equator that one chromatid of each chromosome faces one pole and the other faces the opposite pole. Two spindle fibres are attached to the centromere of each chromosome, one on either side of it. These connect the chromosome to the two opposite poles of the spindle and are called *chromosomal* or *tactile fibres*. Other fibres of the spindle extend from pole to pole and are not attached to the chromosomes. These are called the *continuous spindle fibres*. The spindle fibres are composed mainly of protein, some ribonucleic acid (RNA) and lipids.

Form and arrangement of the chromosomes

The arrangement of chromosomes at the equator of spindle

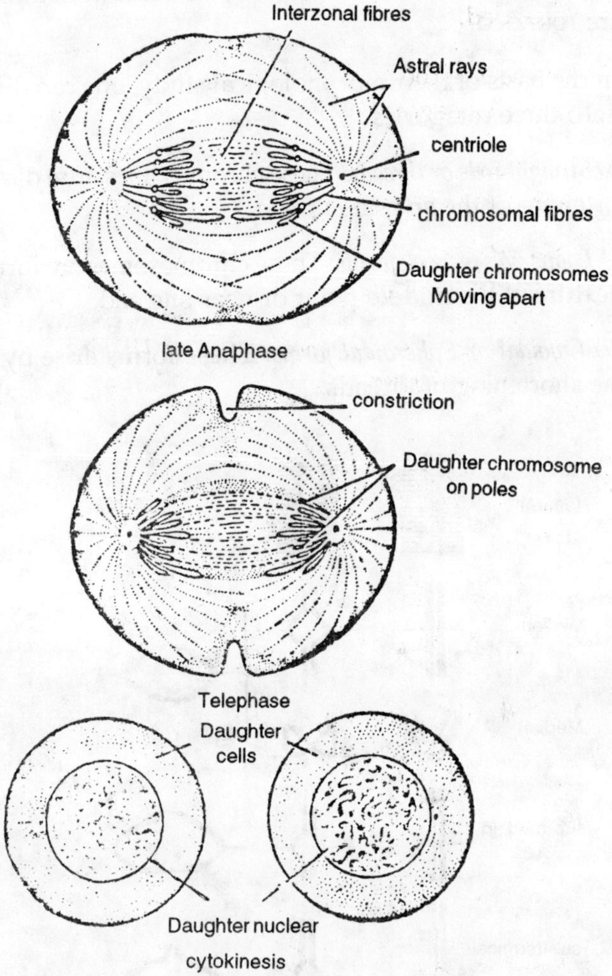

Fig. 4.5. Division of cell cytoplasma.

is not of one type in all organisms, as the arrangement depends upon the shape, size and number of chromosomes that differ in different organisms, because in some organisms they (chromosomes) are thread like [Urodela and in insects

(Orthoptera and Diptera)]. In Odonata, Coleoptera and Hemiptera they are shorter or rod shaped while in Orthropod they are rounded.

On the basis of their morphological study, one can divide them into three catagories :

(i) Straight rods or threads: These chromosomes arise directly by shortening of the spireme thread.

(ii) Loops, V's or hook forms: These chromosomes are formed by a flexture at the middle point or near one end.

(iii) Ovoidal or Spheroidal forms: These forms arise by the extreme shortening of threads.

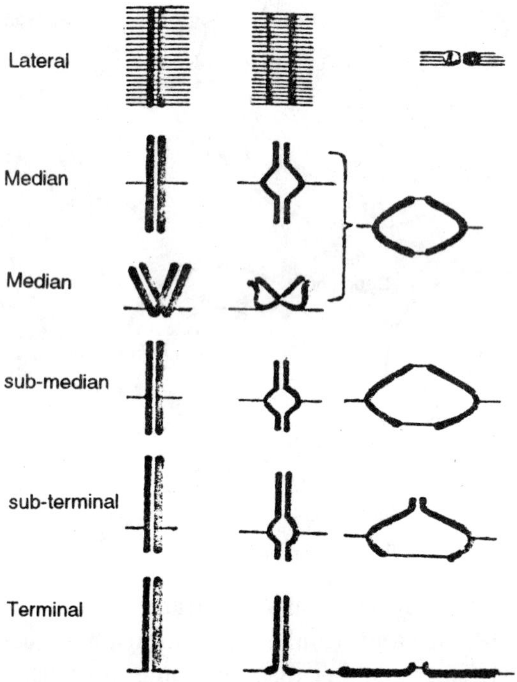

Fig. 4.6. Showing the movement of chromosomes according to their mode of attachment during anaphase.

All the three forms mentioned above are double at metaphase due to their longitudinal splitting. The longitudinal splitting among rod shaped and thread shaped chromosomes, can be traced clearly but in spheroidal forms, it often appears as an apparently transverse constriction due to the extreme shortening of chromosomes.

The Spindle attachment of the chromosomes : The mode of attachment of the chromosomes to the spindle fibres based on the structure of chromosomes: But even though, the arrangement and the way of attachment is constant for every particular chromosome and that it is inherited from generation to generation. Chromosomes are attached to spindle fibres by centromeres. The chromosomes lacking centromere fails to attach with spindle fibres. The way of their attachment may be of two types :

(i) Terminal or telocentric
(ii) Non-terminal or atelocentric

In terminal attachment the chromosomes may have their connection with spindle fibres at the free end. Non-terminal attachment may be at the middle point (*median*) or at an intermediate point *submedian* or sub *terminal.*

Anaphase: In anaphase the pairs of centromeres move apart along the spindle and carry one daughter chromosome of each pair to opposite poles. The spindle eventually grows longer. The movement of chromatids is complex. First separation of chromatid from centromere takes place and then a flow of current along the spindle governs their poleward movement. This stage lasts for very short duration, varying between 5 to 12 minutes. In the late anaphase the zone between the two sets of chromosomes or the equatorial region gradually increases. It seems that the fibers stretch and are named *interzonal fibres.* The expansion of middle part has been referred as the stemmkorper or pushing body by *Bear.* The pushing body looks like a gel which pushes the chromosomal sets towards respective poles. During the poleward movements the chromosomes assume peculiar 'J' or 'V' shapes, depending upon the position of centromere. At this stage, 'J' are called *heterobranchials* and the 'V' chromosomes as *isobranchials*.

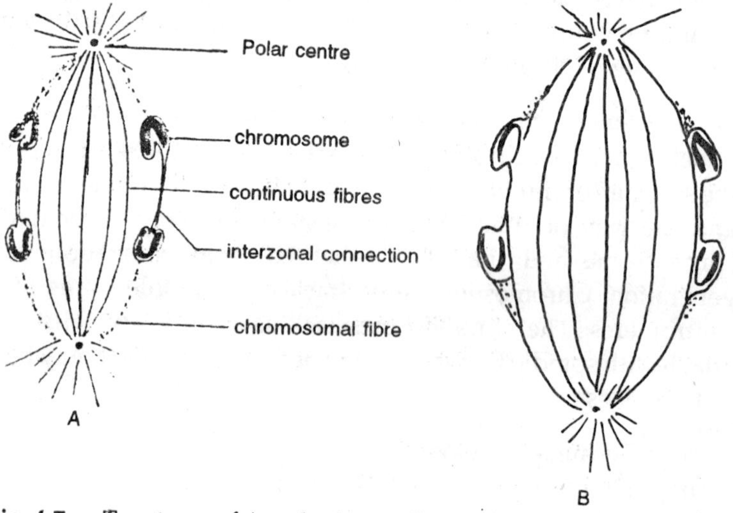

Fig. 4.7. Two types of Anaphasic spindles. A-Direct types B. indirect type.

Movement of chromosomes: The movement of the chromosomes are controlled by spindle fibres. Actually two processes may be at work; a continuous enlargement and elongation of the spindle and the shortening of the chromosomal fibres. Since the fibres shorten without becoming thicker, the process probably involves the removal of water or other molecules from fibres. A band of 'interzonal fibres' is often seen for a time after the separation has been accomplished, connecting the chromosomes which have pulled apart, and often including a remnant of the spindle.

Forces involved in the movement of Chromosoes: Spindle fibres are responsible for the movement of chromosomes during anaphase i.e., from equator to the pole. Several models have been proposed by different workers to explain about the forces involved in the movement. Some of the them are given below:

(i) Simple contraction model: Van Benden (1883) suggested that chromosomes in the dividing cell are pulled towards the poles by the contraction of spindle fibres. *Swann* (1962) described that centromeres secrete some substance which causes the contraction of spindle fibres. The main objection to the theory comes from the direct observations on chromosomes being carried even beyond the centriole to which the spindle fibres are attached. Further during cell division the whole cell elongates which is just opposite to the contraction model.

(ii) Expansion or pushing model: Watase (1981) suggested that the spindle fibres exert pressure on the nucleus and chromosome become flattened on the metaphase plate. The chromosomal fibres become attached to the chromosomes, which are now pushed towards the opposite poles.

(iii) Contraction and expansion model (Traction model): Belar proposed that initial separation of daughter chromosomes is an autonomous process but further separation is the result of both contraction or expansion of various parts of spindle. The chromosomal fibres which extend from the centromeres of chromosomes to the poles of spindle contract and pull the attached chromosomes to the poles. The interzonal fibres present between the separating daughter chromosomes expand and push the daughter chromosomes towards opposite poles.

(iv) Equilibrium dynamic model: This provides the most convincing explanation about the possible mechanism of chromosome movement. *Inove* and *Sato* (1967) have described the occurrence of an equilibrium between a large pool of monomeres which constitute the proteins of microtubules. During polymerization some hydrophobic interactions occur in between the non-polar groups of protein monomeres and some structured water is dissociated.

During chromosomes movement both contraction and elongation of spindle fibres occurs by the substraction or addition of new monomeres to the fibres. During anaphase chromosomal

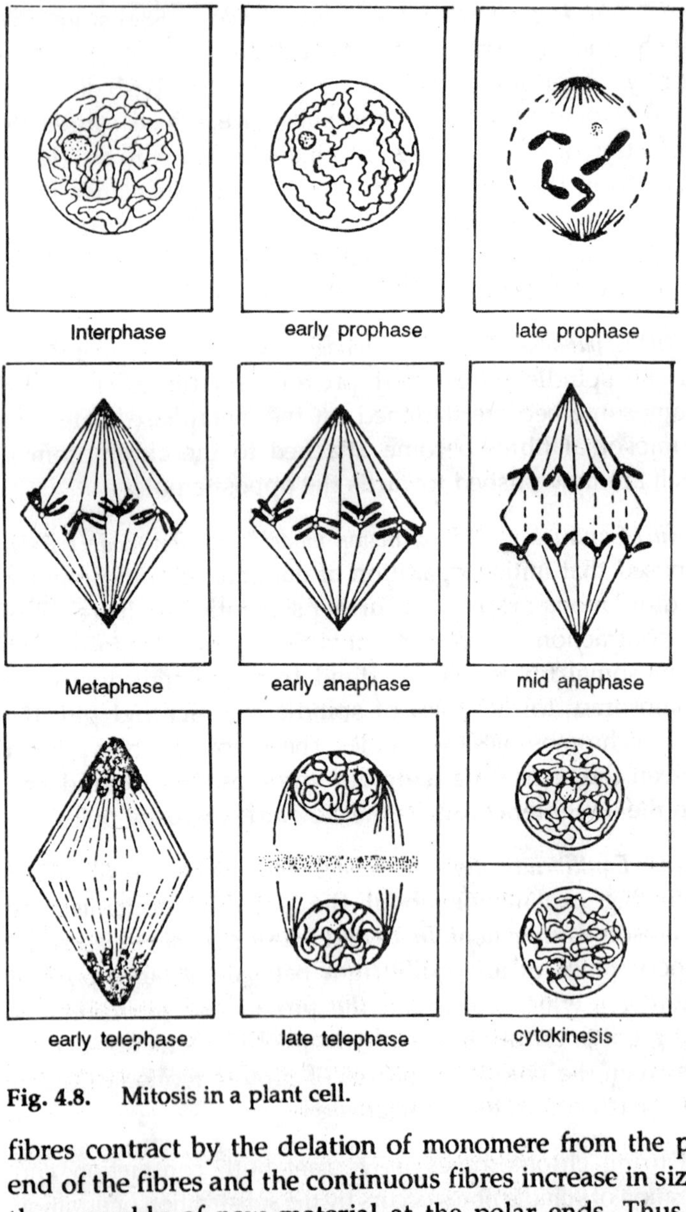

Fig. 4.8. Mitosis in a plant cell.

fibres contract by the delation of monomere from the polar end of the fibres and the continuous fibres increase in size by the assembly of new material at the polar ends. Thus, the spindle elongates and distance between the poles increases.

(v) Gliding model: Bajer, Ostergen and other are of the opinion that the movement of chromosomes is an active process in which chromosomal fibres glide as sail boats between the continuous fibres. *Ambrose* has suggested that motive force for these movements could be electroosmosis and electrohoresis. Breakdown of ATP is supposed to provide the magnitude required for spindle movements.

(vi) Ratchet model.: Mclnuosh, Helper, Van Wie and others have described the presence of mechanical bridges between the microtubules similar to the ratchet connections in the muscles fibres. The chromosome fibres glide-between the continuous fibres and draw the chromosomes apart. The energy is supplied by the break-down of ATP.

(vii) Electrical model: Lillie and *Coller* (1936) suggested that the change in the membrane potential is due to local changes in permeability occurs near the poles and around the nuclear membrane and produces electric field. The chromosomes are negatively charged in prophase and the charged chromosomes easily migrate in the electric field.

(viii) Push model: According to this concept chromosomes are pushed apart by the colloidal component of cell cytoplasm. The colloid imbibes water, swells up and pushes the two chromatids of the chromosomes apart. The spindle fibres act as tracks that guide the movement of chromosomes to the poles, preventing their scattering throughout the cell.

Telophase: The two groups of chromosomes aggregate at the poles and on arrival they progressively lose their chromaticity. Decondensation takes place. Reformation of the nuclear membrane by certain unknown process takes place. Perhaps new material is synthesized by RNA during telophase or it may be possible that the endoplasmic reticular system may be giving rise to a new membrane round the nucleus. All chromosomes become uncoiled. The spindle fibers are absorbed in the cytoplasm. The nucleolar organizer or SAT zone again forms the nucleolus. Finally, these changes result into two daughter nuclei which are equivalent to the parent nucleus in all respects.

Cytokinesis: Nuclear division or mitosis, as it is called, is followed by the division of the cytoplasm. While the daughter nuclei are being organised at the poles, the mitotic spindle disappears except at the eqator, where the continuous spindle fibres become more dense. This region is now called the *phragmoplast*. According to *Porter* and *Machade* (1960) the formation of cell plate is initiated by migration of tubular elements of the endoplasmic reticulum toward the interzonal region of the spindle where they spread out to form a close lattice along the mid-line. The droplets appear in phragmoplast and contain pectic substances which fuse to form in the middle of dividing cell a partition known as *cell plate*. More droplets are added to the cell plate to form the middle lamella which now begins to extend outwards until it reaches the outer walls of the original cell. Primary cell wall material is now deposited on either side of the middle lamella. This divides the cytoplasm into two roughly equal parts. Cell division is now complete and two cells are formed. The division of the cytoplasm is called *cytokinesis*.

The cytokinesis varies in plants and animals. In animal cells, the separation takes place by a process of invagination of the cell membrane in the eqatorial region. The cytoplasmic division takes place at the late anaphase stage. The invagination furrow originates as an indentation, gradually depens to cleave the cells into two daughter cells which may be of equal size or may be quite unequal. *Gray* (1931) reported about two types of cleavages and named them as *disjunctive* and *astral cleavages*. The disjunctive cleavage occurs in those cells where the daughter cells proliferate, e.g., leucocytes or in cells of the tissue culture. Astral cleavage is characteristic of the marine eggs where the zygote cells go on dividing to result in the formation of large number of bastomeres intact.

Role of Cetriole During Mitosis: The centriole acts as an

epicentre for the development of the central spindle. Whether the centriole produces the spindle fibres or serves to direct their formation or on the other hand is entirely passive in the process, cannot be unquestionable deduced from all the informations at hand. According to *Cleveland* (1971) the spindle fibres and astral and chromosomal fibres are recognizable first in the close vicinity of the centrioles. The coompletely developed achromatic appparatus then functions to separate daughter chromatids towards respective poles. With relation to the mitosis, *Cleveland* concluded two important points: (i) relative temporary independence of the achromatic and chromatic apparatus as regards apparent duplication and function; (ii) final dependence of the cell upon the integrity of these two systems and their united function.

TABLE 4.1

Duration of Mitotic Stages (in minutes)

Example	*Temp *C*	*Prop-hase*	*Meta-phase*	*Ana-phase*	*Telo-phase*
Mouse spleen	38	21	13	5	4
Chicken mesen-chime cells in tissue culture)	39	30-60	2-10	3-7	2-10
Frog (Fibroblasts in culture)	20-24	32	20-29	6-11	
Grosshopper (Neuroblasts)	38	102	13	9	57
Sea Urchin Embryo	12	19	17	2	18
Onion (root tip)	20	71	6.5	2.4	3.8
Pea (root tip)	20	78	14.4	4.2	13.2

Duration of Mitosis: The time required for mitosis differs with species and environment. Temperature and nutrition, in particular, are important factors. The entire sequence of phases may be completed in 6 minutes to many hours. Normally the entire cycle of cell division takes approximately 18 hours; about 17 hours for the interphase. Different phases of mitosis are of different duration. Anaphase is the shortest, the prophase and telophase the most prolonged, and the metaphase of intermediate duration.

Singnificance of Mitosis

1. *Equal distribution of chromosomes:* The essential feature of mitosis is that chromosomes are distributed equally among the two daughter cells. With every cell division there is a division of chromosomes. The constant number of chromosomes in all the cells of the body in because of mitosis.

2. *Surface-volume ratio:* Mitosis restores the surface-volume ratio of the cell. A small cell has a greater amount of surface available in relation to volume that a large cell. As the cell increases in size the available surface area in relation to increased volume becomes less. By undergoing division the cell becomes smaller in size and the surface-volume ratio is restored.

3. *Nucleoplasmic index:* The growth of a multicellulr organism is due to mitosis. A cell cannot grow in size to a large extent without disturbing the ratio between the nucleus and the cytoplam. After a particular size has been reached the cell divides to restore the nucleoplasmic index. Thus growth takes place by an increase in the number of cells, rather than by increase in the size of the cells.

4. *Repair:* The repair of the body takes place because of addition of cells by mitosis. Dead cells of the upper layer of the epidermis, cells of the lining of the gut, and red blood corpuscles are constantly being replaced. It is estimated that in the human body about 500,000,000,000 cells are lost daily.

Abnormalities in Mitosis

Under unfavounable conditions the mitotic divisions become defective and this various abnormalities are formed when exposed to physical and chemical agents like temperature, radiations, nacrotics and enzyme inhibitors etc. Some abnormal mitotic divisions are given under the following heads:

1. *C-Mitosis (abnormal spindle formation). Brachet* (1975) has described that colchicine inhibits mitosis by disorganizing spindle formation. The results are the formation of polyploid cells, and reduplication of chromosomes. Besides, during mitosis, sometime cell nucleus is either deformed into a single shperoid compact mass (pycnosis) or becomes broken down (karyorrhexis).

2. *Cytasteral mitosis: Wilson* (1901) noticed the formation of numerous asters in the cytoplasm of unfertilized eggs called cytasters by placing them in hypertonic sea water. In the eggs of marine invertebrates, production to many small cytoplasmic asters is of frequent occurrence. The intermixing of nucleoplasm and hyaloplasm seems to be necessary condition for the aster formation. *Costello* (1940) showed that asters can be produced only if germinal vesicle has broken down.

3. *Multipolar and catenar mitosis:* Mitotic division showing several spindles and centrosomes are common in many protozoans and fish eggs. Multipolarity is usually caused by uneven division of centrosomes as well as irregular distribution of chromatids of the different spindles. It results in the formation of cells as aneuploidy (uneven chromosome number) Catenar mitosis have been described by *Dalcq* and *Simon* (1932) in amphibian eggs. During this, dividing cell forms a large number of asters without spindles for chromosomes to link them. It suggests that asters are capable of dividing in the absence of spindle or nucleus in an autonomous way.

4. *Achrosomal motosis:* Sometimes, mitosis is induced by various agents without use of nucleus or chromosomes. *Briggs et al* (1951), in fertilized eggs with heavily x-rayed sperms, removed the egg's maturation spindle by pricking and sucking.

5. *Anastra mitosis:* In plant cells and many oocytes, generally asters around the centrosomes are absent, thus producing

anastral mitosis. *Bataillon and Tchou Su* (1933) described in details the anastral mitosis in amphibian interspecific hybrids.

Achromatic Figure

Achromatic figure consists of spindle fibers and astral rays. Astral rays are developed just before the disintegration of nuclear envelope during late prophase, while the formation of spindle rays can be traced only after the disappearance of nuclear membrane.

The spindle: In fixed preparations of dividing cells, fibers are seen forming a spindle shaped body lying between the centrioles. It is generally believed that spindle fibers are not visible in normal living cells. They can be produced by adding acid to the medium, and, on removing the acid, the fibres disappear; as most fixatives contain acid, it is stated that the fibers are artifacts. There has been considerable controversy as to the reality of fibers and various theories have been put forward to explain their presence in fixed material. *Schrader* in 1944 observed the spindle fibers in living cleavage cells of mite and certain flagellates.

The spindle is superficially similar to the spindle shaped area found between the poles in the magnetic field. The spindle consists of two kinds of fibers which have been supposed to differ widely in functional significance.

Those spindle fibers extending from the poles to the chromosomes are called half spindle, incompete fibers or discontinuous fibers or chromosomal fibers. The remaining spindle fibers extend without interruption from pole to pole and are called the continuous fibers or direct or complete fibers.

It is believed that the spindle is usually formed in two parts. A part arises between the centrosomes before the disappearance of the nuclear membrane; it is cytopolasmic in origin and is called the central spindle. As the nuclear membrane disappears the central spindle moves into nuclear region. Then the nuclear material undergoes gelation around the central spindle to from the 1/2 spindle fibers or chromosomal fibers. In the metaphase the chromosomes are attached to the 1/2 spindle fibers at the equator. *Schrade* suggested that the spindle may consist of protein molecules arranged in parallel lines.

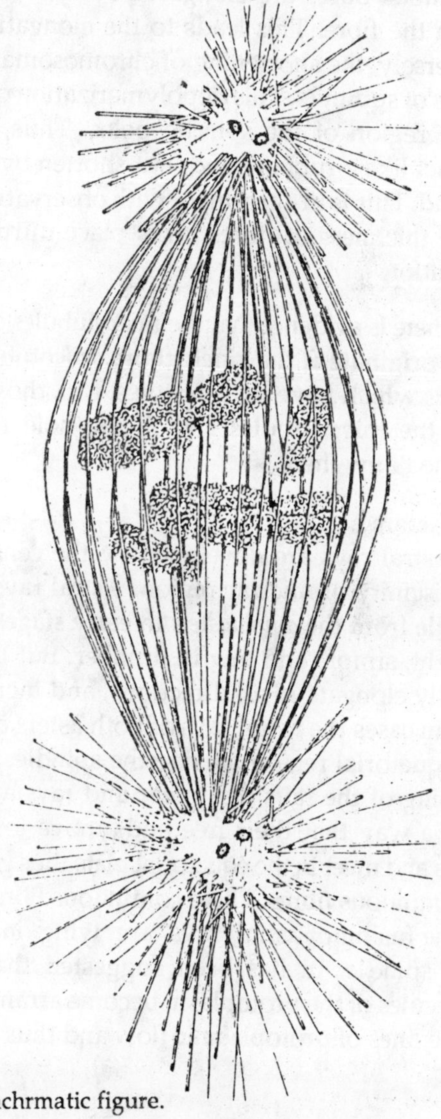

Fig. 4.9. Showing achrmatic figure.

It appears that the spindle is organized from a pre-existing pool of protein. It is postulated that protein monomers aggregate to from microtubules of the spindle fibres. The spindle fibres are thus oriented polymers. It has been proposed that the continuous fibres are elongated by the insertion of subunits within the fibre. This leads to the elongation of the spindle. Conversely, the shortening of chromosomal fibres takes place by loss of subunits. This depolymerization of fibres takes place in the region of the spindle poles. Thus, the fibres do not contract like a rubber band, but shorten by loss of material at one end. This is in agreement with observations that no change in the thickness of fibres takes place during shortening and elongation.

There is evidence that the microtubules of the chromosomal fibres originate at the kinetochore. Micotubules are also found in cells which have no centrioles, e.g., those of higher plants. Here the microtubules end in the pole region and are not attached anywhere.

The Asters

Astral rays are developed by the centrosome present in the viscinity of nuclear region. Astral rays together with the spindle from the *amphiaster*. In early stages the astral rays are straight, simple and few in number. But in latter stages they rapidly elongate in all directions, and increase in number. In certain cases the astral rays of both asters cross one another in the equatorial region outside the spindle. Sooner or later, the crossing of the rays disappear and rays adjust themselves in such a way that their free ends move towards the opposite asters and meet in the equatorial plate thus giving the impression of continuous fibres. These continuous fibres which are formed by the fusion of astral rays are lying in the region outside the spindle. In has been suggested that the long protein molecules of the protoplasm become arranged in rows parallel to the lines of protoplasmic flow and thus constitute the astral rays.

MEIOSIS

Meitosis is a device of increasing the number of cells. During a sexual reproduction it is the means of multiplying the individual. Therefore, in a sexually reproducing organisms, set of characters are transferred unchanged from parents to offspring, except of course accidents. There is uniparental inheritance due to mitosis. But in sexually reproducing organisms, gametes are produced as a result of meiosis. Two gametes unite to give rise to a zygote which divides mitotically to from the individual. More often than gametes from different parents unite to form a zygote, thus ensuring biparental inheritance. This results into a blending of characters from two parents into their offsprings and ensures variation which is important for evolution. Gametes contain half the number of chromosomes, as compared to somatic cells. Union of two gametes restores the somatic number of chromosomes in the zygote. Gametic union results into the mixing of cytoplasms of two cells which is followed by the fusion of two nuclei. During meiosis or at the time of gamete formation, therefore, halving of the number of chromosomes or the amount of DNA is achieved by two successive divisions of cytoplasm and nuclei (meiosis I and meiosis II) accompanied by only one replication of chromosomes. For example, each sometic cells of man has 46 chromosomes. Somatic cells are derived from the zygote due to mitotic divisions. Zygote is the result of a fusion of two gametes, each of which contain only 23 chromosomes. Both the male and female gametes contain the same amount of DNA and equal number or morphologically similar chromosomes. The zygote and somatic cells derived therefrom, therefore, contain 23 pairs of homologous chromosomes, one of each pair derived from the male parent through the male gamete and the other from the female through the female gamete. The members of each pair are homologous to each other and non-homologous to the other chromosomes. In some organisms, the chromosome complements of male and female cells are slightly different and this forms the basis of sex differentiation. Those cells which possess two sets of chromosomes are called *diploid* and they have 2n or dipoid or unreduced or zygotic number of chromosomes. Gamets, on the other have n, or *haploid* or reduced or gametic number of chromosomes. In sexual reproducing organisms the diploid and haploid phases of the life cycle alternate and this phenomenon in called *alternation* of *generation*. In plants, there is an alternation of sporophytic (2n) and gametophytic (n) generation. Reduction division heralds the beginning of gametophytic generation and fusion of gametes that of the sporaphytic phase. Thus, meiosis can be studied at the time of gamete formation which occurs in specialised organs i.e., the embryo sac and the pollen sac. A single cell undergoes meiosis in the embryo but in the pollen sac a large number of spore mother cells divide meiotically, almost synchronously, to produce hapoid pollen grains. It is, therefore, easier to study meiosis is anthers, similarly in testes in animals.

Meiosis consists of two divisions which take place one after the other, during which the number of chromosomes is halved. The two divisions are known as the first meiotic division or heterotypic division and the second meiotic division or homotypic division. The first meiotic division has the same stages as mitosis. These stages are known as prophase I, prometaphase I, metaphase I, anaphase I and telophase I. Except for prophase, the other stages are similar to those of mitosis. Following telophase I is a period of interkinesis or rest. The second meiotic division consists of prophase II, metaphase II, anaphase II and telophase II.

I. Heterotypic or Ist Meiotic Division

This is the first half of meiosis and is true reductional in nature. Here the diploid cell divides into two haploid dauhter cell. As stated already that the phases of this division one called Prophase I, Metaphase I, Anaphase I and Telophase I.

Prophase I: It is the longest stage and its duration makes from a few hours to a few days. The nucleus starts increasing in volume and this increase is much than the mitotic increase to some extent. This is due to an increase in hydration.

Nobel and *Ruttle* (1936) have called this stage as the *premeiotic spiral prophase*. Prophase I is Further divided into six substages.

(i) Pre-leptotene	(iv) Pachytene
(ii) Leptotene	(v) Diplotene
(iii) Zygotene	(vi) Diakinesis

(i) Preleptotene: As noted previously, replication of DNA has already taken place during interphase. The chromosomes at the beginning of meiosis are, therefore, not single but double strands. The beginning of meiotic prophase is marked by an increase in nuclear volume. During preleptotene the chromosomes are very thin and cannot be easily seen, except sometimes for the sex chromosomes.

(ii) Leptotene ("slender thread") or *Leptonema.* During this stage the chromosomes become more distinct, and their double nature is seen in many organisms. Leptotene extends from the time single unpaired chromosomes are first seen to the time when they begin to undergo homologous pairing. Under the light microscope the chromosomes appear as slender threads bearing a series of granule-like structures called *chromomeres.* Whole mount electron microscope preparations show that leptotene chromosomes have an axial filament to which chromatin fibres are attached as a series of lateral loops. In certain regions these loops appear in bunches, and from the chromomeres of light microscopy. The axial filaments are apparently composed of two independent protein fibrils, one for each chromatid. They later become transformed into the lateral elements of the synaptonemal complex. In the leptotene cells of the quail there is evidence of premeiotic pairing. Two homologous chromosomes become apposed to each other, chromomere for chromomore, even though no synaptonemal complex is formed. In leptotene cells there is a tendency for the ends of homologous chromosomes to be associated with each other at the nuclear membrane, where they are attached.

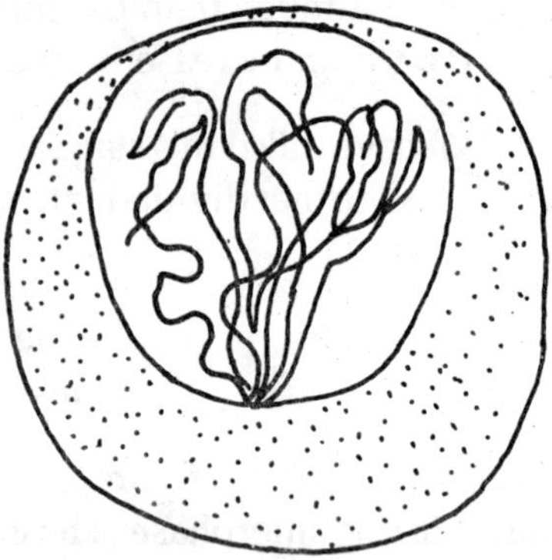

Fig. 4.10. leptotene chromosomes arranged in a 'bouquet'

Leptotene chromosomes may be irregularly arranged, or may be polarized toward the centrioles forming a *'bouquet.'* Electron microscopic studies have shown that bouquet formation results when a group of chromosomes is attached close together on the nuclear membrane.

During leptotene the cytoplasm has many polyribosomes, but endoplasmic vesicles are few. In maize *(Zea mays)* the nucleus increases in size during leptotene, reaching its maximum size during pachytene. This increase in size is correlated with synthesis of RNA and protein.

(iii) Zygotene ("mating thread") or Zygonema. The chromosomes becomes much shorter and thick and number of coils is also increased. Homologous chromosomes begin to pair by force of attraction. In synapsis or pairing, two homologous chromosomes come close to each other precisely chromomere to chromomere. The pairing starts from one point and proceeds onward in a zipper like fashion. Out of the two synaptic chromosomes, one is of paternal origin while the other is maternal. These are called *bivalents.* Since each chromosome is made of two chromatids, each pair of chromosomes consists of four chromatids thus representing *'tetrad'* of chromatids at this stage.

The pairing of zygonemal stage is completed in three different ways, as follows:

(i) *Proterminal synapsis:* The two homologous chromosomes start pairing at the terminals, which gradually then progresses towards the centromere.

(ii) *Procentric synapsis:* The pairing starts near the centromere and progresses onwards on either side.

(iii) *Random synapsis:* The pairing may be at random at many points which then finally fuse together.

Two views have been advocated to explain the mode of initiation of synapsis. *Darlington* (1937) proposed precocity theory which explains that the chromosomes undergo synapsis due to singleness, but this theory does not explain the extra synthesis of DNA and duplication at leptotene stage. Another view is proposed by *Sax* (1932) which is referred as the retardation theory. It explains that pairing takes place due to the retardation or cessation of metabolic activities of the cell. During zygonema, the nucleus remains intact and the centriole moves apart initiating the aster formation.

(iv) Pachytene "thick thread" or pachiynema: It generally begins when the pairing of chromosomes ends. It represents one of the longer substages of prophase I and is characterised by the following important events :

> When the pairing has completed, the aposition of homologues becomes so tight that is difficult to identify two separate chromosomes. Within the major coils, and at right angles to them smaller minor coils appear. At the height of pachytene, the pairing chromosomes twist about one another as relational coils and each soon longitudinally splits if no splitting had occurred. This coiling is further complicated by a coiling of two chromatids of each chromosome around each other. This coiling put the chromosomes under considerable strain. This means that

at this stage, each pacytence element consists of four chromatids. The chromatids of each homologue are called *sister chromatids.* Formerly this element of four chromatids was called a *tetrad*-a name which has now been replaced by *bivalent.* When the chromosomes have longitudinally splitted, transverse breaks occur at the same level in two of the adjacent homologous chromatids. Soon both segments of the chromatids interchange and fuse together with those of homologues. In this process which is called *crossing over,* portions of the two chromatids with their genes are interchanged and two remain intact. The new chromatids thus interchanged will be mixed.

After this interchange, the parts begin to separate, repalling each other. However, this separation is not at all complete since the homologous chromosomes remain joined at their point of interchange. This point where the two are joined is called the *chiasma* (pl. chiasmata). This chiasma is an immediate consequence of crossing over and are found in all plants and animal with few exceptions (e.g., males and females silkworm). The number of chiasmata is variable but for one bivalent atleast one chiasma is found. Bivalent with 2,3 and 4 chiasmata are rare. The highest number of chiasmata has been observed in the long chromosome of the broad bean, *Vicia faba,* where there are probably 12 chiasmata in a bivalent. The average number of chiasmata is known as *chiasma frequency.* This number seems to depend upon the length of the chromosome. Short acrocentric chromosomes have a greater number of chiasmata than metacentric chromosomes. In small chromosomes, chiasmata are localized at the distal ends. Nucleus remains as such, being somewhat larger than the proceeding substage.

Cytomixis: It is a botanical term which refers to the passage of chromatin from one microsporocyte (pollen mother cell) to another during early prophase. This phenomenon is so widespread in the normal species of hybrids that one wonders if it is a general phenomenon.

(v) Diplotene or Diplonema: The separations of homologous chromosomes initiates diplotene. The synaptic forces of attraction between them lapse due to breakage at one or more point, so that the homologous chromosomes uncoil and start separating. But the separation is non-etheless, incomplete since the homologues are in, contact at one or more points where the crossing over has already taken place. These points of contact are known as chiasmata (sing-chiasma; meaning cross) which present crosshaped appearance. The chiasma is the morphological equivalent of genetic crossing over. Their number and position varies in the same pair of chromosomes and in different cells of the same individual.

By the end of diplotene, the chiasmata begin to move along the length of chromosomes from the centromere towards the end. This displacement of chiasmata is termed as terminalization. When the terminalization of chiasmata is complete the homologous chromosomes are kept in contact by terminal chiasmata. The degree of terminalization is generally expressed as a coefficient of terminalization (T).

$$T = \frac{\text{Number of terminal chiasmata}}{\text{Total number of chiasmata}}$$

The average number of chiasmatata in bivalents is known as frequency of chiasmata (fq).

$$\text{Frequency (fq)} = \frac{\text{Total number of chiasmata}}{\text{Total number of bivalents}}$$

According to *Darlingtion* two types of repelling forces operate on the chromosomes at diplotene. One of the forces is electroegatively charged and operates on the surface of the chromosome throughout its length and the other with electropositive charge is localized on the centromere. The former controls the repulsion of the chromosomes and the later causes distal movement of the chiasmata.

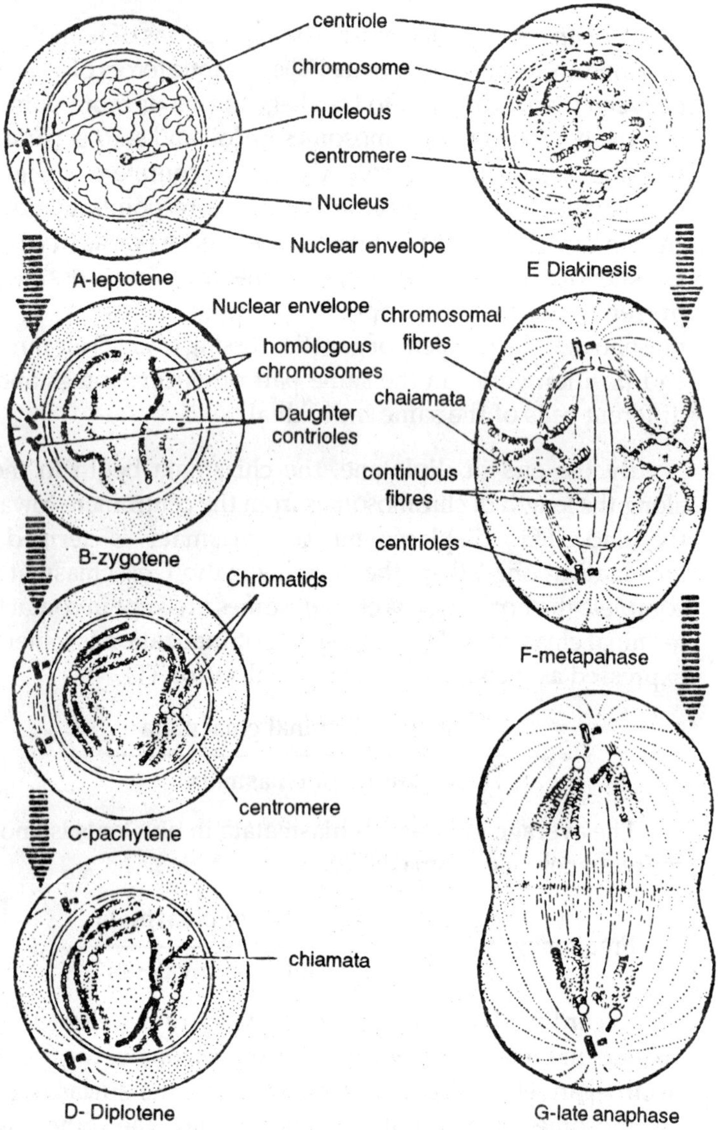

Fig. 4.11. Different stages of first meiotic division.

Diakinesis: At this stage the chromosomes are best seen as very short and thick bivalents. The contraction is nearly

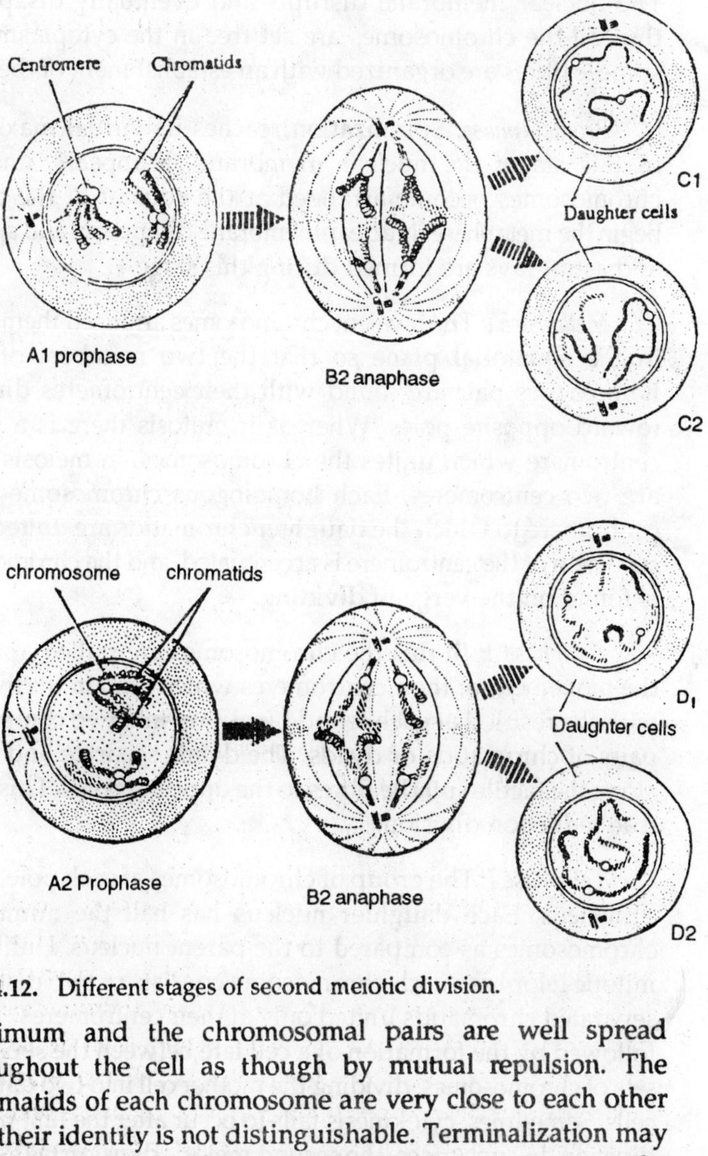

Fig. 4.12. Different stages of second meiotic division.

maximum and the chromosomal pairs are well spread throughout the cell as though by mutual repulsion. The chromatids of each chromosome are very close to each other and their identity is not distinguishable. Terminalization may be completed in the substage. The bivalents migrate to the periphery of the nucleus. The nucleus generally disappears during this stage but may persist, in reduced size until anaphase.

The nuclear membrane disrupts and eventually disappears thus all the chromosomes are set free in the cytoplasm. The spindle fibres are organized with an establishment of the pole.

Prometaphase: Spiralization reaches its virtual maximum in this stage; the nuclear membrane disappears, and the chromosomes become arranged at the equator of the cell to begin the metaphase. Nuclear membrane disappears and spindle and astral rays are formed during this stage.

Metaphase I: The bivalent chromosomes arranged themselves in the equatorial plane so that the two members of each homologues pair are found with their centromeres directed toward opposite poles. Whereas in mitosis there is a single centromere which unites the chromosomes, in meiosis there are two centromeres. Each homologous chromosome has a centromere to which the daughter chromatids are united. The repulsion of the centromere is accentuated, and the chromosome is found on the verge of dividing.

Anaphase I: In this the chromosomes are pulled apart by the movement of their centromeres which remain undivided, with the result that each tetrad breaks up into two component pairs of chromatids or dyads. The dyads separate and move along the tactile spindle fibres to the opposite poles. This is the true reduction division.

Telophase I: The group of chromosomes at each pole forms a nucleus. Each daughter nucleus has half the number of chromosomes as compared to the parent nucleus. Unlike the mitotic telophase each chromosome consists of two distinctly separated chromatids united only at their centromeres. This is followed by the formation of a cell late between the separated sets of chromosomes, dividing the mother cell into two daughter cells. Sometimes, cytokinesis fails to occur after the first meiotic division. In such cases, the second meiotic division takes place in the common cytoplasm of the mother cell.

Interkinesis: After telophase, two haploid daughter nuclei or cells sometimes undergo typical resting stage as in mitosis. This intervening stage between first telophase and the beginning of second prophase is termed as interphase or interkinesis. This stage is either very short or may be entirely absent, if present, the coiling of the chromosomes persists. No increase in DNA takes place between the I and II meiotic divisions. The occurrence of an interphase between the two division suggests that additional RNA and protein sythesis is required to carry meiosis through to completion.

II. Homotypic or 2nd Meiotic Division

Before the start of this division, most of the activities of heterotypic division cease for sometime. This resting stage is termed as the interkinesis. 2nd Meiotic division involves following staps:

Prophase II: This stage superficially resembles that of the mitosis prophase with following exceptions :

At the early prophase II, the two chromatids of each dyad looks like X, since they are conjoined by a common centromere. The four arms are widely separated. There is no rational coilling. The X-shaped dyads are quite linger than at telophase. The chromonemata are still not completely coiled. The genetic constitution of the two chromatids of each dyad depends upon the kind and number of cross overs which took place in the phase I. If there is no crossing over, the dyads would consist of two identical sister chromatids of either paternal or maternal origin. At the end of the prophase II the nucleolus disappears. The nuclear membrane disappears and the acromatic figure is developed.

Metaphase II: As usual, all the chromosomes arrange themselves on the equator for a short duration. The centromeres touch the equator but their arms radiate out in different directions. Later on the centromere in each dyad divides into two sister centromeres.

Anaphase II: Sister centromeres separate to the poles, pulling with them the chromatids to which they are attached. The chromatids, however, become much thick and stout.

Telophase II: The chromosomes at each pole uncoil and thin out to form the nuclear net. Each group gets surrounded by a nuclear membrane. Nucleolus reappears. Thus two nuclei are recognised in each cell. This is soon followed by cytokinesis and two cells are formed from each haploid daughter cell. Thus as a result of meiosis four cells are produced, each with a haploid set of chromosomes, i.e., each contains just one members of each homologous pair.

Cytokinesis: Sometimes meiosis I is followed by cytokinesis and sometimes it is deferred until the end of mMeiosis II. The details of the process are same as during mitosis. Cell plate formation results into a tetrad of cells with reduced chromosome numbers. The arrangement of cells in a tetrad is different in different organisms, but is characteristic of the species.

Like mitosis, meiosis too is a dynamic process and its different stages merge into subsequent ones. There is no sharp demarcation between different stages. DNA synthesis takes place during interphase, prior to prophase I. Duplication of chromosomes takes place during interphase itself and the chromosomes that appear during leptotene are double and not single. Number of chromosomes is reduced during meiosis I. If the reduction process is examined more closely, it becomes clear that there is qualitative reduction during meiosis I of those parts of chromosomes which are not involved in exchange. The homologous chromosomes or the maternal and paternal chromosomes segregate or undergo qualitative reduction during meiosis I. Their quantitative reduction takes place during meiosis II, when their centromeres divide and the sister chromatids separate. But the reverse is true of those parts of chromatids which are exchanged due to the formation of chiasmata. For these parts quantitative reduction takes place during meiosis I and qualitative reduction during meiosis II. Thus, complete of both qualitative and quantitative reduction takes place as a result of meiosis I and II and none of them alone can bring about complete reduction. It is, therefore, not proper to say that meiosis I is reductional and meiosis II equational. Duration of meiosis varies a great deal in different organisms. So does the duration of different stages in an organism. Prophase I lasts for the longest duration.

Types of Meiosis

There are three types of meiosis, *zygotic or initial meiosis, gametic or terminal meiosis and sporic or indeterminate meiosis.*

1. Zygotic or initial meiosis (haplotone pattern) occurs in lower plants. Fertilization is immediately followed by meiosis, giving rise to cells with the haploid chromosome number. The zygote is the only diploid stage in the life cycle.

2. Gametic or terminal meiosis (diplotonic pattern) occurs in animals. The cells resulting from the two meiotic division are transformed into sperms or eggs without further cell division. The process is as follows. Primordial germ cells divide repeatedly to give rise to many generations of daughter cells, spermatogonia (in males) or oogonia (in females). The spermatogonia and oogonia are transformed into primary spermatocytes and primary oocytes, respectively. Ench primary spermatocyte undergoes the first maturation (metiotic) division and froms two secondary spermatocytes. Each secondary spermatocyte undergoes the second matration (meiotic) division to give rise to two spermatides, which are then transformed into spermatozoa. In the female during the first meiotic division the primary oocyte, into a large secondary oocyte and a smaller cell, the first polar body. The secondary oocyte undergoes the second meiotic division and divides into an ootid and a second polar body. The first polar body may or may not divide into two. The ootid is the ovum proper.

3. Sporic or indeterminate meiosis (diplohaplotonic pattern) is characteristic of higher plants and some thallophytes, but is not known in animals. In the diploid sporophyte, meiosis

takes place at some indeterminate point between the zygote and gamete formation. As a result of meiosis haploid megaspores and microspores are formed. The megaspores and microspores undergo repeated mitotic divisions and form the male and female gametophyte, respectively. Certain cells of the gametophyte differentiate and from the gametes. In plants the cells that undergo meiosis are the microspore mother cell and the megaspore mother cell.

Significance of Meiosis

The process of meiosis keeps the chromosome number constant from one generation to other. This also involves reshuffling the hereditary characteristic which arises from reassortment of the parental chromosomes though crossing over. *Carothers* (1933) and others mentioned that in the metaphase I of meiosis it is usually a metter of chance whether the centromere of the paternal or the bivalent lies towards the particular pole. This results in random origin of new set on speration. There may be many new combinations. In fact, since each chromosome can go to either pole, the total possible number of arrangement will be 2n where n is the haploid number of chromosomes.

Darlington (1931) has suggested that the meiosis can be regarded as a precocious mitotic, and that the force which retains daughter chromatids together in mitotic prophase also retains the homologous chromosomes pairs in the earlier stages *(Ambrose)*.

Role of Centriole During Meiosis

The centrioles perform precisely controlled synchronous function during meiosis. Where meiosis occurs in two steps, that is, two nuclear divisions take place in sequence, the centriole reproduces and functions to from two achromatic figures. In the first division of meiosis, the centromeres do not duplicate but rather homologous chromosomes pair together and move to opposite poles, while in the second meiotic division the centromeres of chromatids are separated. A scheme for relation various key events is presented briefly as under :

(i) Non-duplication of centromeres in meiosis-I;
(ii) Non-duplication of chromosomes in meiosis-II; and
(iii) Duplication and function of centrioles in both divisions.

By a complete repression of the duplication, simultaneous function of the achromatic apparatus brings about the reduction. The cetriolar function is super-imposed upon the duplication process of the chromatic apparatus in a sequence which leads to the loss of a generation of chromatic apparatus. When meiosis is gametic and fertilization follows after, the immediate formation of zygote as a cell might contain the maternal as well as paternal centrioles, having endowed as it were, with the extra centriole generation. According to *Daniel Mazia* the kinetochore is an essential part of the chromosomes so far as the mitosis is concerned. In gametogenesis there is a gain in total chromatic material, since these types of cell behaviour lead to duplication and subsequent reamassing of chromosomes. The extra set of centriole is rejected normally in fertilization or autogamy. Thus a cell results in a chromatic apparatus. This is accomplished by meiosis: one division meiosis gives the centriole a chance to 'catch up' in a singel duplication.

DIFFERENCES BETWEEN MITOSIS AND MEIOSIS

Mitosis	*Meliosis*
1. Prolonged resting stage.	1. This stage is very short.
2. The prophase is short without any substages.	2. The prophase is subdivided into six stages, viz., Proleptoten Leptonema, Zygonema, Pachynema, Diplonema and Diakinesis.
3. The chromosomes split longitudinally into sister chromatids during the preceding resting stage. From early prophase they appear as double tetrads.	3. The chromosomes do not split longitudinally.

4.	The split homolgues are not attracted towards each other	4.	At zygonema or succeeding stages, in prophase of first division, the homologous chromosomes are attracted towards each other and they form bivalents.
5.	Synapsis and tetrad stages do not occur.	5.	The chromosomes split after synapsis to from sister chromatids which are arranged in tetrads.
6.	No chromosomal crossing over takes place.	6.	The crossing over is a prominent feature which results in exchange of genic material.
7.	The chromatids are coiled in plectonemic manner.	7.	Coiling is paranemic type, and much evident.
8.	Doubling of chromosomes takes place early in prophase.	8.	Doubling of chromosomes occurs in pachytene stage of prophase.
9.	Nucleolus and nuclear membrane disappear later on and the formation of nuclear spindle is completed with the commencement of metaphase.	9.	The nuclear membrane and nucleolus disappear and the formation of acromatic figure is completed with diakinesis simultaneously.
10.	The nucleus enlarges to some extent.	10.	The nucleus is greatly enlarged.
11.	At metaphase, the centrosome of each chromosome divides and the sister chromatids move towards the opposite poles. The somatic number of chromosomes is maintained without any reduction.	11.	At metaphase I, the centrosome does not divide but the homologous chromosoms move towards opposite poles. The chromosome number is reduced to half by reduction divison.
12.	The chromosomes are seen in dyad consisting of two chromatids.	12.	The chromosomes are in the form of tetrads having four chromatids.
13.	Each chromosome of patemal and matemal origin is longitudinally divided and the separation to opposite side results in identical cells at the end of mitosis.	13.	At metaphase I, the orientation of chromosomes at the equatorial plate is in such pattern that either paternal or maternal chromosome goes to either pole. The orientation of one pair is independent of the other.

14.	During orientation of dyads the centromeres face towards the equator and the arms directed towards the poles.	14.	During the orientation of tetrads, the centromeres are placed equidistant and face the pole while arms are directed towards the equator.
15.	The centromere of each bivalent undergoes division at metaphase.	15.	No division of centromere occurs, but homologour chromosomes simply separate to form two dyads.
16.	The initial movement of chromosomes at late metaphase is due to repulsion between the centromeric parts.	16.	The main repulsive force lies between the centromeres of homologous chromosomes.
17.	At anaphase the chromosomes are less thick and short.	17.	The chromosomes are very thick and short.
18.	At anaphase further separation of chromosomes is due to the formation of stem body.	18.	Most of the changes are similar to mitosis.
19.	At telophase two dauther nuclei, each with diploid number of chromosomes, are reconstituted.	19.	At this phase two daughter nuclei are haploid.
20.	There is no evidence of second division in mitosis.	20.	There are two division for each meriotic division The second division follows after a short gap, interkinesis. The second division is homotypic by nature, and results into four cells, in the form of tetrads.
21.	In the mitosis neither the chromosomal number nor the quality is changed and the dauther cells carry 2n chromosomes.	21.	At the end of meiosis each of tertrads is haploid and quality is changed.
22.	This division takes place in growing somatic cells.	22.	This division takes place only in gonodial cells and result in the formation of gametes.

5

Nucleic Acids

Friedrich Miescher in (1868) isolated a material from pus cells by digesting them for weeks with dilute HCl and called this material as *nuclein. Hoppe Segler* and his co-workers confirmed the work of *Miescher* and also proved the presence of nuclein in yeast and the erythrocytes of the birds and reptiles and various other tissues. The term *nucleic acid* was introduced by *Altmann. Albrecht Kossel, P.A. Levine and Walter Jones* described the chemical nature of nucleic acid. According to them the nucleic acid is composed of phosphoric acid, a sugar and nitrogenous bases. *Franklin W. Stahl* presented first evidence that nucleic acid forms the genetic material. *Chargaff* (1951) described the occurrence of nitrogenous bases in equal proportions. *Dotty* (1961) has emphasized much on the physical properties of DNA. In 1962 *Willkins, Klatson* and *Crick* proposed a model to explain the double helicular structure of DNA. They also shared the Nobel Prize for the same.

Deoxyribonucleic Acid or DNA

DNA is found in the cells of all living organisms except plant viruses, where RNA forms the genetic material and DNA is absent. In bacteriophages and viruses there is a single molecule of DNA, which remains coiled and is enclosed in the protein coat. In bacteria, mitochondria and plastids of eukaryotic cells DNA is circular and lies naked in the cytoplasm. In the nuclei of eukaryotic cells DNA occurs in the form of long spirally coiled and unbranched threads. The number of DNA molecules is equivalent to the number of chromosomes per cell. In them DNA is found in combination with proteins forming nucleoproteins- the *chromatin* material and is enclosed in the nucleus.

DNA is a polymer, the monomeres of which are the deoxyribonucleoside monophosphates. Most DNA have high molecular weight, varying from 10^6 to 10^9 or more. Chemical analysis of DNA molecules from different kinds of organisms suggest that in one kind of DNA the number of adenine (A) nucleotides was always the same as the number of thymine (T) nucleotides. Similarly the number of guanine (G) nucleotides was always the same as the number of cytosine (C) nucleotides. This evidence suggests that nucleotides occur in pairs in the DNA molecules. Further, quantity of A present in one organism's DNA (and it varies from species to species) is equal precisely to the amount of T. Similarly the amount of G present is always equal to the amount of C.

Physical Structure of DNA

According to the model proposed by *Watson* and *Crick*, the DNA contains two polynucleotide chains running antiparallel and twisted around each other in the form of a regular double helix. The diameter of the helix is about 20A° and each chain makes a complete turn every 34A°. Each chain is polynucleotide in which the sugar of each nucleotide is linked by a phosphate group to the sugar of the adjacent nucleotide. There are 10 nucleotides on each chain in every turn of the helix. The two helices are coiled in such a way that they cannot be separated without unwinding. The two chains of a helix are of opposite polanty. It one chain ruins in 3'-5' direction, then the other will run as 5'-3' direction.

The molecular weight for purified DNA is found in the range 1000,000 to 120×10^6. The amount of DNA in a mammalian cell corresponds to about 25,000 molecules of molecular weight 120×10^6, a total of about 10,000 million nucleotides.

Molecular Structure of DNA

The DNA molecule is a polymer consisting of several thousand pairs of nucleotide monomers. Each nucleotide consists of the pentose sugar, *deoxyribose,* a *phosphate* group and a *nitrogenous base,* which may be either a *purine* or a *pyrimidine.*

1. *Deoxyribose suguar: Levine* (1909) identified ribose sugar as a main constituent of nucleic acid structure. *Mori* (1929) isolated sugar from the guanine nucleoside and showed that it is deoxyribose suguar. It is pentose type which in DNA molecule shows the absence of one oxygen molecule from carbon-2 position of ribose sugar. Both deoxyribose and ribose (pentose sugars of nucleic acids) have a pentagonal ring with five carbons, among which two (i.e. 3' and 5') are attached to phosphoric acid and three (1') to the base. This sugar is called *deoxyribose* and it simply acts as a support column to which bases are attached.

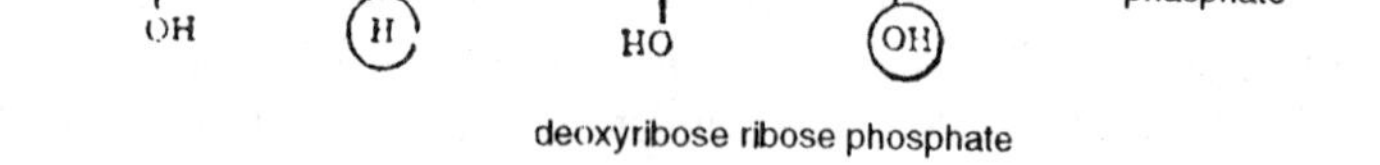

Fig. 5.1. Pentose Sugar.

2. *Phosphate:* In the DNA strand the phosphate groups alternate with deoxyribose. Each phosphate group is joined to carbon atom 3' of one deoxyribose and to carbon atom 5' of another. Thus each strand has a 3' end and a 5' end. The two strands are oriented in opposite directions. The 3' end of one strand corresponds to the 5' end of the other. Consequently the oxygen atoms of deoxyribose point in opposite directions in the two strands. *Phosphodiester bonds* are formed between the sugars of two different nucleotides and a phosphate group.

3. *Nitrogenous Bases:* The nitrogenous bases of nucleic acid are of two types : *Purine* and *Pyrimidine.* Purine bases comprise mainly adenine (A) and guanine (G) which are common to both DNA and RNA while pyrimidine bases comprise cytosine (C) and thymine (T) in DNA and in the cases of RNA, thymine (T) is replaced by uracil (U).

These bases are arranged linearly as the back bones of chains which (back bones) are composed of alternating sugar and phosphate. The bases have specific way of combining, if C is found in one chain G will be opposite to it and in other if A is on one T must be on the other. Or in other words adenine (A) is always paired with thymine (T) and guanine (G) with cytosine (C).

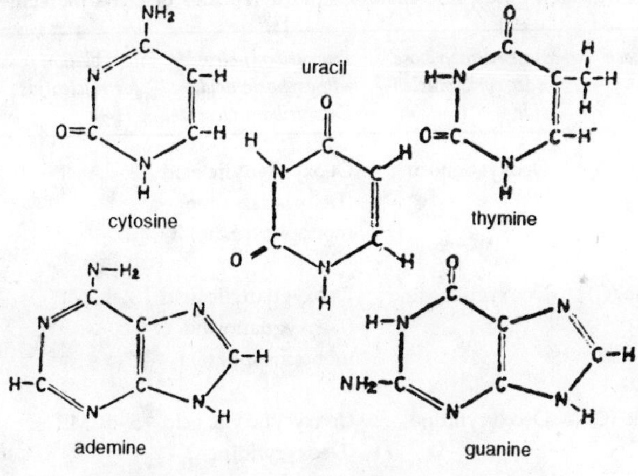

Fig. 5.2. Common nitrogenous bases of nucleic acids.

Among bases A-T and G-C, two hydrogen bonds are formed between A and T and three hydrogen bonds are formed between C and G. This hydrogen bond formation precludes A-C or G-T pairs. Only these arrangements A-T and G-C are possible for two purines would occupy two much space to allow a regular helix and correspondingly, two pyrimidines would occupy too little. The stickness of these pairing rules result in a complementary relation between the sequences of bases on the two interwined chains. For example, if we have a sequence TAGTC on one chain, the opposite chain must have a sequence TACAG. The hydrogen atoms with its positive charge is shared between an oxygen atom and a nitrogen atom, both with slight negative charges. Although hydrogen bonds are weak, the fact that there are so many gives stability to the DNA molecule. The weak hydrogen bonding enables the two strands of the DNA to separate during replication.

The bases are joined to the pentoses by N-C *glycosidic bonds.* For the purine, the glycosidic bond is between the C_1 position of the pentosoe sugar and the N_9 position of the bases. For the pyrimidines the linkage joins the C_1 and N_3 positions.

Four nitrogen bases, nucleosides and nucleotides of DNA molecule

Nitrogen base	*Base+deoxyribose =deoxyribonucleo-side*	*Deoxyribonucleoside +Phosphoric acid= Deoxyribonucleotide*	*Abbreviation for nucleotide*
1. Adenine (A)	Deoxyadenosine	Deoxyadenylic acid (Deoxyadenosine monophosphate)	3'-dAMP
2. Guanine (G)	Deoxyguanosine	Deoxyguanylic acid (Deoxyguanosine monophosphate)	3'-dGMP
3. Cytosine (C)	Deoxycytidine	Deoxycytidylic acid (Deoxycytidine monophosphate)	5'-dCMP
4. Thymine (T)	Deoxyhymidine	Thymidylic acid (Deoxythymidine monophosphate)	5'-dTMP

Nucleosides and nucleotides: A sugar molecule and a nitrogenous base form a *nucleoside,* and a nucleoside plus a phosphate group form a *nucleotide.* In other words a nucleoside is a base-sugar combination and a nucleotide is a nucleoside phosphate. The nucleotides of RNA are called ribonucleotides,

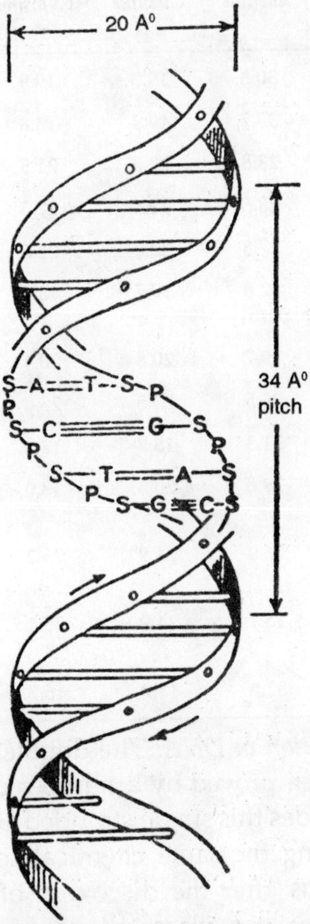

Fig. 5.3. The watson-Crick base pair in DNA, Adenine-Thymine, and Guanine-Cytosine.

and those of DNA deoxyribonucleotides. Ribonucleotides contain the sugar ribose, and deoxyribonucleotides the sugar deoxyribose.

The nucleosides are usually abbreviated as A,T,G or C, with d - as a prefix for deoxynucleosides (e.g. dG- is the abbreviation for deoxy gnanosine).

Showing amount of various bases in different tissues

S.No.	Source	Adenine	Guanine	Cytosine	Thymine	$\frac{A+T}{G+C}$
1.	Human liver	30.3	19.5	19.9	30.3	1.53
2.	Human sperm	30.7	19.3	18.8	31.2	1.62
3.	Hen red cells	28.8	20.5	21.5	29.2	1.38
4.	Rat bone marrow	28.6	21.4	21.5	28.4	1.33
5.	Herring sperm	27.8	22.2	22.6	27.5	1.23
6.	*Paracentrotus lividus* (sea urchin) sperm	32.8	17.7	18.4	32.1	1.85
7.	Salmon	29.7	20.8	20.4	29.1	1.43
8.	Wheat germ	26.5	23.5	23.0	27.0	1.19
9.	Yeast	31.3	18.7	17.1	32.9	1.79
10.	*Diplococcus pneumoniae*	29.8	20.5	18.0	31.6	1.59
11.	K-12 *Escherichia coli*	26.0	24.9	25.2	23.9	1.00
12.	*Mycobacterium tuberculosis*	15.1	34.9	35.4	14.6	0.42
13.	Bacteriophate T_2	32.5	18.2	16.7	32.6	1.86

Structural Variation in DNA: The double helicular nature of the DNA has been proved by a number of workers using the isotope and besides this single stranded DNA and circular shaped DNA, having the same chemical composition have been observed. Thus after the discovery of such abnormal variety of DNA, one can place them under the following headings:

(1) Double stranded DNA
(2) Single stranded DNA
(3) Circular shaped DNA

(1) Double Stranded DNA: This is the common DNA present in all the living organisms. These contain two polynucleotide chain running antiparallel and twisited around each other in

the form of a regular double helix. The detailed structure of this type of DNA has already been explained.

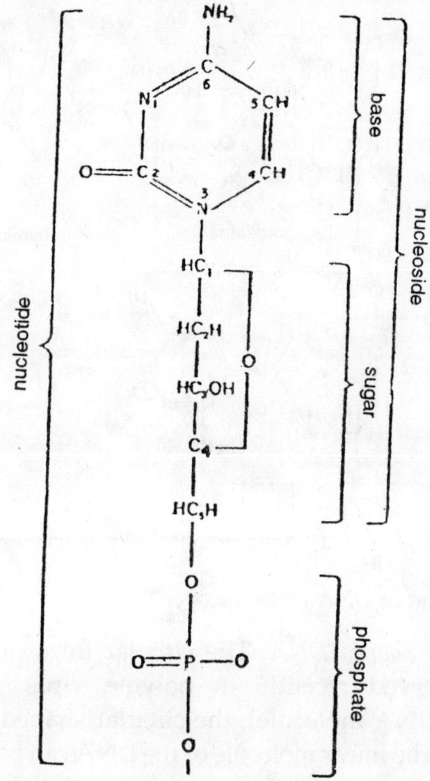

Fig. 5.4. Chemical structure of a nucleotide and a nucleoside

(2) Single Stranded DNA: At first it was thought that all DNA molecules are double stranded except during replication, when small portion near the Y regions are temporarily in a non-hydrogen bonded, single stranded form. Many people were surprised, therefore, when conclusive experiments revealed that the DNA of several groups of small bacterial viruses exists normally as single stranded molecules. This single stranded DNA is having the same chemical composition as we find in the double stranded DNA.

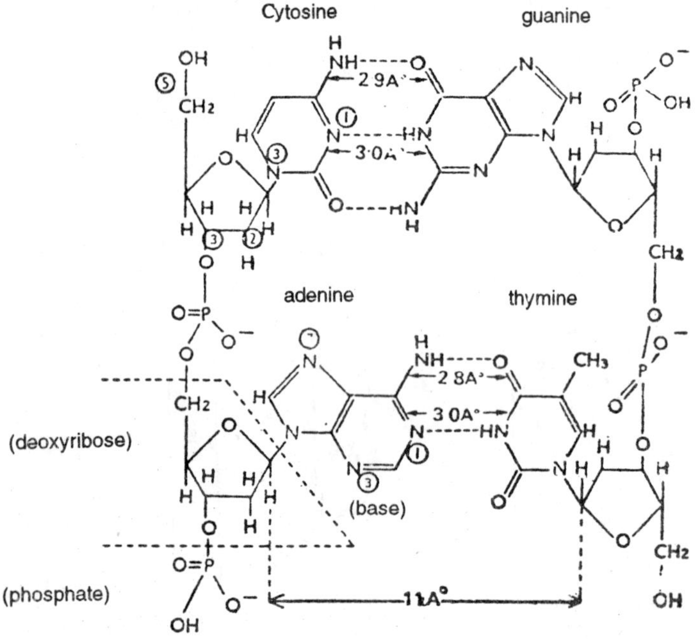

Fig. 5.5. Segment of DNA molecule.

3. *Circular Shaped DNA:* The circular form of the DNA has been observed recently in polyma virus. In *E. coli* chromosome (DNA molecule), the circular shaped DNA has also been seen. The linear molecule of the DNA can be converted in the circular form but we do not know the chemistry of converting a linear molecule of DNA into circular form. When the phage - DNA molecule is isolated from the free phase, it is linear in shape. However, by controlled heating and recooling it can be reversibly converted to a circular form.

Z-DNA or Left Handed DNA

On the basis of the arrangement of pentose sugar forming chains of DNA two forms of DNA have been reported. Usually the sugar chain has the pentose sugar with ascending sequence of carbon atoms. This is the right handed DNA or B-DNA. Recently in 1979, another form of DNA having pentose sugar with descending arrangement of carbon atom has been reported and named Z-DNA or left handed DNA.

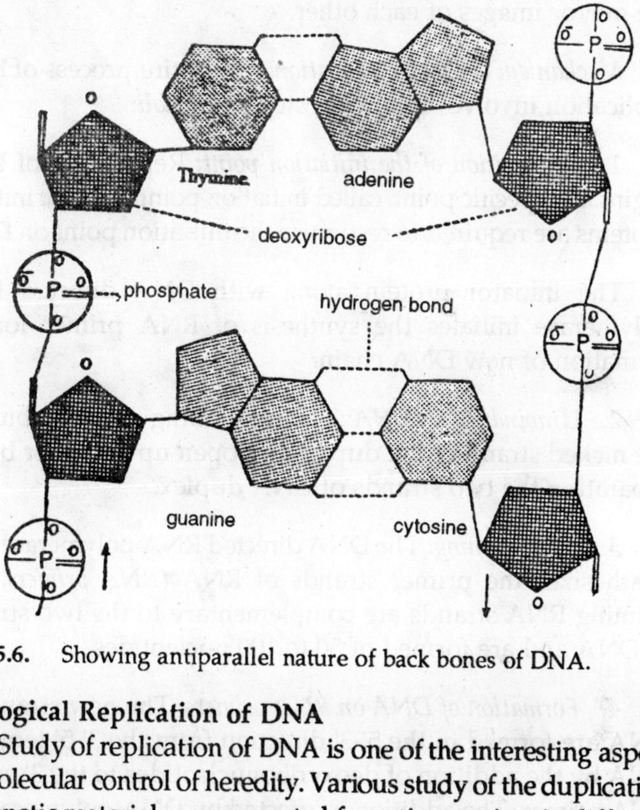

Fig. 5.6. Showing antiparallel nature of back bones of DNA.

Biological Replication of DNA

Study of replication of DNA is one of the interesting aspect of molecular control of heredity. Various study of the duplication of genetic material was presented from time to time in different organisms. But most of them are discarded because no concrete ideas have been proposed till *Watson* and *Crick* proposed a detailed scheme for the molecular structure of DNA. Later on, *Korenberg, Stahl* and *Taylor* have confirmed the Watson-Crick hypothesis. They also made a suggestion for its mode of replication which follows in a natural and attractive fashion from the proposed structure. During the replication, the DNA duplicates exactly into two chains. In the beginning both the strands of double stranded DNA, apart from each other by the disolution of hydrogen bonds and thus uncoiling of the chains occur. And then new chains, "the complementary chains" are built on them so that the newly formed daughter chains are the mirror images of each other.

Mechanism of DNA Replication: The entire process of DNA replication involves following steps in *E. coli*:

1. *Recognition of the initiation point:* Replication of DNA begins at a specific point, called initiation point. Specific initiator proteins are required to recognise the initiation point on DNA.

The initiator protein along with DNA directed RNA polymerase initiates the synthesis of RNA primer for the formation of new DNA chain.

2. *Unwinding of DNA:* The unwinding proteins bind to the nicked strand of the duplex and open up a loop or buble, separating the two strands of DNA duplex.

3. *RNA priming:* The DNA directed RNA polymerase now synthesizes the primer strands of RNA *(RNA primer)*. The priming RNA strands are complementary to the two strands of DNA and are formed of 50 to 100 nucleotides.

4. *Formation of DNA on RNA primers:* The new strands of DNA are formed in the 5'-3' direction from the 3'-5' template DNA by the addition of deoxyribonucleotides to the 3' end of RNA primer. The addition is affected by DNA polymerase III in the presence of ATP. The unwinding proteins separate the duplex strands ahead of the replication fork.

The leading strand of DNA synthesized in 5'-3' direction. The following strand of DNA is formed in its opposite direction. The DNA strands are synthesized the short segments, consisting of 1,000 to 2,000 nucleotides. These segments are called *Okazaki fragments*.

5. *Excision of RNA primers:* Once a small segment of an Okazaki fragment has been formed, the nucleotides of RNA

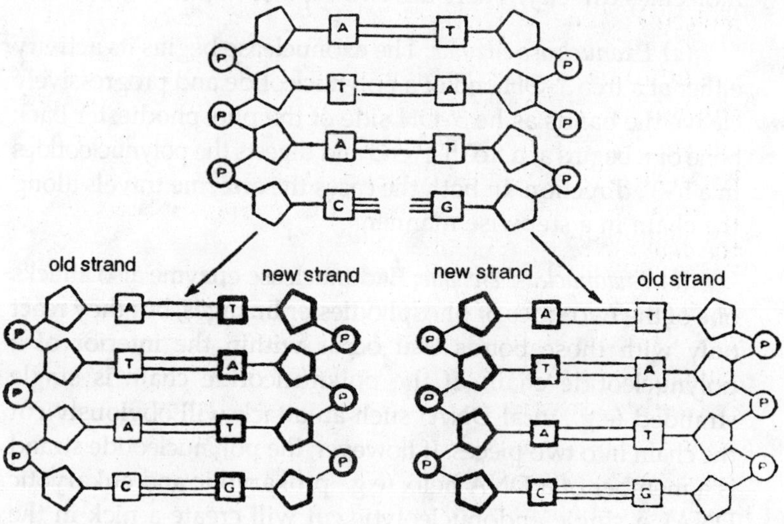

Fig. 5.7. Process of DNA duplication.

primer are removed from the 5' end one by one by the action of 5'-3' exonuclease activity of DNA polymerase-I.

6. *Joining of Okazaki fragments:* The gaps left between Okazaki fragments are filled with complementary deoxyribonucleotide residues by DNA polymerase-I. Finally, the adjacent 5' and 3' ends are joined by *DNA-ligase.*

In eukaryotic cells, the mechanism of DNA replication is expected to be more complex than in prokaryotes. The replication of eukaryotic DNA begins at multiple points of origin. Their DNA primer is formed of about 10 nucleotides and the Okazaki fragments are much shorter (about 100 to 150 nucleotides).

Enzymes for DNA synthesis: Different prokaryotic and eukaryotic cells have been formed to contain three kinds of enzymes that act on DNA namely nucleases, polymerases and ligases.

1. *Nuclease Enzymes:* These enzymes attack on the DNA molecules directly. There are two type of nuclease enzyme:

(a) Exonuclease enzyme: The exonuclease begins its activity either at a free 3'-OH end of a polynucleotide and progressively cleave the bands as he 3'-OH side of the phosphodiester back bone or it begins at a free 5'-P end and digests the polynucleotides in a 5'-3', direction. In both the cases the enzyme travels along the chain in a stepwise manner.

(b) Endonuclease enzyme: Endonuclease enzyme also attacks one of the two sides of phosphodiester linkages, but they react only with those bonds that occur within the interior of a polynucleotide chain. If the polynucleotide chain is single stranded (e.g., viral DNA) such an attack will obviously cut the chain into two pieces. If however, the polynucleotide strand is a member of a DNA helix (e.g., prokaryotic and eukaryotic DNA), a single endonucleolytic cut will create a nick in the helix; the helix remain in one piece but it now possesses a gap that contains two free ends, which can serve as substrates for exonucleases. A nicked double helix suffers a localized disruption of its secondary structure. The molecules becomes free to bend or rotate around its intact strand and the two broken ends are free to dangle. The increased molecular motion in the region of the nick will very likely interrupt hydrogen bonding between bases in the vicinity of the nick, thus effecting a limited "unraveling" of the helix.

2. *DNA Polymerase:* For many years, *Korenberg polymerase* was believed to be the enzyme responsible for the replication of the *E. coli* chromosome. But recently *Delucia* and *Cairus* isolated a true replicase from *E.coli* and named as *DNA polymerase*. The polymerase catalyses the formation of a polymer. The synthesis of a new complementary chain becomes possible in the presence of polymerase enzyme. There are three categories of DNA polymerase enzymes:

DNA polymerase I: Discovered by *Korenberg* in 1955 and has been studied in *E.coli*. It is believed to take part in the repair of DNA. DNA polymerase I is a single polypeptide chain with a molecular weight of about 109,000. There may be as many as 400 molecule of DNA polymerase I in *E.coli.*

DNA polymerase II: It is also made up of a single polypeptide chain having a molecular weight of about 90,000. The number of DNA polymerase II molecule is about 40 in *E.coli.* This enzyme can elongate Okazaki segments even in the absence of polymerase I.

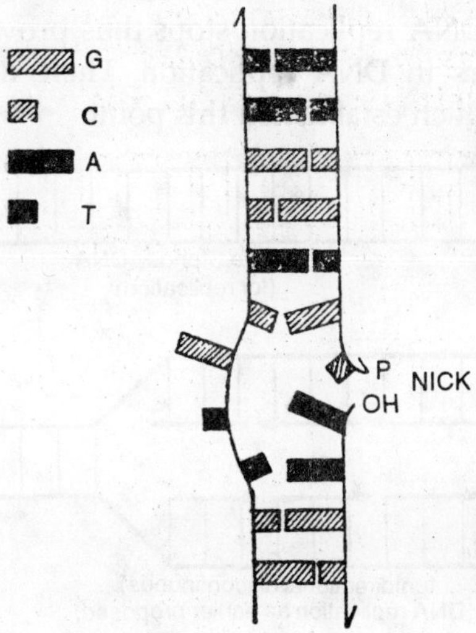

Fig. 5.8. A nicked duplex DNA molecule such as is formed by an endonuclease enzyme.

DNA polymerase III: It is considered to be a true replicase of *E.coli.* It consists of two polypeptide chains with molecular weight of 140,000 and 40,000. It is most effective and can polymersize about 15000 nucleotides per minute in *E.coli.*

3. *DNA Ligases:* DNA ligase enzymes are capable of catalyzing phosphodiester bond formation between free 3'-OH and free 5'-P group of a nick of DNA which is created by endonuclease enzyme, thereby restoring an intact DNA duplex. Many DNA ligases have already been discovered. The ligase

enzyme from *E.coli* requires the presence of oxidized nicotinamide adenine dinucleotide (NAD^+) a cafactor, whereas, the ligase enzyme specified by T_4 bacteriophage requires ATP to bring about the joining reaction.

RNA Primers in DNA Replication

Short segments of RNA are first synthesized before actual DNA replication starts. These RNA segments are later excised and filled with DNA with the help of enzyme DNA poymerase I. It was found that if RNA synthesis is stopped by a drug rifampicin, DNA replication stops thus providing the role of RNA primers in DNA replication. There are several other evidences which established this point.

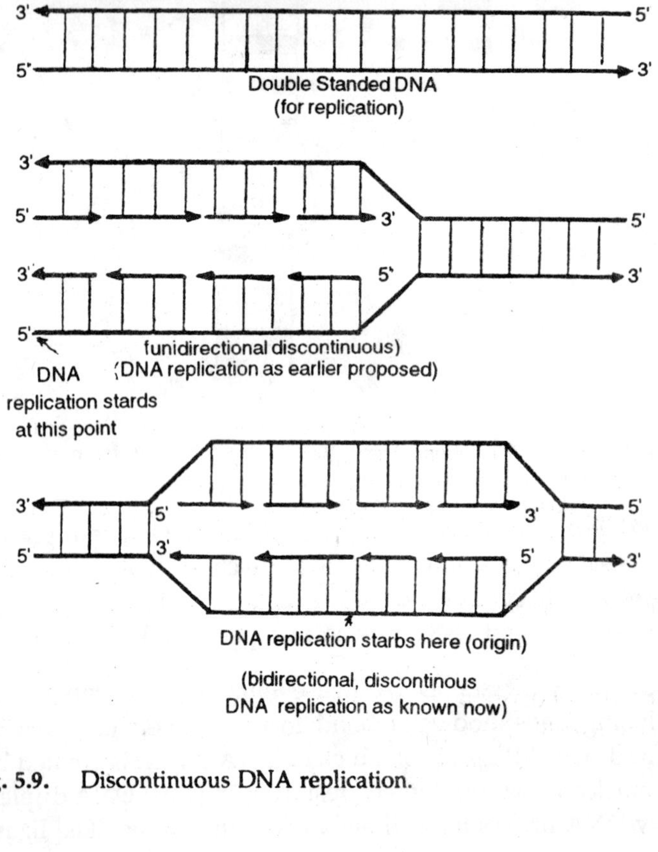

Fig. 5.9. Discontinuous DNA replication.

Types of replication: Three different types of replication of DNA have been discovered :

(1) Semiconservative
(2) Conservative
(3) Dispersive

(1) Semiconservative: Here the replication of DNA involves the unwinding of the double helix, but no rupture of the separated polyuclotide strands. Thus the primary DNA is causevned during the replication process. According to this mechanism, the unwinding of the original strands would be accompanied or followed by the guided synthesis of the complementary strands of each of the two separated polyuclotide strands. The linearly arsembly, catalyzed by enzymes, of the nucleotides of the complementary sequence would be guided by their paining with the Watson-crick partanres on the original strands. In this manner, the proper nucleotide sequence would be automatically assumed. The DNA helix would thereby regenerated each of the strand of the original helical duplex uniting with its freshly synthesized complement to form a total of two new double stranded structures. The semi-convervative model thus predicts that entire polynucleotide strands of the parental DNA will be passed on to the progeny in intact form. Each molecule of progeny DNA will receive either one half or none of its nucleotides from the parental DNA.

(2) Conservative: According to this mechanism both the primary and secondary structure of the parental DNA are conserved. Thus the replication process would involve neither the rupture of the individual strands nor the unwinding of the helical duplex. The details of these propose that the sequences of hydrogen bonded base pairs, rather than individual bases, serves the genetic determinant. According to this picture, particular base pairs might preferentially bind specific nucleotides and determine their sequential position in the daughter polynueleotide.

The convervative model allows for no redistribution of parental DNA among the progeny. The original DNA is transferred as an intact unit or not at all.

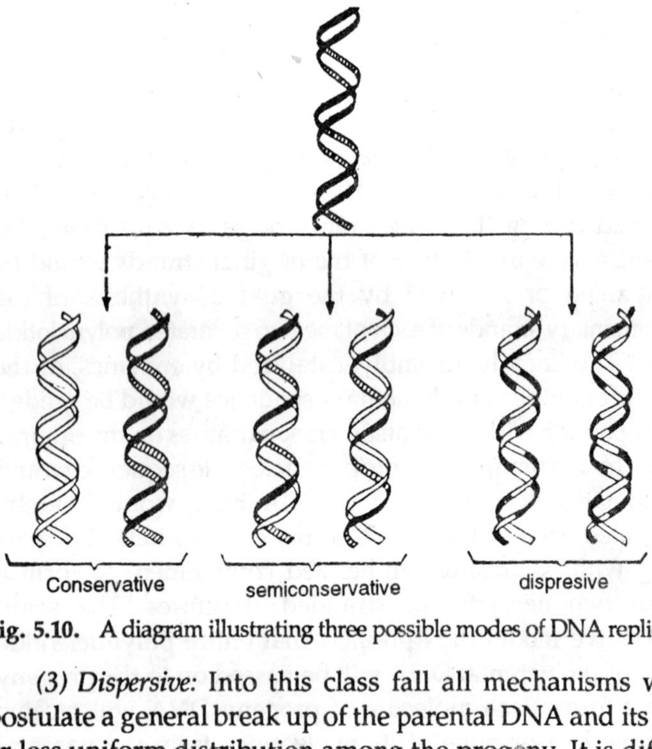

Fig. 5.10. A diagram illustrating three possible modes of DNA replication.

(3) Dispersive: Into this class fall all mechanisms which postulate a general break up of the parental DNA and its more or less uniform distribution among the progeny. It is difficult to propose a model of this type which can possibly allow the preservation of nucleotide sequence. Moreover, the existing experimental evidence is uniformly inconsistent with this kind of scheme.

Hybrid DNA: When double helical DNA molecules are heated above physiological temperature (to near 100°C), their hydrogen bonds break and the complementary strands often separate from each other *(DNA denaturation or melting)*. If cooled slowly, the two complementary stands can meet and reform the normal double helix. This process of reunion is called *denaturation* or *annealing*. This ability to renature DNA molecules

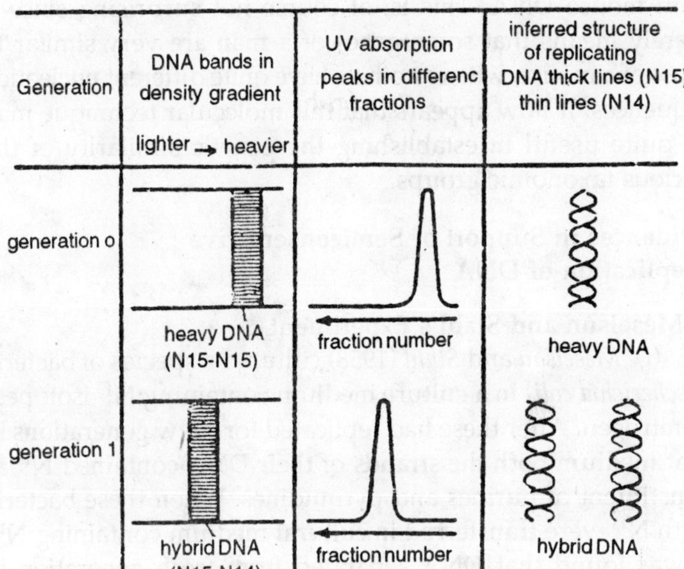

Fig. 5.11. Experiment of Meselson and State in Support of Semiconservative Method of DNA Replication.

permits as to show that artificial hyrid DNA molecules can be formed by slowly cooling mixture of denatured DNA from two species. For example, hybrid molecules can be formed containing on strand from man and one from a mouse. Only a fraction (25%) of the DNA strands from a man can form hybrids with mouse DNA. This is, of course not surprising since it merely means that some genes of a man are very similar to those of a mouse, whereas others have quite different nucleotide sequences. It now appears that this molecular technique may be quite useful in establishing the genetic similarity of the various taxonomic groups.

Evidences in Support of Semiconservative Replication of DNA

1. Meselson and Stahl's Experiment

(i) *Meselson* and *Stahl* (1958) cultured a species of bacteria *(Escherichia coli)* in a culture medium containing N^{15} isotopess of nitrogen. After these had replicated for a few generations in that medium both the strands of their DNA contained N^{15} as constituent of purines and pyrimidines. When these bacteria with N^{15} were transferred in cultural medium containing N^{14}, it was found that DNA separated from fresh generation of bacteria possesses one strand heavier than the other. The heavier strand represents the parental strand and lighter one is the new one synthesized from the culture indicating semiconservative method of DNA replication.

(ii) *Density determination experiments were also conducted:* The DNA of parental generation (labelled with N^{15}) was found to be heaviest. In first generation it was lighter than the first one. In second generation two bands were observed indicating the presence of DNA with two different densities, as is expected according to semiconservative method of DNA replication. These two bands were of equal intensity in the second generation. In the subsequent generations the intensity of light density bands gradually increased and the intensity of hybrid density band gradually decreased.

2. Cairn's Autoradiography Experiment

J. Cairns demonstrated the semiconservative type of replication by using a technique of autoradiography as given in the following lines:

In autoradiography technique, the material is first supplied with a suitable radioactive material like tritiated thymidine (H^3-TdR; H^3 is heavy isotope of hydrogen and it replaces normal hydrogen in thymidine to give rise to tritiated thymidine.). Tritiated thymiline is used since this will selectively label only DNA and will not label RNA, since thymine base is absent in RNA. The tritiated thymidine gets incorporated into DNA and replaces ordinary thymidine. The material is then sectioned or else the cells may be broken down to release the intact bacterial chromosome on slides. These slides are then covered by photographic emulsion and stored in the dark. During this storage the particles emitted by tritiated thymidine will expose the film, which can be developed. This photograph will then show the regions of the presence of tritium and thus indirectly the presence of labelled DNA.

Using this above technique, the replication of DNA could be easily followed by *J. Cairns* and the results were reported in 1963. During incorporation of tritiated thymidine, autoradiographs could be prepared at regular known intervals. Autoradiographs from this replicating material prepared at regular known intervals demeonstated semi-conservative mode of replication. It would be expected that in regions, where one cycle of replication of DNA has been completed, the density of dots will be higher than in the region where the replication has not taken place. The lighter density of dots is considered to indicate that only one of the two strands is labelled, while a heavier density of dots will indicate that both strands are labelled. Such a situation was actually observed. The rate at which the replication proceeds could also be worked out by measuring the length of DNA undergoing replication in a known interval of time. The generation time was worked out by *Cairns* as 30 minutes in *E.coli.* The length of the chromosome was worked out to be about 1 mm. The rate of replication as obvious would be approximately 30μ to 40μ per minute (1 mm = 1000μ).

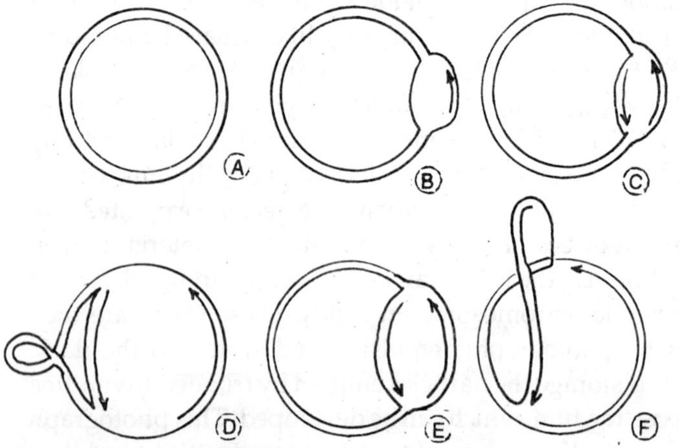

Fig. 5.12. Caim's model for circular DNA replication.

One of the two strands in the daughter DNA molecules is derived from the parent molecule and the other is synthesized afresh. In θ shaped figure, which is obtained in the second cycle of replication in presence of label, the two arcs in the split region would never be equally labelled. For instance, one arc would be twice as heavily labelled as the other arc. This is what was actually observed by *Cairns*. Thus the observation support the semiconservative type of replication of DNA.

3. Taylor's Experiment on *Vicia faba* Root Tips

J.H. Taylor (1957) and his coworkers also demonstrated the semiconservative method of DNA duplication in the root tip cells of *Vicia faba* by autoradiography. The roots were grown in a medium containing radioactive thymidine, so that the radioactivity is incorporated in the DNA of these cells. The outline of this labelled chromosome on a photographic film appears in the form of scattered black dots of silver grains When these root tips with labelled chromosomes were transferred to the unlabelled medium containing colchicine and studied for radioactivity, the following observations were made:

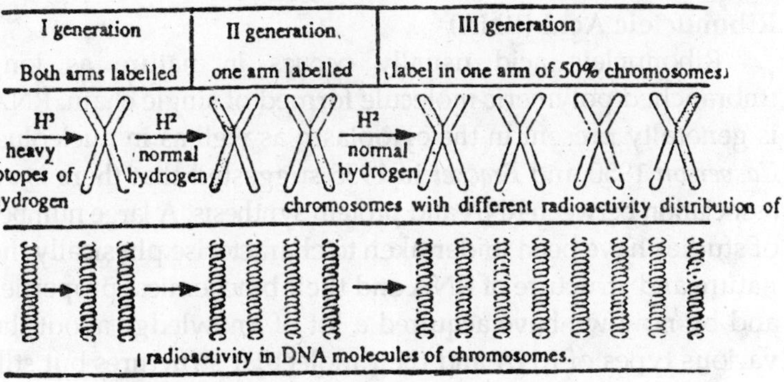

Fig. 5.13. Taylor's experiment on *Vicia faba.*

1. In the chromosomes of first generation the radioactivity was found to be uniformly distributed in both the chromatids, because in them the original strand of DNA double helix was labelled with radioactivity and the new one was non-labelled.

2. In the chromosomes of second generation only one of the two chromatids in each chromosome was radioactive.

This experiment demonstrates the semiconservative method of chromosome replication. But as it is known that each chromatid is composed of only one double helical molecule of DNA, it is the replication of DNA.

Biological Significance of DNA

There are four general requirements that genetic material can be expected to fulfil.

(i) That it replicates itself accurately during cell growth and duplication.

(ii) That its structure be sufficiently stable so that heritable change, mutation occur only very rarely.

(iii) That it has the potentiality to carry all kinds of necessary biological information.

(iv) That it transmits the informations to the cell.

Ribonucleic Acid (RNA)

Ribonucleic acid usually occurs in nature as long unbranched polymeric molecule formed of single chain. RNA is generally present in the cytoplasm as well as in nucleolus. *Casperson* 1950 and *Brachet* in 1955 suggested that there is an association between RNA and protein synthesis. A large number of studies have been undertaken to characterise physically the nature and structure of RNA and their biochemical properties and by now we have acquired a lot of knowledge about the various types of RNA and their molecular structures but still there are various gaps to be filled to make a complete picture.

Some plant viruses (e.g., TMV, turnip yellow mosaic viruses, wound tumour viruses, etc.,), animal viruses (e.g., influenza viruses, foot and mouth viruses; rous sarcoma viruses, poliomyelitis viruses reoviruses, etc.) and bacteriophages (e.g., MS2, etc.) contain ribonucleic acid (RNA) as their genetic material. That is why in these cases it is called the *Genetic RNA*.

Genetic RNA in Different Viruses

Type of Viruses	*Name of Viruses*	*Type of RNA*
1. Plant viruses	1. TMV	1. Single stranded
	2. Wound tumour	2. Double stranded
2. Aimal viruses	3. Influenza virus	3. Single stranded
	4. Poliomyelitis	4. Single stranded
	5. Reo virus	5. Double stranded
3. Bacteriophages	6. MS_2	6. Single stranded
	7. F_2	7. Single stranded
	8. r_{17}	8. Single stranded

Primary Structure of RNA

RNA usually occur in nature as long unbranched polymeric molecule made up of single chain or helix of polynucleotide strand Ribnucleic acid is composed of pentose sugars, phosphoric acid and nitrogenous bases. The pentose sugar of

RNA is ribose sugar. The sugar forms the back bone along with the phosphoric acid just in a manner studied in DNA.

Phosphoric acid is also one of the constituent of the RNA which along with the pentose sugar forms the back bone, on which the nitrogenous bases are linearly arranged, as shown in the Figure. On the basis of phosphate or phosphoric acid, the chemical constituents of RNA can be written in two ways:

Nitrogenous Base + Ribose Sugar + Phosphate = Nucleotides

Nitrogenous Base + Ribose Sugar = Nucleoside

The nitrogenous bases of ribonucleic acid are of two types:

(i) Purine

(ii) Pyrimidine.

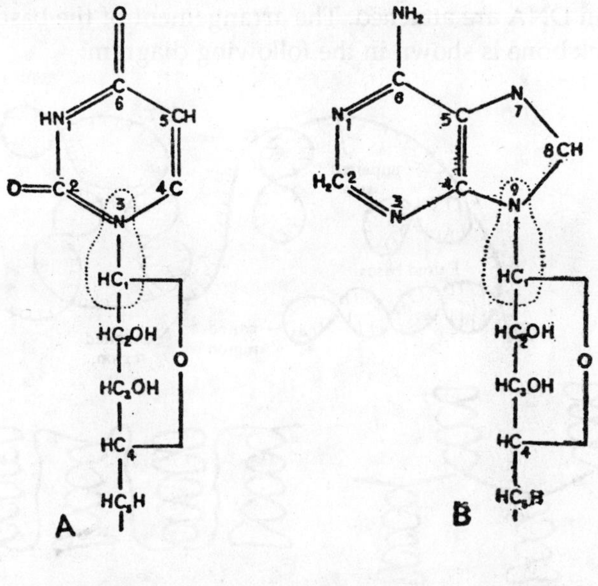

Fig. 5.14. (A) Glycosidic bond between (sugar) C1 and (Base) N_3 (for Pyrimidine base) (B) Glycosidic bond between (Sugar) C1 and (Base) N9 (for Purine base).

Purine bases comprise mainly adenine (A) and guanine (G) while pyrimidine bases comprise cytosine (C) and Uracil (U).

These bases are joined to the pentoses by N-C *glycosidic bonds*. For the purine bases the glycosidic bond is between the C_1 position of the pentose sugar and the N_9 position of the base. For the pyrimidines, the linkage joins the C_1 and N_3 positions as shown in figure given below :

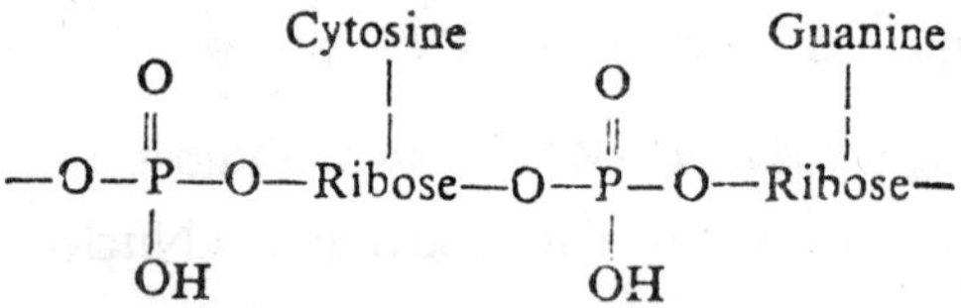

The bases are arranged on the back bone of the chain which is composed of alternating sugar and phosphate. The bases are attached to the ribose sugar in the same way, the bases in DNA are attached. The arrangement of the bases on the back bone is shown in the following diagram:

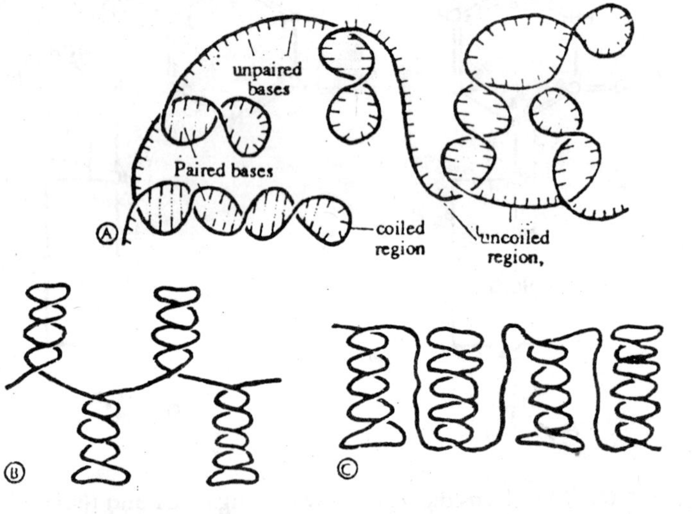

Fig. 5.15. Showing the Secondary structure of RNA. (A) an extended strand (B) with absence of positive enteraction there is mutual repulsion of helicel regions. (C) With presence of positive interaction there is association between helical region.

Secondary Structure of Ribonucleic Acid

Generalization as to the secondary structure of RNA is probably less justified than in the case of DNA, as the available information is much less extensive. The available information does permit the following conclusions :

(1) There is no evidence for the existence of more than one polyribonucleotide strand in the natural RNA's

(2) Hence the helical regions of natural RNA arise from the hair pin like bending of the single strand back upon itself.

The helical regions are stabilized by A-U and C-G interactions similar to those which stabilize the confirmation of DNA. However, the irregular base composition of RNA does not allow the incorporation of the entire molecule into double perfect helic of this type, and the helical zone co-exist with the random segments.

(3) Two alternative models for the structure of RNA may be postulated : (a) The non-helical fraction of RNA may take the form of extended amorphous zones separating segments of nearly perfect helical organization. This model pre-suppose that extensive complementary sequences exist within the single strand. These tend to seek each other out to form perfect or nearly perfect, helical regions. (b) Alternatively, nucleotides which do not participate in the helical organization may occur largely as numerous small loops external to the helically paired structure.

Types of Ribonucleic Acids

Four types of robonucleic acids have been studied and they are as follows :

(1) m RNA
(2) t RNA
(3) r RNA
(4) v RNA

1. *Messenger (mRNA)* : Studies by *Huxley* and later by *Volkin* and *Astrachan* with T_2 bacteriophage infected (bacterial virus, *E.coli)* in 1956 led to the first existence for the unique kind of the RNA, later called messenger RNA. *Jacob* and *Monod* (1961) who first proposed this name, believed "messenger" to be necessary to convey the message of the genetic code from the DNA of the nucleus to the cytoplasm. This messenger is know as mRNA and it acts as a template for the translation of the DNA code into a specific protein. Thus, it helps in determining the structure of protein.

mRNA molecules exist in a large variety of lengths, depending on the length of the polypeptide chain for which they code. Its molecular weight and sedimention value vary considerably. Generally the life span of mRNA is relatively short, although in some eukaryotic cells there is evidence of some degree of stability.

Further mRNA is always complementry to a portion of strand of DNA, with the except that uracil replaces thymine. It should therefore hybridize with the DNA from which it was transcribed.

It is now an established fact that when the code has been transcribed from DNA on to mRNA, the later leaves the nucleus passes through the nuclear membrane into the cytoplasm. Here it moves to the ribosomes, the site of protein synthesis. mRNA molecules attaches reversibly to the surface of ribosome always binding to the smaller sub-units. The magnesium cation, Mg^{++} is involved in the formation of a complex between mRNA and ribosomes.

Types of mRNA

On the basis of size and the coding two types of mRNA has been described :

(a) *Monocistronic mRNA:* Mostly the mRNA carries the codons of single cistron (i.e., codes for one complete protein molecule) of the DNA. Such mRNA molecule is called monocistronic mRNA. For example, for the synthesis of a polypeptide chain of 300 to 500 amino acid residues, a monocistronic mRNA of *E.coli* contains 900 to 1500 nucleotides in its molecule.

(b) *Polygenic or polycistronic mRNA* : Sometimes a mRNA molecule carries the codes from several adjacent DNA cistrons and become much longer in size. This type of mRNA is called polygenic or polycistronic mRNA. For example, for the metabolism of the histidine protein the cell synthesizes about 10 specific enzymatic proteins and a mRNA in this case may carry codons for all the 10 enzymes.

Informosomes

In eukaryons mRNA is associated with protein forming ribonucleo-protein complexes. Some of these complexes remain free in the cytoplasm without being attached to the polyribosomes. These are named as informosomes by *Spirin*. These are very stable and can remain in the cell cytoplasm for several days. The stability is attributed by the presence of protein sheath around mRNA molecule. In informosomes proteins and mRNA are found in the ratio of 4 : 1 . These occur in the form of particles of 80S or less.

Since informosomes are very stable, these are used by the cell in protein synthesis only when there is delay in translation. In embryonic cells, genetic apparatus starts functioning only late during organogenesis. Therefore, in early stages of cleavage when there is no rRNA, informosomes govern the process of protein synthesis.

Stability: The cell does not contain large quantities of mRNA. This is because mRNA, unlike other RNAs is constantly undergoing breakdown. It is broken down to its constituent ribonucleotides by ribonucleases. In bacteria, mRNA may be so short-lived that while one end is translating protein the other end may be undergoing breakdown. In *E.coli* the average half-life of some mRNA is about two minutes.

Eukaryotic cells, on the other hand, contain metabolically stable mRNA. Mammalian reticulocytes are immature RBC which have lost their nuclei but have retained ribosomes. These cells can synthesize haemoglobin for many hours and even days by utilizing the mRNA transcribed by the nuclei when they were present.

2. *Transfer Ribonucleic Acid:* In the literature, t-RNA molecules have been variously termed tRNA, sRNA for soluble RNA, or adaptor RNA. It forms about 10-15% of the total RNA in most of the existence. Each tRNA contains approximate 80 nucleotides. The soluble or t-RNA forms large varieties of different cells, is homogeneous in its molecular size having a molecular weight of about 25000.

The nucleotide sequence of tRNA was first worked out by *Holley et al* (1965) for yeast alanine tRNA. Since then the sequence of about 75 different tRNAs has been established. According to the Holley's leaf model the tRNA is a polynucleotide chain which is folded upon itself to form 5 arms. As a result of the folding the 3' and 5' ends of the chain come near each other. An arm consists of a stem and a loop. Bases are arranged in pairs in arms. There is no base pairing in the loops.

The 3' end of the polynuceotide chain ends in CCA base sequence. This represents site for the attachment of activated amino acid. The end of the chain terminates with guanine base. The bent in the chain of each tRNA molecule contains a definite sequence of three nitrogenous bases which constitute the anticodon. It recognizes the codon on mRNA.

Four different regions or special sites can be recognized in the molecule of tRNA. These are :

(I) Amino acid attachment site: It occurs at the 3' end of the tRNA chain and has -OH group of ribose molecule free to pair. This -OH group combines with the specific amino acid in the presence of ATP forming amino-acyl tRNA. It is common to all the tRNA molecules.

(II) Recognition site: It contains a specific base sequence which dictates the attachment of correct amino acid to the tRNA molecule. It matches with the amino acid activating enzyme through which attachment of amino acid to tRNA takes place.

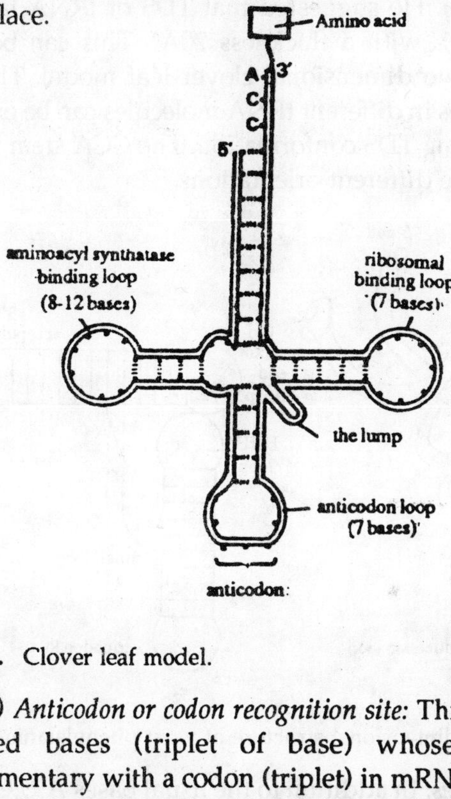

Fig. 5.16. Clover leaf model.

(III) Anticodon or codon recognition site: This site has three unpaired bases (triplet of base) whose sequence is complementary with a codon (triplet) in mRNA. Therefore, it determines the pairing of tRNA with the specific codon (triplet) of mRNA. It is, therefore, the most specific region of tRNA molecule.

(IV) Ribosome recognition site: This helps in the attachment of tRNA to the ribosome. This site is common to all the molecules of tRNA.

Three dimensional structure of tRNA: In order to understand the structure-function relationship of tRNA, its three dimensional structure (TDS) should be known. With the availability of tRNA in crystalline form in 1968, TDS of tRNA could be studied through X-ray crystallography as was done of DNA double helix. About a dozen models of TDS were proposed, but the latest and most acceptable model was proposed in 1973 by *S.H.Kim.* He suggested that TDS of tRNA takes the shape of letter L with a thickness 20A°. This can be easily derived from two dimensional clover leaf model. The extra arm which varies in different tRNA molecules can be extended without distorting TDS conformation. The CCA stem projects out and can take different orientations.

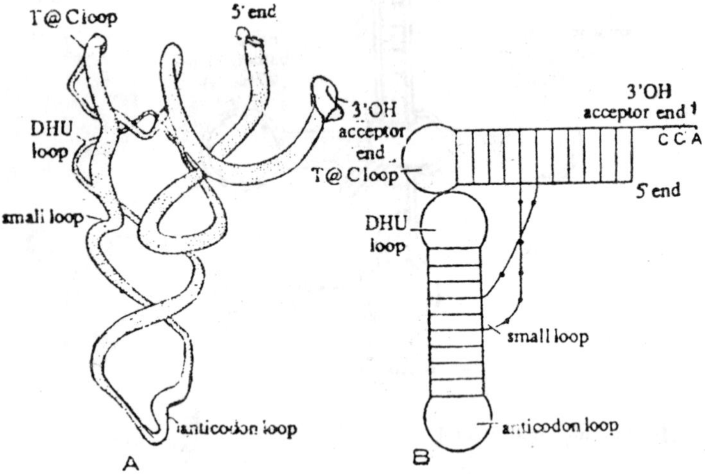

Fig. 5.17. Three dimensional structure of yeast phenylalanine tRNA.

Unusual bases: In addition to the usual bases A,U,G and C, tRNA contain a number of unusual bases, and in this respect differs from mRNA and rRNA. The unusual bases of tRNA account for 15-20% of the total RNA of the cell. Most of the unusual bases are formed by methylation (addition of-CH3 or methyl group to the usual bases), e.g., cytosine and guanine on methylation yield methyclotosine and methylguanine, respectively. Precursor tRNA molecules transcribed on the DNA template contain the usual bases. These are then modified to unusual bases. The unusual bases are important because they protect the tRNA molecule against degradation by RNAse. This protection in necessary because RNA is found floating freely in the cell.

Some of the unusual bases of tRNA are methylguanine (GMe), dimethylguanine (GMe_2), methyicytosine (Me), ribothymine (T) pseudouribine (ψ), dihydrouridine (DHU_2, H_2 UH_2), inosine (I) and methylinosine (IMe, MeI).

3. *Ribsomal Ribonucleic acid rRNA:* Ribosomes are small particles of about 200A° to 250A° in diameter, and are generally present in the cytoplasm either lying free in the cytoplasm or attached to the outer membrane of endoplasmic reticulum or "nuclear membrane." These particles are ribosomes containing about 60% ribonucleic acid or RNA which is ribosomal RNA or rRNA and 40% protein. It is generally agreed that cytoplasmic ribosomes or their subunits are made in the nucleus and are subsequently transferred to the cytoplasm.

However, the mode of their passages across the nuclear membrane is still a matter of controversy.

Even from the morphological evidence, it would appear that the reticular or microvesicular spheres of the nucleus could be engaged in either the synthesis of rRNA, or its assembly, or both.

The finding that hybridization of rRNA takes lace with the chromatin associated with the nucleolar region suggests that rRNA is synthesized on the part of the chromosome traversing the nucleolus (the nucleolar organizer) or, in some cases, on DNA detached from the chromosome and associated with the nucleolus.

Ribosomal RNA differs in size and base contents from t-RNA and from other RNA classes of most cells. r-RNA from most organism is relatively rich in guanine and cytosine, and it lacks many of the rare bases found in tRNA, although some methylated bases are present.

The function of rRNA in protein synthesis is not yet completely clear. The unpaired bases in the molecule may bind mRNA and tRNA to ribosomes possibly by Mg^{2+} linkages between phosphate groups on the two molecules.

Ribosomal RNA also gives some evidence for developing the double helical structure.

4. *Virul RNA (vRNA):* Viruses may contain either DNA or RNA. However, each particular virus generally contains nucleic acid of only one type. Some examples of RNA viruses are influenza, turnip yellow mosaic, tomato bushy stunt, tobacco necrosis, and the particularly well studies tobacco mosaic virus (TMV).

Replication of Ribonucleic Acid

At first it was difficult to apply the rules of DNA replication to RNA replication, because less is known about the macromolecular structure of DNA, but with the basic principles of the template theory laid down, the synthesis of RNA became accessible to further investigation.

As discussed above the four various kinds of RNAs, which are structurally alike, but differ considerably in molecular weight and are distributed in different parts of the cell, are replicated or biosynthesized as follows:

Biosynthesis of rRNA: rRNA although present in ribosomes is formed inside the nucleus. DNA associated with the nucleolus is responsible for coding rRNA. This part of DNA is known as nucleolar organizer. In bacteria about 10-200 cistrons are concerned with the rRNA synthesis, whereas in hither organisms 200-2000 tightly clustered rDNA cistrons are involved in rRNA synthesis.

The two types of rRNA (28S and 18S) are transcribed from the nucleolar DNA as a single elongated unit of 45S. Inside the nucleus, the 45S RNA is methylated and complexed with protein. Finally, via a number steps of it is cleaved into 32S and 18S segments. The 18S RNA gets associated with basic proteins to from the small subunit of ribosome. The 32S segment is further severed and finally change to 28S RNA. This 28S RNA also gets associated with proteins and forms the large subunit of ribosome. These units then come out into the cytoplasm.

The 5S RNA has a extrancleolar origin, transcribed from genes located outside the nucleolar organizer. It gets associated with each large ribsomal subunit. The function of 5S RNA is unknown.

Biosynthesis of mRNA. The existence of a short lived type of RNA was predicted on theoratical grounds by *F. Jacob, J. Monod,* and associates. The name messenger RNA was used to describe a form of RNA that would be made in the cell nucleus, but would be transferred to the cytoplasm during protein synthesis. During protein synthesis, the amino acid sequence was determined by DNA which is mainly in the nucleus. The information from DNA in the form of a specific sequence of bases had to be transferred some how to the cytoplasm.

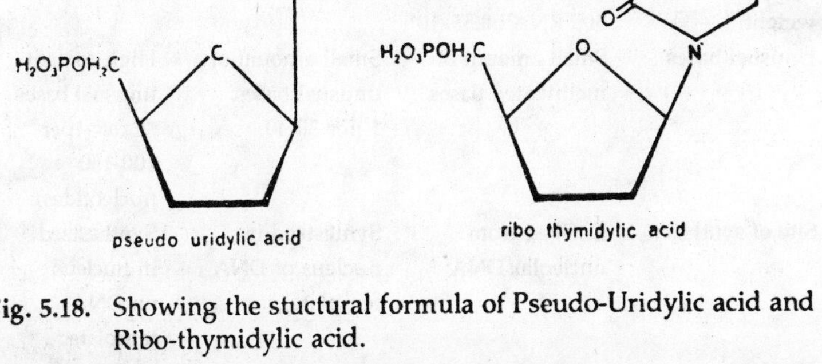

Fig. 5.18. Showing the stuctural formula of Pseudo-Uridylic acid and Ribo-thymidylic acid.

The notion of a mRNA demanded that DNA acts as template for RNA synthesis. The principle of base pairs still applies here, because uracil in RNA has the same dimension as thymine in DNA, and one can substitute for the other. In this manner a DNA strand could function as template for RNA synthesis.

```
A — T — T — G — A — C DNA
|   |   |   |   |   |
U — A — A — C — U — G m RNA
```

Biosynthesis of Transfer RNA: It is of low molecular weight. The structure of tRNA has been explained already. Although the details of its synthesis are not known. It is known that tRNA is synthesized on DNA templates by steps similar to those of the other types of RNA.

Comparison of the different types of RNA

	Ribosomala RNA (rRNA)	*Messenger RNA (mRNA)*	*Transfer RNA (tRNA)*
Percentage of total RNA of cell	80%	3 to 5%	10 to 20%
Sedimentation Coefficient	28S, 18S, 5.8S, 5S 23S, 16S 5S	8S	3.8S
Number of nucleotides	5S RNA:120 nucleo-otides 16-18S RNA : 1,600 to 2,500 nucleotides. 23-28S RNA: 3,200 to 5,500 mucleotides	*E.coli:* 900 to 1,500 nucieotides	73 to 93 nucleotides.
Molecular weight	23S RNA: 1.1×10^4 30S RNA : 0.55×10^4	500,00	25,000-30,00
Unusual bases	Small amount of methylated bases.	Small amount of unusual bases. 1 per 30-40	High content unusual bases **E.coli:* 1per 100-150 nucleotides).
Site of synthesis	Derived from nucleolar DNA	Syntesized in nucleus of DNA template.	Synthesized in nucleus on DNA template.

Beginning of Synthesis	Synthesis begins at gastrulation, and increases as development proceeds.	Some mRNA is found in the ovum. New mRNA is synthesized during early cleavage.	tRNA synthesis occurs at the end of cleavage stages.
Base of relation ship DNA	no obvious base relationship to rDNA. rRNA is from only small section of DNA.	mRNA shows base relationship to DNA. It is formed from all sections of DNA.	Same as in mRNA
function	Unpaired bases may bind mRNAand tRNA to ribosomes.	Conveys genetic information from DNA ofchromosomeswhere it takes part in protein synthesis	Adaptor for attaching amino acids to MRNA template.

Comparison between DNA and RNA

Deoxyribonuclelc acid	*Ribonucleic acid*
1. DNA is the usual genetic material	1. RNA is the genetic material of some viruses.
2. DNA is usually double-stranded, (In certain viruses DNA in single stranded, $\phi\chi$ 174).	2. Most cellar RNA is single stranded. (Some viruses e.g., revirus, have double stranded RNA).
3. The pentose sugar is deoxyribose	3. The pentose sugar is ribose.
4. The common organic bases are adenine, guanine, cytosine and thymine.	4. The common organic base are adenine, guanine, cytosine and uracil.
5. Base pairing : adenine paris with thymine and guanine with cytosine	5. Adenine pairs with uracil and guanine with cytosine.
6. Pairing of bases is throughout the length of the molecule.	6. Pairing of bases is only in the helical region.
7. There are fewer uncommon bases.	7. Three are more uncommon bases.

8. DNA is only of one type	8. Three are three types of RNA, messenger RNA, ribosomal and RNA transfer RNA.
9. Most of the DNA is found in the chromosomes. Some DNA is also found in the cytoplasm e.g. in mitochondria and chloroplasts.	9. Messenger RNA is formed on the chromosomes, and is found in the nucleolus and cytoplasm. rRNA and tRNA are also formed on the chromosomes, and are found in cytoplasm.
10. Denaturation(melting) is partially reversible only under certain conditions of slow cooling (renaturation).	10. Complete and practically instantaneous reversibility of the process of melting.
11. Sharp, narrow temperature interval of transition in melting	11. Board temperature interval of transition in melting.
12. DNA on replication forms DNA, and on transcription forms RNA.	12. Usually RNA does not replicate or transcribe. (In certain viruses RNA can synthesize an RNA chain).
13. Genetic messages are usually encoded in DNA	13. The usual function of RNA is DNA into proteins.
14. DNA consists of a large number of nucleotides, up to 4.3 million.	14. RNA consists of fewer nucleotides, up to 12,000.

Several aspects of tRNA merit some comment. One is that all tRNA types so far isolated contain the same sequence of bases at one end of the molecule, cytosine-cytosine-adenine (C-C-A). Another interesting observation is that in tRNA one invariably finds nucleotides not yet described: Seudo-uridylic acid and sub-othymidylic acid.

The presence of the two bases prsent a problem. If synthesis of tRNA proceeds by the same mechanism as other types of RNA, the enzymes involved in tRNA synthesis must be assumed to have unique specificity for the two new nucleotides. A more attractive idea is that tRNA is formed from the familiar four nucleotides, and by enzymatic modification of selected uridylic acid units in the tRNA, the new nucleotides, are formed. Pseudo-uridylic acid could be the product of an isomerization reaction of tRNA. Ribothymidylic acid is known to be formed by methylation of uridylic acid present in tRNA. The characteristic terminal-C-C-A sequence is believed to be added enzymatically after tRNA is already made.

6

Role of Nucleus in Development

A living cell is made of a nucleus and cytoplasm. There is a partnership between the nucleus and the cytoplasm. The nucleus, with its store of hereditary factors, the *genes,* is the conservative member of the partnership. It controls the syntheses which go on in the cell. Hence, it directs that most amazing of all syntheses. The course of development. Hence also it determines the species and individual characteristics which are the outcome of development. On the other hand, it is primarily the cytoplasm which develops and is the progressive member of the partnership. It begins in relative simplicity and then, by differentiation and molding, becomes the mature organism in all of its structural and functional complexity.

For the smooth functioning of the cell and also for the continuation of life both the nucleus and the cytoplasm are needed. Many experiments have been performed in which the nucleus of a cell has been removed. For example, when an amoeba is cut into two parts, the part containing the nucleus heels and continues to live. The part which lacks a nucleus-may carry on for a time, but ultimately it wears itself out and perishes, for it cannot continue to rebuild its substance. It is reported that a sea urchin egg from which the nucleus has been removed may cleave several times, but it will not continue to develop.

It is even more obvious that a nucleus cannot live and carry on its functions without the cytoplasm. In particular, the nucleus depends upon the cytoplasm for its raw material and energy. It is probable that only in the cytoplasm do the oxidations take place by which energy-rich molecules (ATP) are synthesized.

The Life Cycle of the Nucleus

The story of the nucleus is a repetitive one and begins at fertilization when the chromosomes of the two germ cells gametes), an egg and a sperm, come together in the nucleus of the fertilized egg or zygote. The story ends and is ready to start again, at least as far as the contribution of the individual to the race is concerned, when the new individual has matured and has produced gametes of its own. Between fertilization and the maturing of the germ cells, the story of the nucleus is one of alternating growth and division by mitosis. It is responsible for growth repair and inheritance.

Fertilization

It is the union of the male and female gametes. From the standpoint of the nucleus, fertilization is the coming together of the chromosomes of the egg and the chromosomes of the sperm in the same nucleus, the zygotic nucleus. One set of chromosomes, the haploid number, is present in the egg. Another set is present in the sperm. This makes two sets, the diploid number present in the zygote. In man the haploid number is 23. Hence the human diploid number is 46.

Each chromosome of a haploid set has its own individuality; that is, it is of its own kind. It becomes visible at the beginning of cell division and disappears at the close of cell division in its own characteristic manner. Very significantly, its genes are arranged within it in a definite sequence. Moreover, it affects development in its own specific way. Each chromosome of a diploid set usually has a mate. Yet although the two mates are within the same nucleus, with rare exceptions each remains independent of the other throughout the cell divisions (mitoses) of development.

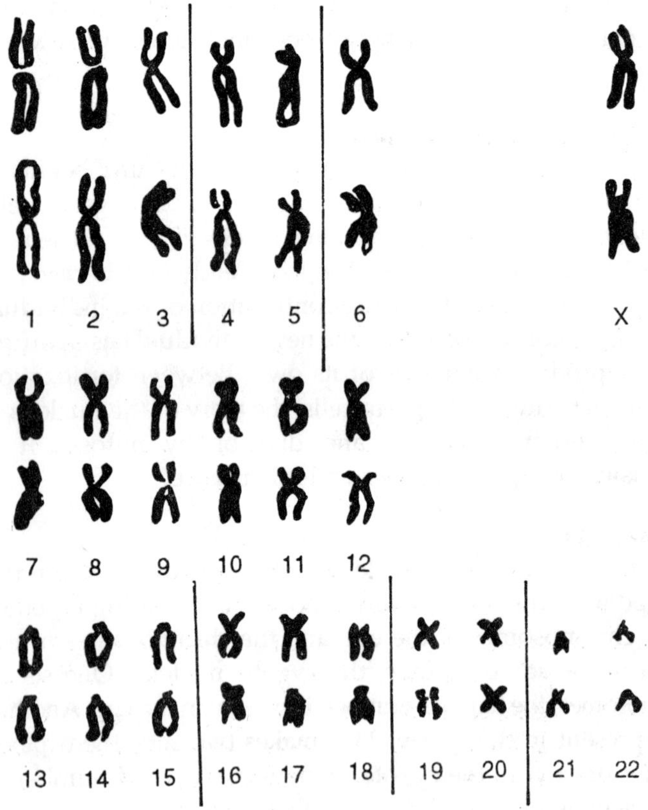

Fig. 6.1. The 46 human female chromosomes.

Mitosis

The coming together of the male and female nuclei in fertilization is followed by *mitosis*, the process by which one cell becomes two cells in so complicated but precise a manner that each daughter cell possesses just the same two sets of chromosomes that the mother cell possessed. It is the mitosis only that convents a single called zygote into a multicellular organism with billion cells.

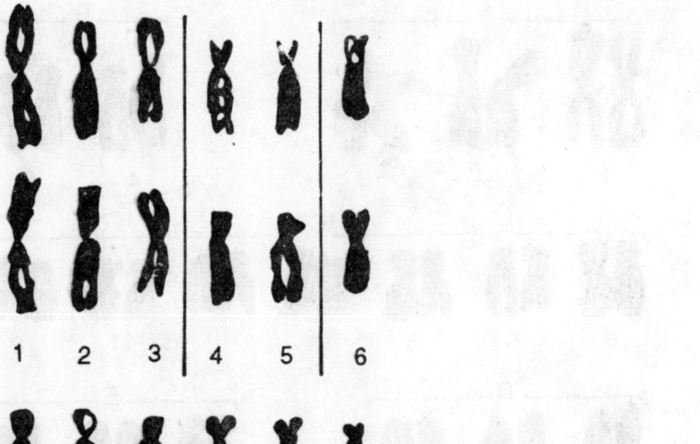

Fig. 6.2. The 46 human male chromosomes.

In most of cases the *Karyokinesis* which the division of nucleus is followed by *cytokmesis*—the division of cytoplasm. But in group unsecta specially during cleavage the nuclear division is not followed by the division of cytoplasm. The abnormal division results a number of chromosomal abnormablities to such as *Turnev's syndrome,* Mangloid idiocy, klinifeltev's syndrome, super female etc.

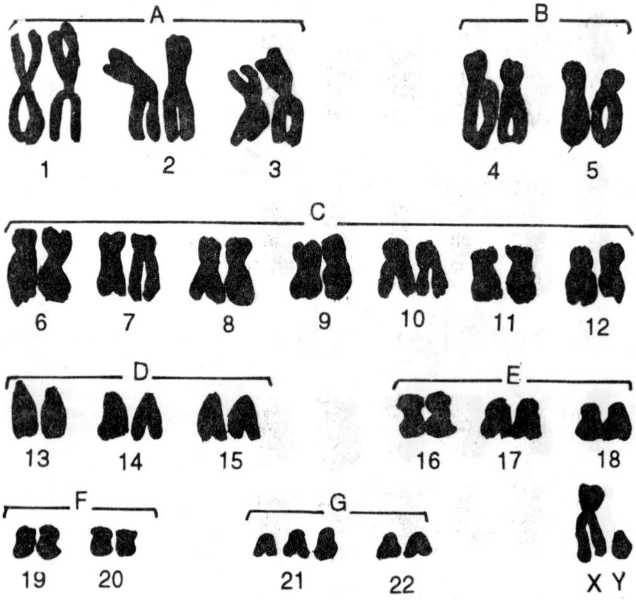

Fig. 6.3. Chromosomes of a mongoloid idiot.

The process of all division is a continuous one but on the basis of certain major changes it can be divided into four phases—prophase, metaphase amaphase and telophase. The interphase is the stage between telophase of one division and the prophase of the next division. All necessary biochemical changes occur during interphase at which the cell is metabolically most active.

It is during the interphase that the genes self-duplicate and carry on their function of supervising syntheses. Following self-duplication, the chromosomes continue to be stretched out. Usually they are so attenuated that their boundaries cannot be seen with the light microscope. The extended chromosomes are able to carry on their chemical functions in an efficient manner.

"Nuclear sap," or karyoplasm fills the interstices between the chromosomes. One or more rounded bodies, the *nucleoli*, are usually present in the interphase nucleus. The visible boundary between the nucleus and the surrounding cytoplasm in the *nuclear membrane*. In the cytoplasm adjacent to the nucleus there is a body, the central body, which consists of two granules or, after each granule has replicated, of two pairs of granules, the *centrioles*.

Prophase: The events in the nucleus and in the cytoplasm which lead to the division of the cell constitute the prophase. The following is a generalized account.

1. In the cytoplasm the pairs of centrioles move apart, and a sphere of gel rays, the *aster*, grows around each pair. Together the two asters constitute the *amphiaster*. It is of interest that the centrioles are short cylindrical bodies each consisting of nine rods (or double or triple rods). In this they are structurally comparable to a cilium or flagellum. (They lack the two central tubules of a cilium.) In a typical case, the centrioles of a pair lie at right angles to each other and to the axis through the pairs.

2. The chromosomes condense into tight spiral coils. Under the right microscope these appear to be rods. But when the material is suitably fixed and strained, each chromosome can be seen to be a double structure, each half of which is a chromatid.

3. A fusion body, the *mitotic spindle*, now forms between the pairs of centrioles, sometimes in the cytoplasm, sometimes in the substance of the nucleus. The spindle's poles are anchored in the region of the centrioles of the cytoplasm. The electron microscope shows that it consists of hollow fibrils (microtubules). It elongates as the centrioles move apart. Possibly this is a result of imbibition of water, or incorporation of other molecules, or change in the shape of its constituent proteins. The amphiaster and the spindle together form the "achromatic figure," so-called because it does not stain deeply with basic dyes.

4. According to some accounts, chromosomal fibers grow out from particular locations in each chromosome until they reach the poles of the spindle. The locations from which they grow out are known as *centromeres* or kinetochores. The chromosomal fibers seem to consist of a contractile fibrous gel.

5. Toward the end of the prophase, the nuclear membrane disappears.

6. The nucleolus also disintegrates, and its substance, along with the nuclear sap, passes into the cytoplasm.

Metaphase: The chromosomal fibers appear to contract and pull the chromosomes to the equator of the spindle (prometaphase). When thus spread out and seen in polar view, their number can be counted.

Anaphase: Continuing, or resuming, what appears to be contraction the chromosomal fibers seem to drag each chromatid toward one pole of the spindle. Actually two processes may be at work a continuing enlargement and elongation of the spindle and a shortening of the chromosomal fibers. Since the fibers shorten without becoming thicker, the process probably involves the removal of water or other molecules from the fibers. As the chromatids move apart, they are often V or J-shaped depending on the location of the centromeres to which the chromosomal fibers are attached. A band of "interzonal fibers" is often seen for a time after the separation has been accomplished, connecting the chromosomes which have pulled apart, and often including a remnant of the spindle.

Telophase: Having reached the poles of the spindle, the chromatids swell and disappear from view, as seen with the light microscope. Although usually called "daughter chromosomes," they are actually single chromatids. The nuclear membrane reappears, and shortly one or more nucleoli re-form. Of the mitotic apparatus, little more than the centrioles of the cytoplasm remain.

Interphase: The daughter calls have now returned to the interphase or "resting stage" except for one feature: each chromosome still consists of only one chromatid. Investigations by a staining technique known as the Feulgen reaction (specific for DNA) have made it clear that the restoration of chromosomal material takes place during the interphase. At that time each chromatid becomes two chromatids and the double nature of each chromosome is restored.

Studies employing radioactive isotopes as tracers have added some important information. It is not the chromatids which replicate. It is half-chromatids, also termed chromonemata, which undergo self-duplication. Taylor (1957) placed bean seedlings in a solution containing radioactive (tritiated) thymidine (a substance which the cell uses to synthesize DNA). He left them in this solution' until those cells which were in early interphase had each replicated once. He then cut the roots off and transferred them to a non-radioactive solution. In this they replicated a second time. Then, having killed the cells, he determined by autoradiography whether or not the chromosomes were radioactive. Now, if whole chromatids had replicated when they were in the first or radioactive thymidine solution, they would have made for themselves radioactive mates. After cell division half of the daughter chromosomes would have been radioactive, and half, namely, the original chromatids, would have been non-radioactive. But Taylor found that *all* the daughter chromosomes were radioactive! Why? The answer is that half-chromatids replicated, so that each chromatid now consisted of a new radioactive half-chromatid and an original non-radioactive half-chromatid.

At the second replication, this time in non-radioactive solution, each half-chromatid, whether radioactive or non-radioactive mate. Therefore following the next cell division, the daughter chromosomes (chromatids) were half radioactive and half non-radioactive.

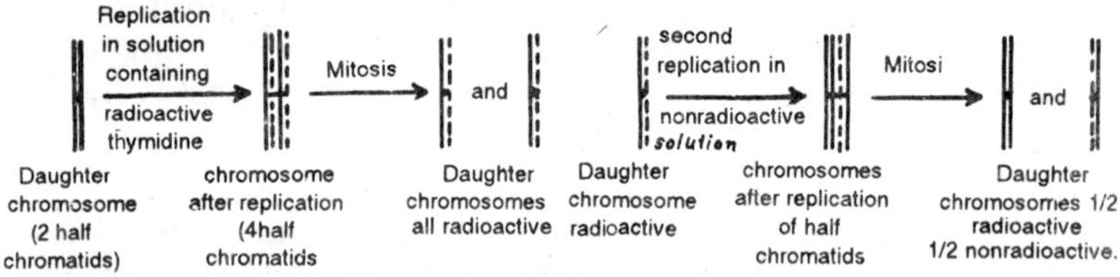

Similar results were obtained by Prescott and Bender, using cells of hamster issues raised in tissue culture and also human leucocytes.

Experiments of this sort emphasize the remarkable stability of the half-chromatids. Although most of macromolecules within the cell are in a state of flux, continually braking down and being replaced, the half-chromatids are passed on intact from one cell generation to the next. And this has been going on for countless generations!

There is evidence that the centrioles may self-duplicate even earlier than the chromosomes, in some cases as early as the preceding anaphase. At this time each centriole (there are two at each pole) makes for itself a partner, not usually by longitudinal splitting, however, as might be expected, but apparently by a process of lateral budding.

The introduction of mitosis in the history of life on the earth was one of the grand innovations of all time. Possibly it took place only once, for, which variations, it occurs in all plants and animals except bacteria and blue-green algae. Even in these there must be a process akin to mitosis, for without some such process the transmission of hereditary characters to each new generation in a balanced fashion would not be possible.

Meiosis

In 1884 Van Beneden described the processes of maturation, fertilization, and cleavage as they take place in the parasitic roundworm of the horse. *Ascaris megalocephala.* He demonstrated that the egg and sperm contribute an equal number of chromosomes to the offspring. This led, a few years later, to the discovery of *meiosis,* the process by which the number of chromosomes is reduced to one half when the germ cells ripen.

Each body cell and each unripe germ cell or *gonium* has two sets of chromosomes, the *diploid number*. Each of the chromosomes has a mate (except in the case of the unpaired sex chromosomes). When meiosis takes place, an oogonium or spermatogonium, as the case may be, gives away one set of its two sets of chromosomes; that is, it gives away one chromosome of each pair of chromosomes. The result is that each ripe egg or sperm has left only one set of chromosomes, the *haploid number*.

Now, meiosis, unlike mitosis (1) takes place only in gonads, (2) occurs only in those unripe germ cells which at the time are in the process of ripening, and (3) although it begins earlier, reaches completion only during the period of sexual maturity of the plant or animal. In most cases some unripe germ cells (residual gonia) remain in the gonad where they multiply by mitosis and so produce more gonia. In female birds and mammals, however, the period of multiplication of oogonia comes to an end about the time of hatching or birth. At this time several million gonia may be present; yet only a very few ever ripen into ova-about 400 in the case of women.

The gonia are thus lineal descendants, by mitosis, of the original fertilized egg or zygote. Hence, they are cousins of the body cells, and, generally speaking, they possess the same two sets of chromosomes which the body cells possess.

Meiosis has been compared to two mitotic divisions, for

two successive spindles are formed, and two separations of chromosomes take place. Yet there is only one complete prophase, and this differs from a typical prophase in that it involves a pairing of the chromosomes side by side, each with its mate. The process of pairing is known as *synapsis.* In a sense this pairing of homologous chromosomes completes the coming-together of the germ cells at fertilization; for all through the mitotic cell divisions of development the chromosomes of the egg and those which came from the sperm have remained separate. Now each chromosome joins its mate.

The stages of the prophase of meiosis have been given names; leptotene, when the chromosomes become threadlike; synaptene when they come together in pairs; pachytene, when they contract and become tightly coiled; diplotene, when they begin to separate, except for locations when "crossing over" between synapting chromatids has taken place; and diakinesis, when again they become compacted prior to separating.

Note that at the beginning of meiosis each chromosome consists of two chromatids as in ordinary mitosis, so that, when pairing takes place, the result is bundles of four chromatids. These are known as *bivalents* because each consists of two chromosomes. They are also known as tetrads because each is four chromatids.

During this period preparatory to the first meiotic division, the ripening germ cell is called a *primary gametocyte,* primary spermatocyte or oocyte, as the case may be. The diplotene stage of oocytes is of particular interest, not only because it is long drawn out (it may be as long as 40 years in the human female), but because it is a period of great growth. The nucleus enlarges greatly due to the accumulation of nuclear sap and is often referred to as the *germinal vesicle.* The chromosome become fuzzy objects which have been compared to lampbrushes.

First meiotic division: The stages of the prophase which have just been described end as the metaphase approaches. The bivalents move to the equator of the first meiotic spindle.

At the metaphase they split; and at the anaphase each bivalent divides, and *monovalent chromosome* or dyad (pair of chromatids) goes to each pole of the spindle. The monovalents are actually the original chromosomes which united in synapsis, except that some crossing-over has usually taken place. The products of the first meiotic division are known as *secondary gametocytes.*

Second meiotic division: Usually the telophase of the first meiotic division is brief and is quickly followed by the much reduced prophase of the second meiotic division. In some instances, the anaphase of the first meiosis leads directly into the metaphase of the second meiosis. In any case, the monovalent chromosome (dyads, two chromatids each) take a position at the equator of the second metaphase spindle. At the anaphase the chroatids or *daughter chromosomes* (monads) mcve apart to the poles of the spindle. The resulting cells are known as *tids,* ootids or spermatids, as the case may be.

Note that a reduction in the number of chromosomes has taken place. The gonia had two sets (diploid number) of chromosomes. The primary gametocyte had one set (haploid number) of bivalent chromosomes (tetrads). The secondary gametocyts had one set of monovalent chromosomes (dyads). Now the tids have one set of daughter chromosomes, that is, one set of chromatids (monads). But this is important; the set which each tid had is a complete set with one chromosome present to represent each pair of chromosomes of the original diploid set.

In the case of spermatogenesis, each primary spermatocyte divides twice equally and produces four equivalent spermatids. In oogenesis, however, although the meiotic divisions of the nucleus are equal, those of the cytoplasm are grossly unequal. At each division most all the cytoplasm goes to only one of the daughter cells and thus is conserved to supply the substance of the embryo. The result is that the first meiotic division gives rise to one secondary oocyte and one small *first polar body* . The second meiotic division similarly produces one ootid and a small *second polar body*. The spermatids undergo metamorphosis and become sperm cells or *spermatozoa*. The ootids, on the other hand, as a rule need only to burst from the ovary to cause them to ripen and become ready for fertilization.

It is a strange fact that the stage in meiosis which is attained before the fertilizing sperm enters the egg is not the same for all species. In sea urchin eggs meiosis is complete and both polar bodies have been formed before the egg is receptive to the sperm. This is rather rare. In vertebrates the first polar body has been given off and the second meiotic division has progressed to the metaphase before fertilization takes place. The extreme case is the parasitic roundworm, *Ascaris*, in which the sperm enters the cytoplasm of the egg before even the first meiotic spindle has formed. It remains inactive in the center of the cytoplasm while the egg completes meiosis. In most animal species, therefore, the meiosis which closes the nuclear cycle of the egg overlaps to a greater or lesser extent the entrance of the sperm which begins the nuclear cycle of the new individual.

Why does this complicated process of meiosis exist? The answer was pointed out in 1903 by Walter Sutton, who showed the parallelism between the story of hereditary factors (later termed genes), as worked out by Mendel and other students of plant and animal breeding and the behaviour of chromosomes in fertilization, mitosis and meiosis. Meiosis, Sutton showed, is nature's way of reducing the number of chromosomes in each germ cell so that, at fertilization, the normal diploid number of chromosomes will be restored. But it is more than this. It is nature's way of seeing to it that the germ cells are all different. Let us assume that the chromosomes of different pairs are different and also that each chromosome has some mutant genes which make it unlike its mate. In nature this is probably always true. Since all the cell divisions during development are made by mitosis, every gonium will have the same two sets, or diploid number, of chromosomes and genes. Then, when meiosis takes place, each ripe germ cell retains only one chromosome of each pair of chromosomes, one gene of each pair of genes. Moreover, it is a matter of chance which chromosome of a given pair of chromosomes the cell retains.

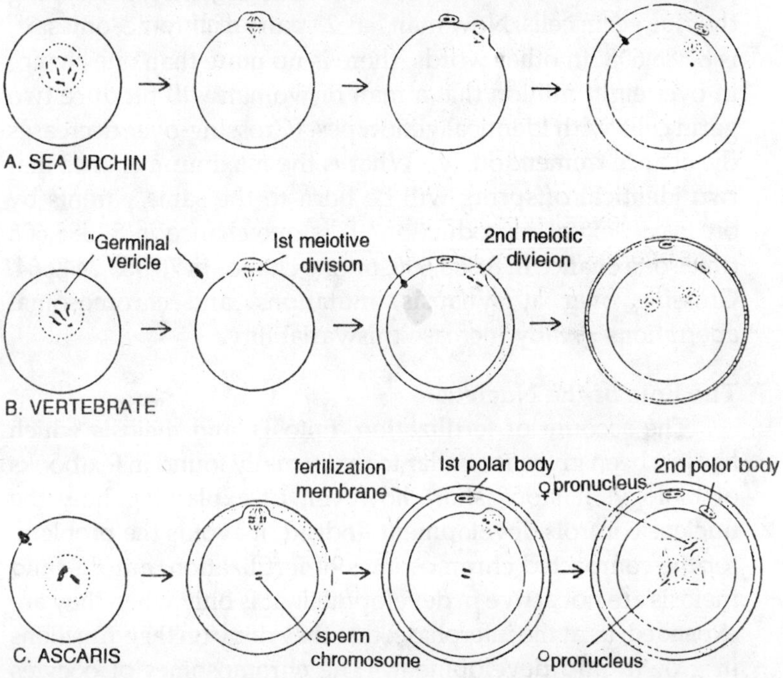

Fig. 6.4. Fertilization (entrance of sperm) takes place at different stages of meiosis in different animals: A, after meiosis is complete in the sea urchin, *B*, midway in the second meiotic division in most vertebrates; and C, before meiosis in the parasitic roundworm *Ascaris*.

Thus, from the standpoint of heredity, there can always be as many different genetic kinds. or *genotypes*, of ripe germ cells as there are possible combinations of chromosomes taking one chromosome from each mating pair of chromosomes. If there would be two kinds of ripe germ cells. If there were just one pair of chromosomes in an unripe germ cell, there wère two pairs to begin with, there would be four possible kinds of germ cells. If there were three pairs, the number would be eight. In general, if there are N pairs of chromosomes before meiosis, there will be 2^N possible kinds of genotypes among the ripe germ cells. Now man has 23 pairs of chromosomes. 2^{23} is 8,388,608. In other words, there is no more than one chance in over eight million that a man or woman will produce two germ cells with identical genotypes. (Crossing-over decreases the chance tremendously.) What is the maximum chance that two identical offspring will be born to the same parents by ordinary sexual reproduction? It is one chance in 8, 388,608 times one chance in 8,388,608, or one chance in 70,368,744,664. Crossing over at synapsis, mutations, and chromosomal aberrations greatly increase this variability.

The Role of the Nucleus

The account of fertilization, mitosis, and meiosis which has just been given is similar to that usually found in textbooks of biology. It stops short however, of explaining how the nucleus controls development. Indeed, it avoids the problem, for the compacted chron osomes of fertilization, mitosis, and meiosis are not active in development. It is only when they are expanded, as at the interphase, that they carry on their functions in growth and development. (The chromosomes of oocytes are expanded also during the long prophase.) Furthermore, the chromosomes presumably are alike in every cell of the body of an embryo. Now, how can chromosomes which are everywhere the same account for the origin of differences within the embryo? In order to approach this problem we must consider the chemistry of the nucleus.

The Chemistry of the Nucleus

The distinctive chemical substances of the nucleus are long chain molecules known as *deoxyribonucleic acid,* or DNA for short. Each such molecule is a gene; more likely it is a string molecule is a gene; more likely it is a string of gene, the material units of heredity. A closely related type of nucleic acid is *ribonucleic acid,* or RNA. Both DNA is in the chromosomes.

RNA is the chromosomes, but especially in the nucleoli, and it is also present in the cytoplasm.

Now, nucleic acids have three unique and remarkable properties:

1. They are self-duplicating molecules. Possibly they are the only truly self-duplicating molecules. Centrioles and plastids are morphologically self-duplicating, but it is likely that, in this case, nucleic acids are involved.

2. Nucleic acids supervise the synthesis of proteins within the cell. The proteins are the principal structural compounds of protoplasm and are the chemical basis of its functioning. They include the enzymes which control the innumerable chemical reactions of the cell and also structural proteins. In fact, a gene (*cistron*) has been defined as the template for the production of one polypeptide chain. It is probably true that, in the absence of nucleic acids, no proteins are ever produced.

These statements are oversimplified. The self-replication of a nucleic acid requires the presence of specific enzymes (polymerases) and numerous other molecules in the protoplasm. The process is therefore a circular one: nucleic acids bring about the synthesis of proteins, and proteins are needed to make possible the replication of the nucleic acids. It is the entire protoplasmic system which endures.

Genes are nor invisible. Crossing-over of parts of genes may take place between synaptic mates during meiosis. The smallest part of a gene which may interchange thus with a mate has been termed a *recon*.

3. Nucleic acids (genes) occasionally mutate; that is, a short segment of a gene may change in composition. When the occurs the gene replicates according to its new or mutant form. If this were not so, life on earth would never have evolved, and the ever-increasing adaptation of living things to their environments would not have been possible. The shortest

segment of a gene which may mutate is probably a single nucleotide unit. It is known as a *muton*.

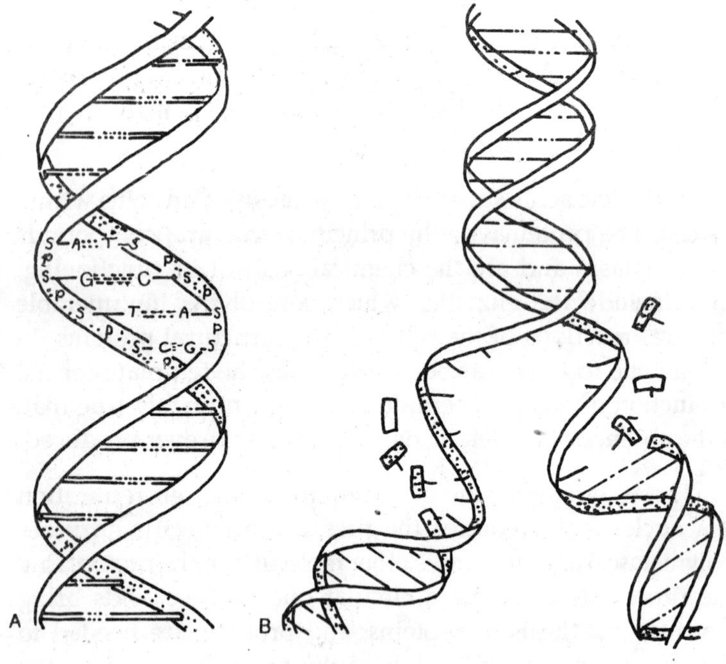

Fig. 6.5. The structure of DNA and the manner of its replication.

Great progress has been made in the last few years in the knowledge of the structure of nucleic acids and of the manner in which they duplicate. According to the Watson-Crick hypothesis, each DNA molecule is a long double chain of units known as nucleotides. It may be compared to a rope ladder twisted into a helix. Each single nucleotide consists of a sugar, deoxyribose (S), a phosphate group (P), and one of four different "bases." The bases are acenine (A, a purine), guanine (G, also a purine), thymine (T, a pyrimidine), and cytosine (C, a pyrimidine). The single chain consists of alternating sugars and phosphates, with a base attached as a side group to each sugar. The precise sequence in which, the nucleotides are arranged in the long nucleic acid chain is the basis of the coded information which is passed on by heredity and which guides the course of development.

In the double chain of the DNA molecule, each base of one chain is linked crosswise, by weak hydrogen bonds, with it mate in the other chain. The relations are precise. Adenine is always bounded with thymine, and guanine with cytosine. There are no other combinations. Hence, the two chains fit each other exactly like a mold fits a model.

When a nucleic acid (DNA) molecule self-duplicates, the chains supposedly separate from each other. Each single chain then makes for itself a replica of its former partner by attracting and assembling to itself appropriate nucleotides from the surrounding protoplasm. The result is that there are now two identical DNA molecules (double chains of nucleotides) where previously there was one. One notes at once the comparison between the replication of the half-chromatids (chromonemata), which takes place during the interphase of mitosis, and the replication of the single chains of DNA molecules.

The nucleus also contains proteins of two sorts: histones (proteins rich in basic amino acids), which are associated with the genes; and nonhistones, which compose the larger amount of protein and may be a part of the metabolic and mitotic machinery of the cell. The relation between histones and DNA is loose, yetit is so consistent that the combination is usually referred to as nucleoprotein. A small amount of nonhistone protein, termed residual protein, presumably holds the genes in order and organizes them into the definite bodies, the chromosomes. Lipids are also present in the nucleus.

The Influence of the Nucleus on Development

The genes (DNA) do not directly influence development. Instead, they "transcribe" their coded information to RNA, which then collects around the chromosomes and in the nucleoli. Apparently one of the two strands of a DNA molecule assembles ribonucleotides and makes for itself a complementary strand of RNA. (The other strand possibly self-duplicates.) The nucleotides (ribonucleotides) which compose the RNA, however, differ from those which compose DNA. The sugars are ribose instead of deoxyribose, and one of the four nucleotides is uracil instead of thymine (it lacks the methyl group of thymine). Although these differences from a chemical standpoint appear to be slight, yet they account for important differences in function. The RNA molecules are single chains and as such are not self-replicating. (RNA is self-replicating in some viruses.) They are dependent on DNA for their production. Moreover, the RNA molecules do not remain in the nucleus. From time to time they migrate into the cytoplasm where they function in the manufacture of proteins. Some of them serve as templates for the production of proteins, notably for the production of the enzymes which catalyze the chemical processes of the cell. Possibly this release of RNA is not a continuous process. An impressive release takes place at the prophase of the first meiotic division of oogenesis when the swollen oocyte nucleus (the so-called germinal vesicle) breaks down and its nuclear sap and the material of the large nucleolus mingle with the cytoplasm. A similar release occurs at the prophase each time a cell divides.

In very active cells, such as growing oocytes and certain cells of larval insects, the nuclear membrane has been shown by the electron microscope to be perforated. Nuclear material has been found to "bleb" through pores into the cytoplasm where it supervises the syntheses of proteins.

There may be a delay between the time of the formation of the RNA in the nucleus and the time when its influence becomes apparent in cytoplasmic processes. The clearest evidence that this is so comes from experiments on species hybrids, that is, on eggs which have been fertilized by sperm of another species. In this case, the early events of development, for example, the pattern and rate of cleavage, are determined by the egg cytoplasm alone, presumably by the RNA that was synthesized while the oocyte grew in the mother's ovary and under the influence of the mother's genes. Generally speaking, the effect of the genes (DNA) brought in by the sperm is not apparent until about the time of gastrulation. But from then on, the genes of both the eggs and the sperm control the course of development.

Even more crucial are certain experiments in which eggs were enucleated before they were fertilized by the sperm of another species, or in which the egg nuclei were destroyed immediately after the foreign sperm had entered. In such haploid eggs (termed androgenetic hybrids), the cytoplasm belongs to one species and the nucleus to another. Such combinations have not lived beyond the early embryonic stages, but the evidence indicates that, from the time of gastrulation onward, the influence of the chromosomes of the stranger nucleus is pronounced.

Do Nuclei Undergo Differentiation?

In the early days, Weismann and Roux believed that the nuclei of body cells become different during cleavage as a result of what were thought to be unequal divisions of the nuclei. The organization of the embryo, so they theorized, is the expression of the inherited organization within the chromosomes of the zygote nucleus.

This view has long since been given up. There is organization within the cytoplasm of the uncleaved egg. There is strong evidence that the nuclei of every cell of the body contain initially the same set of chromosomes and genes, and that, during cleavage at least, mitosis is equal cell division in so far as the chromosomes are concerned. This fact was originally demonstrated in an experiment by Spemann in which he constricted an uncleaved amphibian egg by tightening a hair noose around it. The egg was pinched into two halves, with only a narrow neck of protoplasm between them. One half contained the nucleus; the other half lacked a nucleus. After several cleavages had taken place in the nucleated half, one of the descendant nuclei, migrated across the isthmus of cytoplasm into the non-nucleated half (delayed nucleation). Cleavage followed, and both halves (provided the needed cytoplasm was present) became whole embryos of half size. This proves that a cleavage nucleus can do everything that the original zygotic nucleus can do. It is an exact copy of its original.

The fact that the daughter nuclei are equal during early development has been clinched by some remarkable experiments by Briggs and King. They removed the nucleus of an uncleaved frog's egg, and substituted a nucleus from a cell of a blastula or an early gastrula. With refined techniques, many such operations were performed and were successful. Normal embryos developed. Here again, it is demonstrated beyond question that at this early stage of development when the future fates of the regions of the embryo are being determined. The nuclei, i.e., the genes of the cells of the different tissues, are equivalent. Since the nuclei are alike, it therefore must be the cytoplasms which are different.

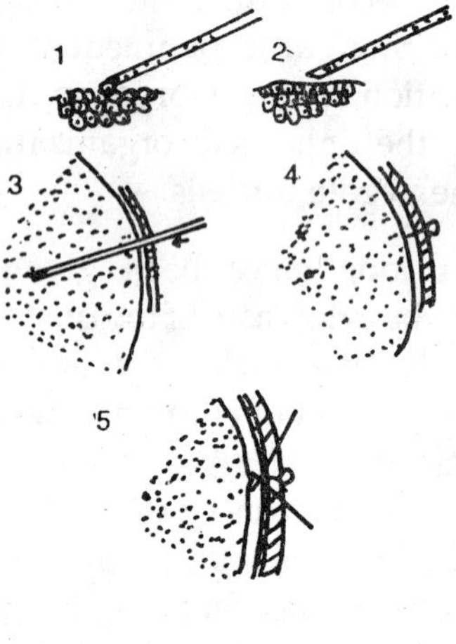

Fig. 6.6. Method used by Briggs and King when they substituted nuclei from older embryos for the nuclei of uncleaved eggs of the frog.

Gurdon and others have extended this work and have found that some nuclei of stages as fate as swimming larvae, when transplanted to enucleated eggs, are capabie of supporting a complete development. Some of the tadpoles even live to metamorphose into frogs and become sexually mature.

This, however, is not the whole story. It is a common observation that the nuclei of cells of different tissues do differ in size and shape. Somehow during development nuclei do change. The cytoplasm, which supplies them with substance, energy, and necessary enzymes, also affects the visible characteristics which distinguish them. It does more than this it affects their activity; and, at least in some cases, it affects their composition.

The nuclear changes which arise during development are of three sorts, and they arise in three distinct ways: (1) by the diminution of the chromatin, (2) by stable changes in the chromosomes, and (3) by the epigenetic control of gene activity by the cytoplasm.

1. Diminution of the chromatin: This is a rare phenomenon. Boveri observed many years ago that a reduction of the substance of the chromosomes (diminution of the chromatin) takes place during the early cleavages of the parasitic roundworm, *Ascaris.* At the two, four; eight, and 16-cell stages only one of all the blastomeres retains the typical four chromosomes of the zygote. In each case, this cell is in the line of descent of the germ cells (germ line). The other cells cast off a part of their chromatin into the cytoplasm; that is, the tips of the chromosomes swell and disappear. At the same time, the central part of each original chromosome breaks into several small segments which remain as separate chromosomes.

Other examples of the diminution of the chromatin in the development of body cells have been described, especially among the insects. Sometimes this process takes place early in development, sometimes late. But it never occurs in the germ line. We may conclude that, in the body cells of these embryos, the chromatin (DNA) has already performed its function by producing RNA. The RNA is stable and functions during development. The DNA (chromatin), no longer needed in the development of the body cells, is discarded.

Experiments have shown in *Ascaris* that a certain substance localized in the cytoplasm prevents the diminution of the chromatin. Normally, this material is passed on to only one blastomere during each of the first four cleavage divisions. But if, as a result of manipulation (centrifuging), this material becomes divided between two daughter cells, then diminution of chromatin does not take place in either of them. This is a clear case in which substances in the cytoplasm control the nucleus.

2. Changes in the chromosomes: Diminution of the chromatin is not known to occur in vertebrate development; yet there is evidence that changes in the genome or sum total of genes may take place. In their studies of transplanted nuclei, Briggs and King found that when a nucleus from a late gastrula or neurula of an amphibian is substituted for the nucleus of an uncleaved egg, normal development does not usually take place, although it *may* take lace. When examined under the microscope, the cells often do not have the normal chromosomal equipment. It is not clear whether changes of this sort take place in normal development. Possibly the nuclei of differentiating cells become increasingly sensitive to experimental manipulation. Possibly the transplanted nuclei are forced by the mitotic apparatus to divide before they are fully ready. Possibly, also, when cels differentiate, the mitotic cell divisions become slovenly. In any case, in these experiments the daughter cells did not always receive the normal complement of chromosomes.

The nuclear changes which take place in this manner are stable. Briggs and King have demonstrated that when the nucleus of a defective embryo is transplanted to an uncleave egg, the result is an embryo with a defect similar to the defect of it nuclear parent. This transfer of nuclei may be repeated over and over with the same result (nuclear cloning).

For a time it was suggested that the changes in the genome which Briggs and King had discovered might be related to the normal differentiation of cells. For example, the nuclei derived from the endoderm seemed to favour the development of endodermal organs. But this interpretation has not been borne out. Defects of a similar sort have also been obtained when the nuclei are taken from the neural tube. On the other hand, some nuclei are capable of supporting a normal development even when taken from the gut of a swimming tadpole. However, one thing is certain. Any changes which may take place in the genome of differ entiating cells always occur *after* the cytoplasm has already become specialized and committed to its fate.

3. *The epigenetic control of the genes by the cytoplasm:* Whether or not nuclear substance differentiates during development, it is certain that the same genes are not equally active in every cell and at every place, nor are they equally active at all times. Genes represent potentialities (in the sense of possibilities rather than powers) which may or may not have opportunities to express themselves. Which potentialities are realized depends upon the conditions within the cytoplasm which surrounds a given nucleus; these conditions change epigenetically as development progresses. It has been said that the cytoplasm turns the genes off and on like faucets. To change the metaphor, a nuleus is like a stockroom of a foundry full of patterns. From its vast store, patterns are requisitioned when where, and as they are needed. No casting can be poured at the foundry for which a pattern is not provided, and, of course, the form of the casting depends upon the shape of the pattern. In the same way, no synthetic chemical process can take place in a cell unless a gene is present to supply the pattern (i.e., template) for the enzyme which mediates the process. Genes are templates in stock. But the decision as to which genes are active at a given time and place—this depends upon conditions in the cytoplasm. And these conditions change epigenetically as development goes on. It is a mistake to pass over lightly the epigenetic factors which turn genes on and off, for they, and

not the genes, ultimately control the time and place of developmental processes.

Direct evidence that this is true is found in the behaviour of the giant chromosomes of certain insect larvae as they aproach metamorphosis. Such chromosomes consist to multiple strands of chromatids (so-called polytene structure). They are the result of self-duplication of the chromatids without accompanying separations. For example, the chromatids of the salivary glands of fruit fly larvae multiply nine or ten times without separating, with the result that each giant chromosome consists of 512 to 1,024 chromatids. These compound chromosomes show bands (possibly representing

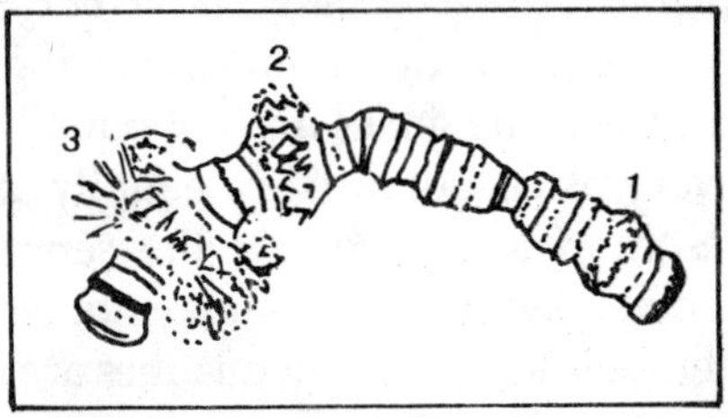

Fig. 6.7. Polytene chromosome of the salivary glands of the midge, *Chironomis,* showing stages of "puffing."

genes) of DNA which stain darkly and are large enough to be seen under the light microscope. They also show swellings, or "puffs," at definite loci. These seem to be the result of RNA and protein accumulating and forcing the strands apart. The puffs may expand until they become what are known as Balbiani rings. Beermann has studied the puffs and rings in the larvae of the midge, *Chironomus,* and has shown clearly that they occur at different locations in the chromosomes of different locations in the chromosomes of different tissues and at different times. The explanation seems to be that the puffs and rings are centers of intense synthetic activity in particular tissues, in preparation for the role which that tissue will play in the metamorphosis..

How are the Genes Controlled

The analogy of a nucleus to a stockroom of pattern at a foundry has one obvious flaw. A stockroom has a stockkeeper who delivers the patterns as they are called for. The nucleus of a cell, however, is automated so that, cued by conditions in the surrounding cytoplasm, it delivers the templates when they are needed.

Lately developmental geneticists have been much interested in the problem of how the genes are activated. We shall make reference to the closely reasoned concepts of Jacob and Monod of the Pasteur institute who studied enzyme induction in the bacterium, *Escherichia coli.* Normally this microbe produces no enzyme to metabolize the nutrient sugar lactose. Yet, when lactose is supplied in the nutrient medium, the enzyme β-galactosidase, which is able to metabolize it, appears in abundance. The presence of the substrate (substance acted upon) serves as an *"inducer"* and leads to the production of the enzyme.

The authors reasoned that no enzyme could have been produced if a "structural gene" for its production had not already been present in the germinal material. (Genes which determine the structure of a protein are called structural genes.) If this is so, then why was the enzyme β-galactosidase not produced until lactose was added? Genetic studies by Jacob and Monod on mutant strains of *Escherichia coli* supplied them with the answer. The gene for the production of β-galactosidase is inactive because another gene, so-called "regulator gene," is present which produces a *repressor substance* that inhibits the activity of the structural gene. When the substrate lactose is supplied, however, it inactivates the regulator gene (or the repressor substance which it produces). The result is that the structural gene, now freed from repression, gives rise to be enzyme β-galactosidase. The structural gene is thus activated by a process of *drepression.*

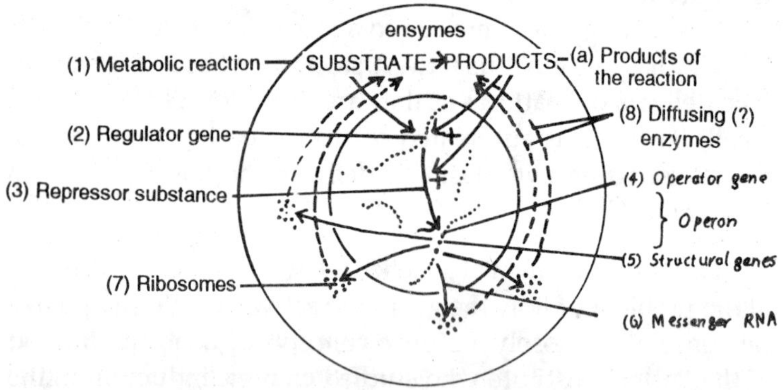

Fig. 6.8. Schema illustrating Jacob and Monod's concept of mucleo-cytoplasmic interaction. 1. A substrate in the cytoplasm inhibits the activity of a regulator gene. 2. A regulator gene, the function of which is to produce a repressor substance. (When inhibited by the substrate in produces less repressor substance). 3. Repressor substance which prevents the functioning of an operator gene in another chromosome. (When less repressor substance is present, the operator gene is "derepressed.") 4. Operator gene which activates a series of structural genes. 5. Structural genes which synthesize messenger RNA (and other sorts of RNA). 6. Messenger RNA which migrates from the nucleus into the cytoplasm where it associates with ribosomes in the production of specific enzymes (or other proteins). 7. Ribosomes scattered through the cytoplasm. 8. Enzymes which catalyze specific chemical reactions in which the substrate is altered. 9. Products of the reaction, which may augment the activity of the regulator gene or activate a repressor substance. In so doing the products act as inhibitors of the operator gene.

This statement is oversimplified. The repressor substance produced by the regulator gene acts upon an "operator gene" within a complex of structural genes. It is the operator gene which directly controls the structural genes, turning them on and off. An operational complex of genes of this sort was called an "operon" by Jacob and Monod. Such a complicated mechanism would be unbelievable if its several steps had not been experimentally verified. Generally speaking, a substrate acts as an inducer. Its presence leads to the production of the enzyme which metabolizes it. In the language of modern communication theory, this is a case of "positive feedback." In a similar manner, the products of a reaction are inhibitors which repress the production of the enzymes which produce them. This is a "negative feedback."

Embryologists are much interested in concepts such as this one of Jacob and Monod because they suggest how epigenetic factors in the cytoplasm (represented in the above case by the substrate lactose) can control the activity of the genes. There is, however, an important difference between the actions of genes in bacteria and in the development of higher organisms. The genes of bacteria are immediately responsive to their environment, and the mRNA which they produce by transcription is immediately active in protein synthesis. It is also short-lived. The genes which control the development of higher organisms, on the other hand, commonly produce "masked" mRNA, which may be stored for a time until the time and place for its activity have arisen.

7

Role of Cytoplasm in Development

Cytoplasm is the part of protoplasm present between nuclear membrane and the plasma membrane. Nucleus alone fails to regulate the activities. It the mutual affords of nucleus and cytoplasm that regulate all the developmental processes.

The genes of the nucleus control the course of development, but it is mainly the cytoplasm which develops. The genes are the "physical basis of heredity," but principally it is the cytoplasm which expresses this heredity insofar as it is expressed. There is interaction between the genes and the cytoplasm, but their roles are different. The genes largely control the species characteristics of the new organism and to a great extent its individual characteristics as well; but the characteristics which distinguish one tissue from another, or one organ from another *in the same individual,* are decided initially by processes which take place in the cytoplasm. The history of the cytoplasm is indeed a true eoigenesis; that is, it is a straightforward coming-into-being , of molecules, structures, and functions which did not at first exist. It starts in the relative simplicity, or at least plasticity, of a single cell, the unripe egg. It moves forward irreversibly in ever-increasing complexity through the stages of the life cycle; embryo; larva (if any), juvenile, adult, and finally the senescent individual. The ultimate end of development is death unless, as in some of the lower organisms which are capable of regeneration, the story is able to begin again. Embryology is therefore, for the most part, a recital of processes which take place within the cytoplasm.

The cytoplasm is a colloidal aqueous system of proteins, lipids, sugars, and numerous other organic molecules plus various mineral salts. It has a structure, an intimat, precise anatomy such as no system of a merely chemical nature ever had. Its proteins and lipids are organized into membranes filaments, and granules, at the surface of which chemical reactions take place. These are the organelles which carry on the many processes of living. The precise patterns of the protein surfaces are the principle basis of the specificity of the cell's chemical reactions. for the proteins include the enzymes which catalyze and control the rates of chemical processes. They also include non-enzyme proteins. A very few years ago cytoplasm was described as a homogeneous, transparent gel or sol which was optically "empty." The various formed bodies which were seem under the microscope were spoken of as being within the "ground cytoplasm"; or they were interpreted as artifacts produced by fixation and staining. An intricate morphologic structure of cytoplasm has been studied with the help of phase-contrast microscope, the electron microscope, and the ultracentrifuge.

There is a variety of cells and the cytoplasms of different kinds of cells have specialized functions (no single type of cell can be said to be "typical"), yet the cytoplasms of most cells posses certain organelles in common. These include: a plasma membrane (cell membrane of plasmalemma , the outer layer of the living cytoplasm); inner membranes, which form vacuoles, tubules, and lamellae; a nuclear membrane (actually a part of the cytoplasm); and formed bodies such as ribosomes and mitochondria. Each of these organelles plays of role in the physiological life processes of the cell. Specialized cells have filaments and granules of various kinds which perform distinctive functions. Muscle cells have contractile fibrils; nerve cells have conductile fibrils; secretory cells have secretory granules; certain other cells produce pigment granules; while still others store reserve food materials such as yolk platelets, glycogen granules, and fat globules. The embryologist sees these various structures appearing in the cytoplasm as development proceeds, each at the proper time and in the proper place. It is believed that they result from the synthesizing activities of distinctive enzymes which are released in the cytoplasm, and that in turn each enzyme obtains the "information" (specific pattern) for its production from a particular gene or genes in the nucleus.

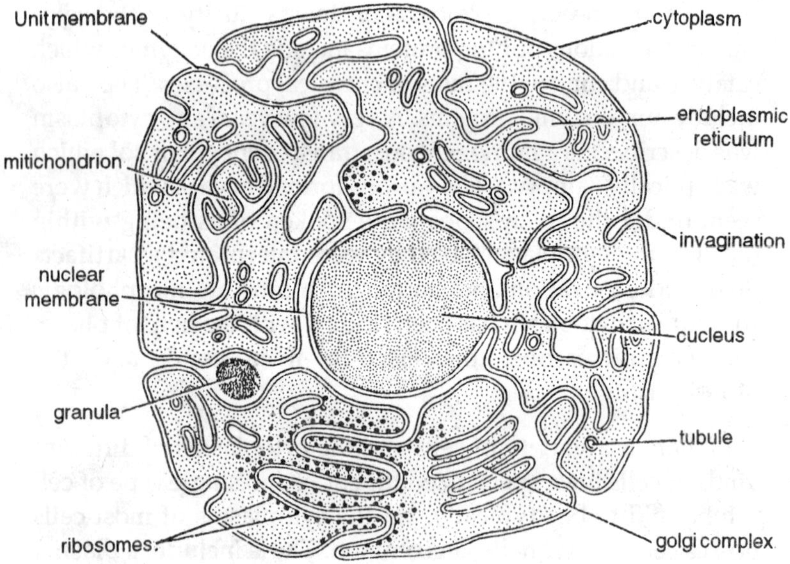

Fig. 7.1. Diagram showing the origin and development of membranous structures of the cell from the unit membrane.

Ribosomes and the Synthesis of Proteins

Protein is the building stone and it is to be synthesized to make the developmental growth possible. Beginning in 1941 and 1950 with the demonstration by Caspersson and Brachet that ribonucleic acid (RNA) is involved in the synthesis of protein, a great deal of information has been accumulated. As noted in the last chapter, the genes of the nucleus (DNA) "transcribe" their specific genetic "information," coded in the sequence of their nucleotides, to RNA molecules. These accumulate in and around the chromosomes and ultimately enter the cytoplasm. Apparently the DNA produces RNA of three sorts: *ribosomal RNA* (rRNA), *messenger RNA* (mRNA), and *transfer* or *soluble RNA* (tRNA). The ribosomal RNA associates basic proteins to itself and accumulates in nucleoli within the cell nuclei. Later this material (rRNA plus protein) enters the cytoplasm where it appears as ultramicroscopic bodies. The *ribosomes*. As much as 30 per cent of the dry weight of cytoplasm may be ribosomes, and about one-half of each ribosome is rRNA.

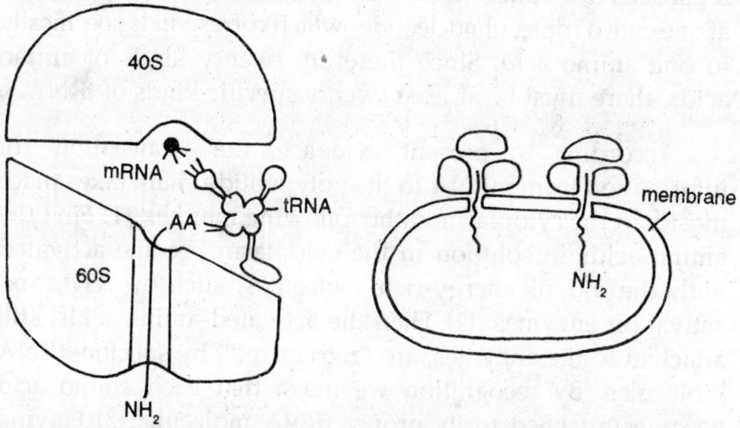

Fig. 7.2. Diagram of ribosome showing the two sub-units and the probable position of mRNA and tRNA.S

From time to time messenger RNA (mRNA) also leaves the nucleus and enters the cytoplasm. There it becomes closely associated with the ribosomes and serves as templates, at the surfaces of which the polypeptide chains of amino acids are build up. These chains become the proteins.

How is this "information" which is brought from the nuclei by the mRNA transferred ("translated") to protein molecules? It is carried in the form of an alphabetic code of four "letters," the ribonucleotides. These differ from the nucleotides of DNA in that the sugar is ribose, instead of deoxyribose, and the bases are abenine, guanine, cytosine, and uracil (which substitutes for thymine). The precise sequence of the nucleotides is the most important feature, for it determines the sequence of the amino acids in the polypeptide chain. But since there are only four kinds of nucleotides and there are twenty or more kinds of amino acids in a polypeptide chain, it must take a "word" or *codon* of at least three nucleotides to pick or choose one amino acid.

The third sort of RNA is transfer or soluble RNA (tRNA). It consists of smaller molecules, each of which bears, possibly at one end, a triplet of nucleotides which corresponds specifically to one amino acid. Since there are twenty kinds of amino acids, there must be at least twenty specific kinds of tRNA.

According to present evidence, the "translation" of information from mRNA to the polypeptide chain takes place mostly in the cytoplasm in the following manner: (1) First the amino acids in solution in the cytoplasm become activated with the aid of energy-rich molecules, such as ATP, and activation enzymes. (2) Then the activated amino acids, still attached to the enzymes, are "recognized" by specific tRNA molecules. By recognition we mean that each amino acid becomes attached to its proper tRNA molecule. (3) Having accomplished this, the tRNA-amino-acid combinations separate from their enzymes, and each adheres one by one in sequence to a specific locus on a messenger RNA molecule as it moves across and past a ribosome. Thus, in turn, the amino acids, having been brought close together, link up, and a growing polypeptide chain is produced. A polymerizing enzyme is required for the linking-up process.

What is the role of the ribosomes in the synthesis of proteins? It seems not to be specific. Cell-free solutions in which mammalian mRNA have been combined with bacterial (!) ribosomes (plus a "pool" of amino acids, transfer RNA, and ATP) are capable of synthesizing mammalian proteins. One writer has referred to ribosomes as "workbenches" on which the polypeptide chains are forged. Ribosomes are said to "read" the message brought in by mRNA as it moves past. It may help us to visualize the process if we think of a ribosome as a sort of automatic machine which is controlled by a punched tape, the mRNA. The machine (ribosome) receives, as raw material a mixture of activated and recognized amino acids (amino acids bound to tRNA molecules) and attaches particular combinations as the mRNA tape dictates. The emino acids link up and an ever-lengthening polypeptide chain is produced. is released to be used again). When the tape runs out, the polypeptide chain is complete and moves away. The entire process takes but little over one minute. The machine is also free to start over again with the same or another tape. Since the tape is long, it may be moving through or past several ribosomes at the same time, each of them in the process of producing a polypeptide chain. Such a series of ribosomes, connected by a thin thread of mRNA molecules, can be demonstrated with the electron microscope. They are known as *polyribosomes* (polysomes).

(There is some evidence that ribosomes are not all the same in their nucleotide composition, even that they differ in the cells of different tissues. Still, they may all be just work-benches.)

The "code" by which mRNA transfers its information to polypeptide chains has been broken. This was accomplished first by students of microorganisms. Bacteria were homogenized without destroying the ribosomes. The resulting cell-free system was supplied with amino acids, a source of energy (ATP), and artificially synthesized RNA chains of known composition. The polypeptide chains which were produced were then analyzed. The breakthrough came in 1961 when Nirenberg "fed" his system an artificial RNA consisting of the nucleotide urocil only. He obtained polypeptide chains composed of the amino acid phenylalanine. From that point on it was only a matter of time until the entire code was broken.

At first the ribosomes in the cytoplasm are free and scattered.

Then, about the time that newly synthesized proteins begin to be secreted away from the cell, the ribosomes be come attached to the endoplasmic reticulum within the cytoplasm.

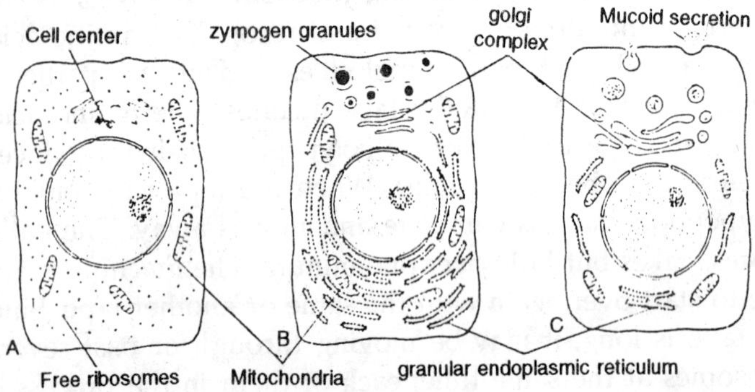

Fig. 7.3. Three types of cells engaged in synthesis (idealized). *A.* A protein-retaining cell with free ribosomes. *B.* A protein-secreting cell with an extensive ndoplasmic reticulum coatd with ribosomes. C. A mucus-secreting cell with a well-developd Golgi complex.

This work on RNA and the ribosomes is of great interest to embroyologists because it is possible that differentiation manifests itself first quantitatively in the distribution of the ribosomes. At any rate, gradients in the arrangement of ribosomes exist early in development, and any manipulation which disturbs.

In cytoplasm there are several cell organelles like mito chandrias, Golgi body, Dendoplosimic reticulun, rebosomes lysogomes etc (see chapter 2). These organells play their alloted roles there by helping the cytoplasm during development. As usual mito chandria provides energy to perform various metabolic processes possible. Golgi body secretes all necessary enzymes and endoplasmic reticulam transports the various products from one place of all to another wherever they are required. Rebosomes play is the synthesis of protein as explained above.

The Cell Cortex

The form and activity of a cell are intimately dependent on the properties of the outer zone (or zones) of the cytoplasm, namely, the *cell cortex*. The outer layer of the cortex is the *plasma membrane*. It directly relates and cell to its environment. Nutrients must enter through it, and unwanted substances must be kept out. Wastes are eliminated by it. Stimuli, except perhaps those of light, affect it first. The plasma membrane is of the order of 100 A thick (=100x10^{-7} mm). It has properties of semipermeability, but is must not be thought of as a passive sieve. On the contrary, it is an extremely active membrane which produces and retracts minute processes (microvilli) from its external surface, and forms microtubules and vesicles that push inwardly into the interior of the cytoplasm. By this means the cell feeds (phagocytosis) and drinks (pinocytosis). The particles and droplets which thus enter the cell do not directly penetrate the plasma membrane. Rather, the membrane moves before them and surrounds them as they advance into the cell. Later the membrane may dissolve, or the contents of the vacuole may be digested and absorbed into the cytoplasm.

In addition to the plasma membrane, the cell cortex commonly includes a somewhat thicker layer of cytoplasm which is from 1 to 5 μ thick (=1 to 5x10^{-3}mm). This layer, sometimes referred to as ectoplasm or plasmagel, is usually gel-like in its physical structure, although it may change state from gel to sol and back again. When photographed with the electron microscope, the plasmagel often differs very little, if at all, from the more liquid internal cytoplasm (endoplasm, plasmasol). Indeed, it may not be present at all. But in other cases, it is specialized by the presence of filaments or granules which are not found elsewhere.

The various responses by which a cell changes its shape are often cortical reactions. Development, also, is to a remarkable degree a matter of cortical response. This is obvious in the case of the extrusion of the polar bodies, the rising-up of the fertilization membrane, and the division of the cytoplasm during cleavage. It is not so obvious that the folding of the cellular layers by which the gut and later the nervous system are formed are also to a large extent cortical reactions.

Protozoologists long ago became aware that the form and much of the activity of one-celled animals are properties of the cell cortex. Consider, for example, the ciliate, *Stentor*. The body of this animal is conical in shape. The sides are marked by a hundred or so longitudinal rows of cilia separated by intervening pigmented stripes. The base of the cone (anterior surface or "head") is a disc with spiral rows of cilia. The head is surrounded by an almost complete circle of membranelles. At the clockwise end of the arc of membranelles are the structures used in feeding, namely. The oral pouch, cytostome, and gullet. Nearby

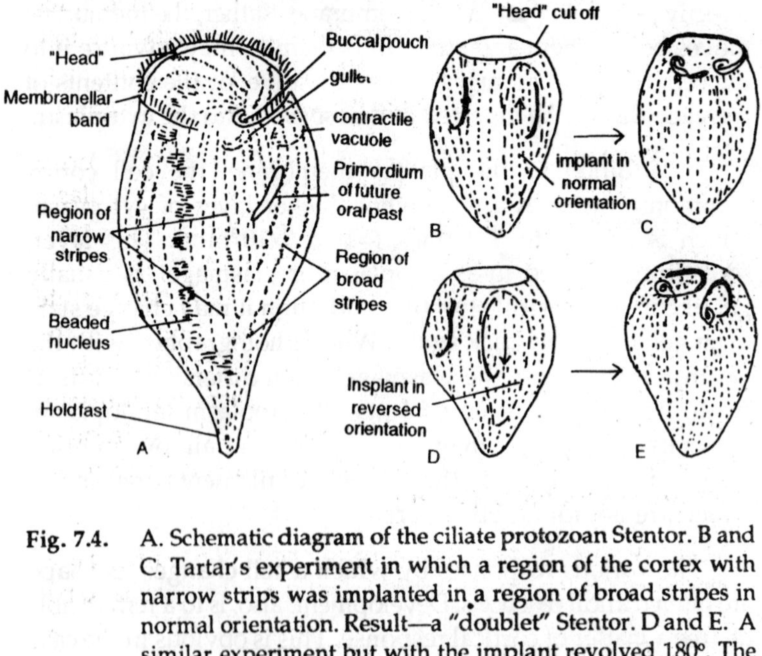

Fig. 7.4. A. Schematic diagram of the ciliate protozoan Stentor. B and C. Tartar's experiment in which a region of the cortex with narrow strips was implanted in a region of broad stripes in normal orientation. Result—a "doublet" Stentor. D and E. A similar experiment but with the implant revolved 180°. The implant developed a set of oral parts with reversed asymmetry.

is the anal pore. The apex of the cone (posterior end) is specialized as an organ of temporary attachment, the holdfast. All these structures are specializations of the cell cortex.

Tartar has performed numerous surgical operations on *Stentor* and has repeatedly observed its extensive capacity for regeneration and reorganization. The controlling feature, he finds, is the presence and orientation of the cortex. A tiny bit of *Stentor,* only 1/123 the size of a normal animal, is capable of regenerating a whole *Stentor* provided that a little of the cortex and some nuclear material are present. But a far larger piece of *Stentor,* lacking a bit of cortex, does not regenerate even though adequate nuclear material is included. Undoubtedly the nucleus is necessary to control the processes by which substances of the cell are synthesized; but, Tartar concludes, the polarity and asymmetry of the body are transmitted from one generation to the next solely by the cortex.

In a somewhat similar fashion, the cortex of a metazoan egg transmits the pattern which largely controls the polarity, symmetry, and asymmetry of the future embryo. For a time these features are labile to varying degrees and subject to modification, by the environment, Raven has studied the role of the cortex in the eggs of mollusks. When such eggs are centrifuged, the visible materials of the cytoplasm are displaced either toward or away from the axis of rotation. They are displaced toward the axis if they are less dense than the fluid cytoplasm which surrounds them, and away from it if they are more dense. In some cases, development proceeds normally except that the granules and droplets remain unnaturally distributed. In other cases, the displaced visible stuffs move back to their original locations. Now, since the internal cytoplasm is quite fluid and subject to flow (witness its movements during meiosis, fertilization and cleavage). It must be the cortex which remains fixed and so transmits the pattern of the developing organism. (Extreme centrifugation liquefies the cortex and results in abnormalities.

The Adhesion of Cells

The cells of an embryo adhere to one another in varying degrees and in manners which are characteristic of the different tissues. At first the adhesions are weak. Cell easily dissociate when early embryos (blastulas and gastrulas) are placed in calcium-free solutions. As development proceeds, however, the adhesions between most cells become stronger. This is especially true in the case of epithelial cells, that is, in the case of cells which form layers and either bound the outside of the embryo or line cavities within it. Adhesion is especially close and strong immediately adjacent to where the contacts between cells border a cavity. In effect, an epitheium presents a continuous outer surface—Holtfreter thought of it as a "coat"—which is semipermeable and which serves to separate the external medium which surrounds an embryo (or fills its cavities) from the internal medium which bathes its inner cells. Actually, it is not a continuous coat, but a network of close contacts between cells (often referred to as *terminal bars*). Except at these outer contacts pockets of liquid may separate the plasma membranes.

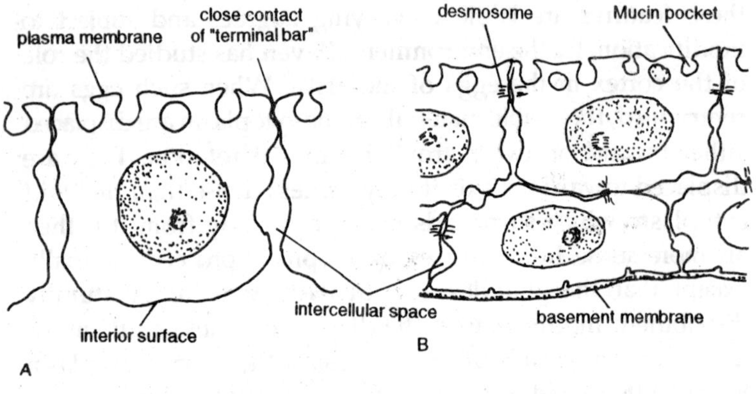

Fig. 7.5. Adhesion of epithelial cells. A early one-layered epithelium in which there is close contact next to the external surface. B. Later two-layered epithelium with desmosomes and a basement membrane.

As development proceeds, adherent patches known as *desmosomes* develop between adjacent cells and strengthen the bonds between them. The electron microscope sometimes shows filaments extending from each desmosome into the cytoplasms on both sides.

Other cells of the embryo, notably those known as mesenchyme, are less adherent to each other. Apparently they are free to wander independently; but they do not wander haphazardly. On the contrary, they move to approptiate positions in the embryo, and there they remain as if trapped by their surroundings.

What guides mesenchyme cells? What finally brings their wanderings to an end? In general, what holds embryonic cells to their places in the embryo?

Holtfreter has attempted to account for the direction of cell migration in terms of "contact guidance." Movement on a surface, he supposes, is controlled by the oriented macromolecular structure of the surface. He explains the final location of cells by the principle of "selective adhesion." That is, cells (or rather, cell surfaces) adhere most strongly to other cells for which they have an "affinity" possibly a comparable surface pattern). When a cell makes contact with a congenial cell, its wanderings cease, Abercrombie terms the behaviour "contact inhibition."

These responses of the cell surfaces are markedly specific. In some cases, they appear to be species-specific, as when an egg cell reacts to sperm cells of its own species, or when lymphocytes react to foreign proteins (antigens) by producing antibodies. In other cases—and this is especially characteristic of embryos—the specificity of the cell surface seems to be tissue-specific. When an embryo is experimentally dissociated into, its constituent cells and then the cells are permitted to reaggregate, they first adhere to each other and then tend to sort out and reform the tissues from which they were taken.

Superfically, it seems as though each cell, when it contacts another cell, recognizes whether the other cell is of its own or another kind.

Steinberg offers a less mysterious explanation of contact guidance and selective adhesion. He explains segregation in terms of quantitative differences in the strength by which cell surfaces adhere. The principle is much the same as that which causes oil and water to separate; namely, the mean of the cohesive forces between like molecules (water to water and oil to oil) is greater than the adhesive forces between unlike molecules (water to oil). Similarly, intermingled cells sort out because like cell surfaces adhere to each other with more average strength than unlike surfaces adhere. Sorting out, therefore, is not so much a property of cells as individuals, as it is a property of cells in populations.

A corollary of this is that those cells of a mixed population which have the strongest mutually adhesive properties tend to pull together and become encapsuled within an outer mass of cells with less adhesive tendency.

The foldings and moldings of tissues during development (about which we shall have much to say in the following pages) have been interpreted in terms of programmed changes in the adhesive characteristics of cell surfaces. For example, a thickening of an epithelial layer of cells involves an increase in the areas of contact between cells. Presumably this results from an increase in the strength of their cohesiveness. A thinning of an epithelium is associated with a decrease in the areas of contact.

The Role of the Cytoplasm in Development

Many embryologists today are engaged in describing in detail the chemical and cellular changes which take place in an embryo as development proceeds. This is the problem of *cytodifferentiation,* Notable progress is being made with the aid of sophisticated physical and chemical techniques such as electron microscopy, radioactive tracers, and immunological methods. As a result, today we know something about the sequence of chemical events in the embryo and when and where new cellular structures appear and disappear. But it is one thing to *describe* the physical and chemical changes which take place during development and quit another thing to give a *causal account* as to how these changes are brought about. The key question is: What controls differentiation?

There is no doubt today that the cytoplasm controls the course of cell differentiation. But there are two "levels" at which this control maybe exerted: (1) it has been generally assumed that control takes place at the level of "transcription": that is, takes place in the nucleus when DNA becomes activated and transfers its information to messenger RNA. Some genes are "turned on" at any given time and place, while other genes (probably most of them) remain repressed. (2) There is, however, increasing evidence that some control by the cytoplasm takes place at the level of "translation," namely, when mRNA supervises the synthesis of protein. In this case the mRNA produced in the nucleus remains "masked" until it reaches the proper time and place in the living system.

The classic example of control of differentiation at the level of translation is seen in the marine alga, *Acetabularia*. This is a giant cell, some two inches long, and shaped like a mushroom or parasol. Its apical end spreads out radially like a hat or cap. Its basal end forms rhizoids (holdfasts), one of which contains the large nucleus. The entire life cycle of *Acetabularia* does not concern us now. Our present interest is in its capacity to regenerate a new cap when the old cap is cut off. Surprisingly, the regeneration of a cap in acetabularia. Hammerling, who studied these matters intensively, interpreted his findings, in terms of gradients in the distribution of morphogenetic substances, i.e., a cap-producing substance with its highest concentration at the apical pole, and a holdfast-producing substances centered at the basal pole. His concept is that these gradients determine the location of the new parts.

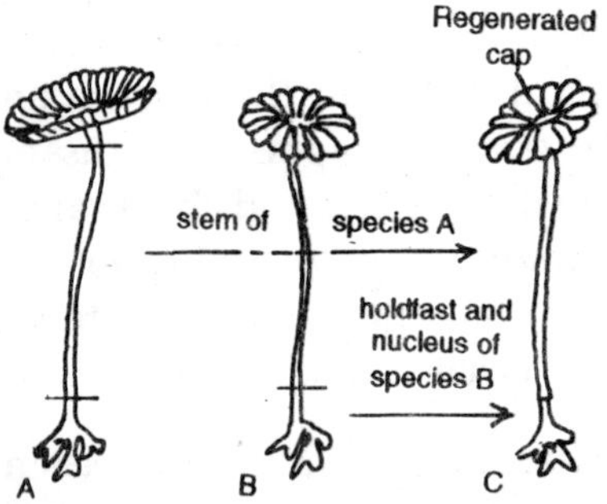

Fig. 7.6. Regeneration in the unicellular alga; Acetabularia. A. Acetabularia mediterranea ("species A"). B. Acetabularia crenulata ("species B"). C. Result of grafting a decapitated stem (no nucleus) of species A to a hold fast (with nucleus) of species B. The first cap which regenerated was cut off. The second cap to regenerate had the characteristics of species B.

The place even when the nucleus has been cut away two or three weeks previously. Clearly the presence of nuclear DNA is not directly responsible for the restoration of lost parts.

The role which the nucleus plays in regeneration must be an indirect one. Presumably the genes of the nucleus produce the morphogenetic substances which then migrate within the cytoplasm and concentrate at the apex or base, as the case may be. There they stimulate regeneration and at the same time control the character of that which regenerates. This interpretation is borne out by Hammerlin's observation that, when a piece of the stem of *Acetabularia mediterranea* (we shall cell this "species A") is grafted by its basal end to a holdfast of *Acetabylaria crenulata* ("species B"), the cap which regenerates resembles the caps of species A. He interprets this to signify that, at the start, regeneration is under the control of the morphophogenetic stuffs which were emitted by nucleus A before it was removed. But if this first cap is cut away, the second cap which regenerates has the character of species B. This confirms the view that the nucleus, which is located at a distance from the apex both in time and space, is the source of the morphogenetic stuffs.

In terms of present theory we may identify the morphogenetic stuffs (as Brachet does) with messenger RNA. They are produced by transcription in the nucleus. Under the influence of the cytoplasm. The cap-forming stuff migrates and concentrates at the apical end of the cell. Here translation takes place; that is the information brought by the mRNA is transferred *in situ* to newly synthesized enzymes and structural proteins. No theory of turning genes on and off is sufficient to account for this localized reaction.

What brings about the concentration of morphogenetic stuff at the apex? Since the inner cytoplasm of *Acetabularia* is engaged in active streaming, some factor in the cortex must be responsible. But it must be admitted that, as yet, we know very little of how this influence of the cortex originates or is exerted.

8

Cytoplasmic Localization

The term-localization is used here to show the specification of cell fate according to the sector of egg cytoplasm inherited by an embryonic cell lineage. As the early blastomeres divide up the egg cytoplasm, they appear, in some organisms, to inherit "instructions" for various kinds of cell differentiation, including specific patterns of macromolecular synthesis. Among the main problems posed by the localization phenomenon are the molecular nature of the morphogenetic agents stored in the cytoplasm; the level(s) of control at which these agents might operate; their subcellular location, and the cytoskeletal structures to which they are anchored; and the means by which they are distributed to given regions of the egg or early embryo.

The description by C.O. Whitman (1878) of cleavage and cell fate in the leech *Clepsine* was the first complete study of this kind, and greatly influenced the field. In this organism, as in many others shortly thereafter the subject of cell lineage analysis, the number of individual blastomeres at the cleavage and blastula stages is relatively small, and they can be distinguished by size, shape, and position, so that the developmental destiny of each cell and its lineal descendents can be followed. Whitman showed that the mesodermal and neural "germ bands" of the leech embryo are budded forth from bilateral stem cells, the teloblasts (annelid teloblasts had first been noticed by Kowalevsky, (1871). Whitman's most important finding was that the ventral nerve cords of *Clepsine* develop from bilateral sets of four individual neuroblast stem cells, themselves deriving from early bilateral blastomeres.

The mesoderm also forms from individual bilateral stem cells, the mesoblasts. Though the discussion in Whitman's 1878 publication is largely focused on the evolutionary affinities implied by his observations, their significance for the cell biology of development was widely recognized. This was the demonstration that highly differentiated structures, such as nerve cords, can be derived lineally from *specific individual blastomeres that arise only in certain regions of the egg, by a precise pattern of cleavage, and that normally inherit certain segments of the egg cytoplasm.* In the following decades elegant and in some cases extremely detailed cell lineage investigations were carried out on animals belonging to diverse phylogenetic groups, including numerous molluscs and annelids, the nematode *ascaris,* and various ascidians, flatworms and ctenophores.

In many cases the fates of *every* cell of the embryo were determined. Sufficient data were obtained for the annelidmollusc group to support comparative analyses of the developmental values of homologous blastomeres by Wilson (1898); by Child (1900) in his study of cell lineage in the polychaete *Arenicola;* and by Conklin (1897), in his classic description of the cell lineage of the mollusc *Crepidula.* Many of the investigators who carried out cell lineage studies were deeply interested in the evolutionary implications of cellular homologies in the development process. Thus a by-product of these studies was the construction of an imposing morphological argument for the phylogenetic relationships of various protostome phyla, particularly polyclad worms, annelids and molluscs. For us, as for contemporaries focused primarily on the problems of developmental mechanism, the main significance of the cell lineage determinations lies in a different direction. Conklin's description of the lineage of *Crepidula* in many ways illustrates the major theoretical impact of these studies. Conklin followed the fate of individual blastomere lineages beyond the 100-cell stage, from their initial appearance to their participation in differentiated structures, and thus defined exactly the embryonic cellular origins of all the organ systems and structures of the advanced veliger larva. Contemporary writers therefore stressed that the determinants in mosaic developmental processes are the maternal cytoplasmic substances, rather than being a particular property of particular embryonic blastomeres.

(i) Localization of in the Egg that Display Determinate

The cases of localization that most impressed classical experimentalists were those in which areas of future cell fate could be visibly mapped out on the uncleaved egg cytoplasm. A spectacular example is Conklin's (1905a) study of development in the ascidian *Cynthia (Styela)*, which greatly expanded and confirmed earlier experiments on ascidian embryos by Chabry (1887). In the egg of *Styela* pigmented areas of cytoplasm corresponding to a morphogenetic fate map for the cells of the early embryo can be distinguished. Some of Conklin's elegant hand-drawn figures are reproduced. These display the relationship between the various pigmented regions of the egg cytoplasm and the tissues ultimately formed from these regions. Five kinds of cytoplasm were noted by Conklin: a dark yellow granular cytoplasm eventually included in the tail muscles of the larva; a light yellow material later segmented into the coelomic mesoderm of the larva; a light gray substance inherited by notochord and neural plate cells; an opaque gray material segregated into the endoderm cell lineage; and a transparent cytoplasm later present only in ectodermal cells. The dark yellow granular cytoplasm referred to by Conklin as the "yellow crescent" is shown in recent photomicrographs of eggs that had been extracted with detergent. As described in the legend, the persistence of the yellow crescent in these eggs indicates that the pigment granules that Conklin saw are embedded in the cytoskeletal matrix of the cytoplasm.

Conklin focused on the striking correspondence between the sets of blastomeres that give rise to specific groups of differentiated larval cells, and the distribution of the five types of egg cytoplasm that he could visually distinguish throughout cleavage. The definitive distribution of the "organ forming substances" only appears after fertilization. It is not a primordial property of the egg. The process by which this distribution develops was observed carefully by Conklin and some of his illustrations are reproduced. One axis of polarity is already present in the unfertilized egg. The establishment of the anterior-posterior axis depends on the acentric movement of the male pronucleus. The fusion nucleus comes to lie near the future posterior pole, and there the yellow crescent cytoplasm later incorporated in the embryonic muscle cells is localized.

Conklin regarded the ascidian embryo as a "mosaic-work," and as had the other observers cited above, he also argued that "the mosaic is one of organ forming substances rather than of cleavage cells." He concept of ascidian development was supported by an extensive series of experiments in which he determined the developmental potentialities of individual blastomeres and sets of blastomeres in embryos in which the other blastomeres had been killed. In general the surviving blastomeres were found to establish cell lineages the differentiate in their respective normal directions, though the overall organization of the embryo is of course affected. The drawings reproduced. That each embryo fraction contains the potentiality of forming certain presumptive tissue types, such as notochord, mesoderm, neural plate, gut, or ectoderm, even though the isolates were often not cultured sufficiently long to reveal completely their developmental capabilities. The ascidian cell lineage includes some corrections of Conklin's original assignments, mainly obtained by the method of horseradish peroxidase injection into individual blastomeres of 8-cell embryos of *Ciona intestinais, Ascidia abodori,* and *Halocynthia roretzi* and into blastomeres of 16-and 32-cell embryos of *Halocynthia* for similar assignments obtained for some lineages of *Phallusia mammaillata* embryos by a different method). This lineage describes diagrammatically the ultimate capacities of the various early blastomeres *in situ,* and in general *in partial embryos as well,* and thus provides a developmental linkage between Conklin's "organ forming substances" and embryonic cell differentiation. However, the relation between embryonic

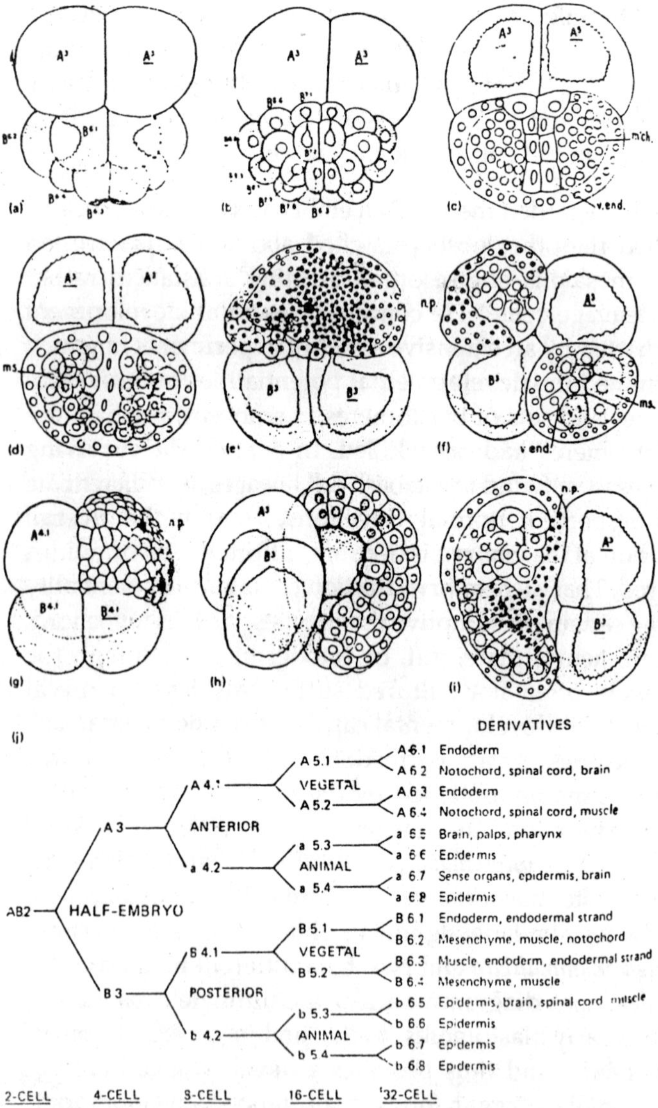

Fig. 8.1. Fate of partial embroyos of *Cynthia (Steyla) parita*.

cell fate and cell lineages is in some areas not as simple as implied by Conklin's (1905a) proposal that *segregation of morphogenetic agents* has been completed by the 64-cell stage, and thus that the progeny of each 64-cell blastomere contribute to only one type of tissue. The lineage chart indicates that some functionally asymmetric divisions must occur later than the 64-cell stage, since certain 32-cell blastomeres contribute to at least three different types of tissue. Examples of clonal determinations by the 64-cell stage, and also of later determination, are to be found in the muscle cell lineages.

The concept that early embryos are a "mosaic-work" of determined blastomeres was widely challenged. Beginning in the late 1880's, blastomeres were isolated and cultured from many different embryos, in order to determine whether their division products indeed develop *in vitro* as do their Normal lineages in the while embryo. Diverse results were obtained with different organisms, and a ferocious controversy over the basic nature of embryological development soon arose. The conflicting theories that followed are reviewed briefly in the next section, while in this we consider further examples of the evidence supporting one side of this controversy, *viz.*, those isolated blastomere experiments that provided powerful evidence for early fixation of morphogenetic potential in localized regions of the embryo. Consistent results were obtained with other isolated cell types and with partial *Patella* embryos. Among these were isolated one-sixteenth embryo macromeres, which produced endodermal gut rudiments, and isolated spical progenitors, which differentiated *in vitro* into apical sensory and ectodermal cells.

The morphogenetic fates of the cell lineages descendant from early embryonic blastomeres were studied in marine annelids, such as the polychaetes *Sabellaria* and *Nereis* and an illustrative series of experiments was carried out by Penners (1926) on the oligochaete annelid *Tubifex rivulorum*. This animal develops essentially as described for *Clepsine* by Whiteman. Thus a neuroblast germ band and mesodermal germ band both derive from the D quadrant of the embryo (blastomere nomenclature here. At the 4-cell stage of the Tubifex embryo the D macromere is the largest, and with respect to both size and rate of division, its products remain distinct from those of the A, B and C quadrants. Penners found that each of these blastomeres would continue its normal course of development even if all the others were killed *in situ* by UV microbeam irradiation. Thus if A,B and C are killed, the D macromere nonethless adheres to its unique cleavage pattern. It gives rise to an ectodermal neuroblast stem cell, 2d and to a primary mesodermal stem cell, 4d. After this columnor ectodermal and mesodermal germ bands are produced by anterior budding. Penners showed that if the 4d cell is individually killed, the neuroblast germ bands still form, but the embryo lacks coelomic mesoderm. Recent investigation has demonstrated that the individual neuroblast stem cells of annelids each give rise to detain specific neurons. In the leech, for instance, ablation of a given germ band neuroblast results ultimately in the absence of one member of each of the three bilateral segmental pairs of larval neurons that contain serotonin, and ablation of a different neuroblast causes the absence of one member of each pair of segmental body wall neurons that contain dopamine.

(ii) The Early History

The initial discovery of determinate, asymmetric cell lineages was among the antecedents of the proposals that morphogenetic substances are localized in the cytoplasm of the very early embryo. There was also a parallel theoretical development, the concept of "precocious segregation" of maternal molecular determinants. The origins of this idea lie in the nineteenth century controversy between preformationist and epigenetic theories of development. The participants in the final stages of this controversy included many eminent biologists engaged personally in research on localization. Huxley (1878), Hertwig (1894), Bourne (1894, and Whitman (1895a) all published discussions of the implications of localization in which the main issue considered was whether the embryo actually increases in biological completely or, on the other hand, merely reveals progressively to the observer an organizational complexity that is already resident in the unfertilized egg. A major accomplishment of mid-19th century biology had been to establish the apparently epigenetic nature of early development. The first investigations directly supporting epigenetic interpretations were those of Wolff (1759, 1768), which showed that embryonic chick blood vessels and gut develop from undifferentiated tissues; Pander's (1817) description of epigenetic development in the chick from primitive germ layers; and the studies of von Baer (1828, 1837), as a result of which the germ layer theory was generalized to other animals. Von Baer showed that skin develops epigenetically from ectoderm, muscular and skeletal systems from mesoderm, etc. Kowalevsky (1867) suggested that the germ layers of many animal phyla are also formed epigenetically. These observations were regarded as crucial refutations of the traditional preformationist theories previously current, such as those of Bonnet (1762 see translation and discussion in Whiteman 1895b). Bonnet had argued that little change incomplexity actually occurs in embryological development, and that "organized bodies preexist from the beginning," According to Bonnet's rather crude ovist theory of preformation (1746) a complete embryo is patterned in every egg, and each such embryo contains an ovary, with eggs that bear embryos within, etc.

The matter seemingly settled by scientific investigation, it was thus a striking event when in the 1870's a novel and sophisticated new form of preformationism was proposed, by Wilhelm His and others. His was the teacher Johann F. Miescher, the discoverer of DNA, and was a proponent of the view that satisfactory explanations of biological phenomena can only be obtained at the molecular level. In 1874, several years before the publication of Whitman's *Clepsine* study. His suggested that the epigenetic character of early chick development is only apparent, the underlying phenomenon being the "coalescence of performed germ." The close relationship between these ideas and the subsequent concept of cytoplasmic localization is obvious. However, these passages were written a decade before it was first realized that the hereditary determinants are confined to the chromosomes of the cell nucleus. To obtain from them the idea of cytoplasmic localization as we now think of it requires the conceptual separation of localized *cytoplasmic* determinants from the *genomic* determinants shared equally among all the blastomere nuclei.

Roux proposed that two kinds of processes leading to embryonic cell differentiation occur in development, *self-differentiation* and *correlative differentiation.* The first of these describes the determinate behaviour of blastomeres and their cell lineages in mosaic forms of embryogenesis, and predicts that when isolated, embryonic blastomeres will continue to display their normal potentialities. Correlative differentiation was invoked by Roux to explain the regulative behaviour of partial embryos. Through their powers of correlative differentiation embryonic cells could secondarily modify their normally determined fate. Roux suggested that correlative differentiation is mediated by means of interactions between cells. Unfortunately, Roux (1883) and Weismann (1885, 1892) initially attempted to explain blastomere self-determination as the consequence of the qualitative partitioning during cleavage of *nuclear* genetic determinants, a proposal that was almost immediately discredited. Experiments of Driesch on sea urchin eggs, Wilson on *Nereis* eggs, and Yatsu on nemertine eggs, among others, showed that treatments which *redistribute* nuclei destined for given blastomeres to other blastomeres nonetheless often result in normal development. The most convincing experiments were carried out by the transient application of pressure to cleaving eggs, e.g., by exposing them to the weight of a cover slip. The result is to inhibit normally occurring transverse cleavages, and to induce extra rounds of vertical cleavage. Three-dimensional development resumes on release of the pressure, and nuclei normally assigned to given blastomeres are not found to be located in different blastomeres. For example, in Wilson's *Nereis* experiments (1896a) the nuclei that would in normal embryos populate the entoblasts were shown to be able to direct the development of first quartet micromeres, and vice versa. In other words, these nuclei must contain genes for endodermal as well as ectodermal differentiation. Generalizing, it follows that any cleavage stage nucleus contain all the zygote genes, and therefore that the differences in cell fate displayed by given lineages must be due to segregation of *cytoplasmic* factors.

The promulgation of Roux's mosaic theory of development—ignoring the nuclear segregation portion—initiated a most interesting intellectual divergence which to some extent is still with us. While tests of the developmental capacities of isolated blastomeres in many cases directly supported the mosaic view of embryogenesis, some regulative capacity was often noted even in blastomeres from eggs that were described overall as "mosaic-works." Most importantly, it was discovered that in some organisms isolated blastomeres could give rise to complete rather than partial larvae. The latter result was obtained with at least some of the early cleavage blastomeres of species of sea urchins, hydroid coelenterates, cephalochordates, nemertean worms, amphibians, and fish (for a complete summary of the classical partial embryo experiments). Proponents of various forms of correlative differentiation regarded the amazing regulatory behaviour revealed in these partial embryo experiments as a decisive blow to the mosaic theory that the embryo is the sum of its self-differentiating blastomeres. The basic difference in attitude this related to the significance of specific substances in specific cells. Wilson and Conklin, for example, considered there to be convincing evidence that the fate of a given blastomere or set of blastomeres may be *determined* by its specific cytoplasmic contents.

The focus on the overall organizational properties of the embryo, and on the role of *position* can be regarded as ancestral

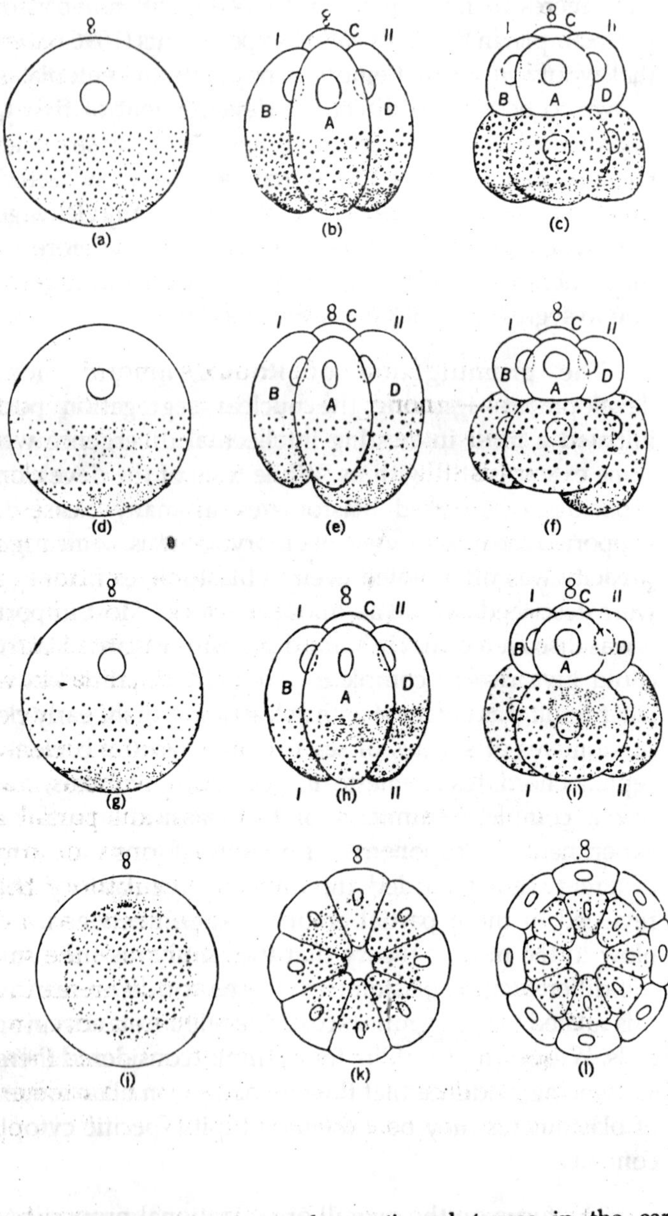

Fig. 8.2. Localization of morphogenetic substances in the early blastomeres of four types of egg. Diagrams of the primary

stratification in the eggs of the sea urchin (a)-(c) and the annelid or gastropod (d)-(f). The first two cleavage planes are designated as I or II. The upper or white zone of each egg is ectoblastic, the middle or granular one the entoblastic, the lower or stippled one the mesoblastic. In (a)-(c) all the zones are equally divided; in (d)-(f) only the two upper zones are thus divided, the lower one being segregated entirely to the D quadrant, (g)-(i). Primary stratification in the ascidian egg, and (j)-(l) in the hydromedusan egg. In the ascidian egg the lower mesoblastic stratum is equally divided between A and D quadrants. In the hydromedusan egg this stratum is absent, and the remaining two are equally distributed up to the time of delamination (l).

schools of embryology oriented toward analysis of organizing centres, morphogenetic fields, and gradients, and that more recently has given rise to the concepts of positional information. The opposing *cellular* focus was soon bolstered by the rediscovery of Mendelism and the general acceptance of the presence in blastomere nuclei of genomic determinants for all the specific traits manifested by differentiated cells. An event of considerable significance at the time was the demonstration of Boveri (1902) that blastomere gene action is specifically required for early development (reviewed in Davidson, 1985). This study concerned the causes of developmental arrest in dispermic sea urchin eggs. Boveri showed that the tetrapolar and tripolar mitoses that occur in these eggs usually result in aneuploidy, so that particular chromosomes are missing or present in greater than diploid number in some blastomeres. He demonstrated that the developmental potential of isolated blastomeres containing partial genomes is invariably incomplete, while in contrast, as Driesch had shown, complete plutei could be obtained from isolated blastomeres of normal embryos.

The classical *cellular* theory of localization, and Boveri's essentially modern assertion of the role of the genome in differentiation, are the traditions that led to the persistent effort to understand embryo cell function, which today is largely molecular in nature. It is an interesting historical note that in the modern pursuit of the mechanism by which initial specification of embryo cell function occurs we once again find ourselves face to face with "the localization problem."

Determination of Specific Cell Lineages

Three examples of embryonic cell lineages that are apparently determined by processes of cytoplasmic localization are considered in this section. One difference between the classical studies just reviewed and the modern investigations we now review is the use of specific ultrastructural and biochemical markers as indices of differentiated cell lineage function. The first of these examples is the determination of the ascidian muscle cell lineage, for which current evidence on the role of localization is probably the most complete; the second is the determination of the gut cell lineage in a nematode; and the third concerns determination of germ cell lineages in certain nematodes, insects, and anuran amphibians.

The Gut Cell Lineage of *Caenorhaditis Elegans*

It is a well-known fact that the development in the nematode *Caenorhabditis elegans* is a highly determinate process. The timing and orientation of each embryonic cell division is programmed genetically, and under normal circumstances the ultimate fate of each cell is rigidly fixed. The simplified diagram in Fig. 8.3 illustrates the origin in the initial cleavages of the six embryonic founder cells. The first cleavage produces an anterior cell, AB, and a posterior cell, P1, and the latter then undergoes several unequal divisions, each resulting in a germ line precursor cell plus a primordial somatic cell. The endoderm derives from the Somatic founder cell EMS, the descendants of which give rise to endoderm, various mesoderm structures, and some neurons. The endoderm lineage is definitively segregated out at the first division of EMS, with the formation of E (endoderm) and MS mesoderm-stomadaeum) founder cells. Since these cells all arise from topologically specific regions of the egg, and since the destiny of the lineage descendant from them is fixed, a possible interpretation of the initial lineage assignments is that each is determined by the localized cytoplasmic elements that it inherits. In this section we consider evidence related to this proposition in respect to the gut cell lineage decendant from the E blastomere, and in the following, the homologous arguments that can be applied to the germ line lineage deriving from the P4 stem cell.

Two markers that specifically indicate gut cell differentiation are the appearance of fluorescent particles that contain tryptophan catabolities, called *rhabditin granules*, and a specific esterase that can be identified by histochemical staining. These appear in the gut cells a few cycles prior to the terminal divisions of their lineage. Laufer *et al.* (1980) and Cowan and McIntosh (1985) used the appearance of rhabditin granules to determine the cellular location of gut cell determinants in embryos whose cleavage had been blocked with cytochalasin and colchicine, and similar experiments monitoring the appearance of esterase activity in cytochalasin treated eggs were reported. The drugs were introduced after bursting the impermeable chorion by application of pressure, or shear forces. The results of these cleavage arrest experiments are essentially similar to those of the ascidian cytochalasin studies just reviewed. When cytokinesis is arrested at the 2-cell stage, rhabditin granules and esterase appear after a number of hours in the PI cell but not in AB; when it is arrested at the 4-cell stage the fluorescent bodies and esterase are observed in EMS, but in no other blastomere: and in arrested 8-cell embryos only the E cell develops these granules or gut esterase. By the time the differentiation reflected by the presence of the cytotypic rhabditin granules has occurred, the arrested blastomeres have accumulated an amount of DNA equivalent to many genomes. It is evident that gut differentiation does not require the appearance of the individual gut cell precursors normally produced during subsequent cleavage of the E lineage, and probably does not depend on intercellular interactions among them, or with other cells. Whether the genomic multiplication is necessary. However, unlike the to those 20th century case in ascidian embryos. cytochalasin-blocked zygotes are unable

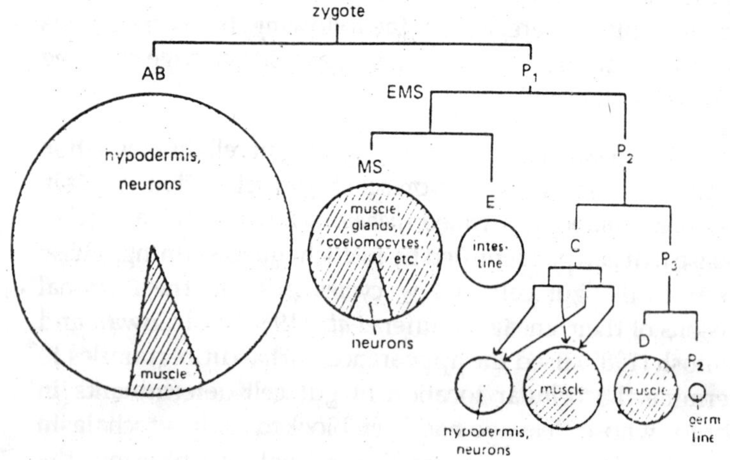

Fig. 8.3. Founder cells of the *Caenorhabditis elegans* embryo, and diagrammatic summary of cell types derived from them. Areas of circles and sectors are proportional to number of descendent cells in the hatched larva. Cell types usually regarded as ectodermal are stippled and those usually regarded as mesodermal are striped. The AB, MSt and C founder cells are multipotential in terms of these germ layer designations.

to produce rhabditin granules, though they do produce hypodermal marker. nor do blocked one-cell embryos produce the gut esterase. In any case, the capacity to synthesize rhabditin granules and gut esterase is shown to be an exclusive and intrinsic property of the E blastomere, and the EMS and PI blastomeres from which it derives. This is confirmed in studies on the development of burst partial embryos in which given blastomeres have lysed. Rhabditin granules are observed only in the E cell derivatives, and not, for example, in isolated AB, PI or P2 cell partial embryos.

These results are consistent with a determinant role for cytoplasmic factors that are segregated into the E cell during cleavage. However, demonstration of such a role would require that gut differentiation occur in cells normally destined to develop into other structures as a consequence of the experimental introduction of E or EMS cytoplasm. One test of this possibility was carried out by fusing AB a blastomeres (i.e., the anterior daughter of the AB cell) to the contiguous EMS cell by puncture of the membranes separating them with a laser microbeam. Following the next cleavage, either the MS cell or the E cell was destroyed by UV irradiation. In 90% of cases when the E cell was intradiated no rhabditin granules formed, though they appeared in the remaining 10% of the experimental embryos, while most control embryos in which the E cell was preserved developed these granules. However, when cleavage was arrested with cytochalasin after the AB.a cell. An alternative experimental design has yielded more positive results. At the 2-cell stage the P1* blastomere was punctured by laser microbeam and the nucleus extruded. In such embryos only the nucleated AB cell divides further and no rhabditin granules are ever formed. However, if nuclei of the progeny of the AB cell are exposed to the enucleate P1 cytoplasm, by laser puncture of the intervening membranes, rhabditin granules form amongst the AB cell progeny in 50% of cases. These blastomere fusion experiments thus provide preliminary support for the causal specification of the gut cell lineage by maternal cytoplasmic factors segregated into the P1, EMS, and then the E blastomeres. The perfect precision, and the predictability of the developmental fate of each blastomere indeed suggest the solution, just as did the same features of many other mosaic embryos to classical writers, However, there is no evidence in this or the foregoing ascidian examples, as to the mode of action, or the directness with which such maternal factors may function. At one extreme. They could immediately bind to and activate gut-specific genes, and at the other, they could serve merely as the initial triggers for a cascade of subsequent regulatory events, an ultimate consequence of which is the activation of these genes.

(iii) Localization of Germ Cell Determinants

In most species that have been examined the definitive

germ cell lineage can be observed after midembryogenesis, and thereafter be traced directly to the gametogenic cells of the larval or juvenile gonads. Primordial stem cells which serve as the exclusive progenitors of the germ line are known to arise in specific embryonic locations at the outset of development in some organisms, but in others, where germ line specification occurs later, the origins of the germ cell lineage remain controversial or obsecure. A unique, localized region of the egg from which the germ line arises can probably be excluded in the case of mammals, for example, since any of the first four or even eight blastomeres can give rise to a complete and fertile embryo. Grafting experiments suggest that in the mouse determination of germ line stem cells occurs in the primitive ectoderm, but not until about 6-7 days of development, as the cells of this region remain totipotent until this time. It is interesting that very early germ cell determination from localized regions of the egg occurs in phylogenetically distant groups, e.g., nematodes, insects, and anuran amphibians. Thus specification of the germ line by means of a cytoplasmic localization process may be an extremely ancient mechanism, since it is found at least occasionally in both deuterostomes and protostomes.

Determination of the Germ Cell Lineage in Nematodes: Boveri and his associates described the Cytoplasmic localization of determinants for the primordial germ cells of the parasitic nematode *Ascaris*. In *Ascaris,* as in *Caenorhabditis,* the definitive germ line stem cell is the product of an early series of unequal cleavages. A special feature of *Ascaris* development not observed in *Caenorhabitis* is the phenomenon of *chromosome diminution.* At the first cleavage one daughter cell retains the complete chromosome complement while the other eliminates a large fraction of the genome in deeply staining pycnotic granules. The stem cell not undergoing chromosome diminution retains the capacity to give rise to the germ cell lineage, while at each of the next several cleavages it produces one daughter which becomes a somatic cell. After fifth cleavage there are 31 somatic

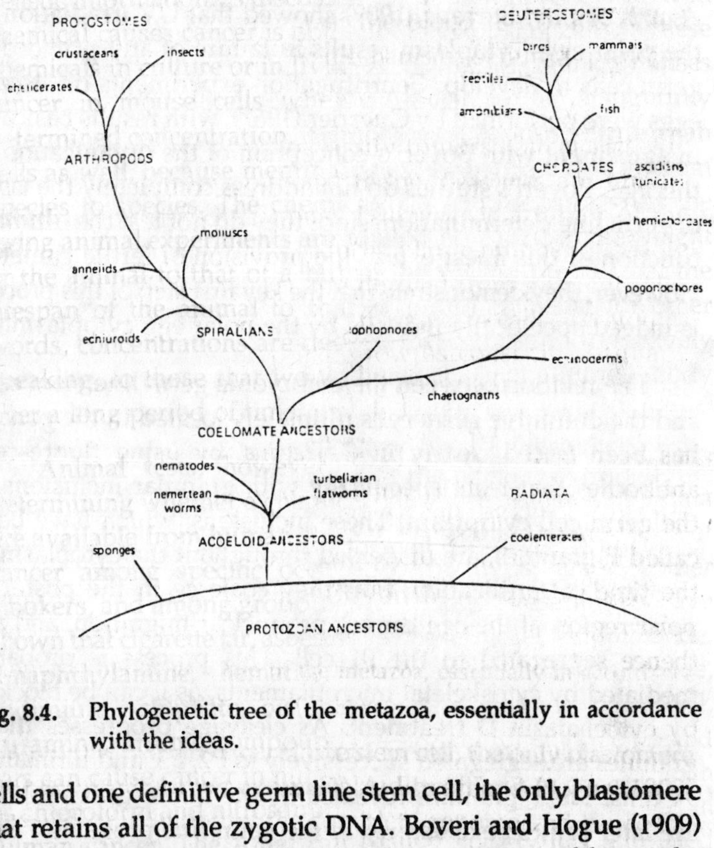

Fig. 8.4. Phylogenetic tree of the metazoa, essentially in accordance with the ideas.

cells and one definitive germ line stem cell, the only blastomere that retains all of the zygotic DNA. Boveri and Hogue (1909) showed that retention of the complete genome and hence the initial step in determination of the germ line, depends on a special sector of polar egg cytoplasm. In eggs undergoing abnormal cleavage induced by polyspermy or centrifugation, nuclei other then the normal germ line stem cell nucleus that are surrounded by the polar cytoplasm now behave like normal germ line nuclei, in that they are spared chromatin loss at mitosis, while all nuclei distributed into the remainder of the cytoplasm undergo diminution. As many as 3 prospective germ line stem cells may form compared to the normal one depending on the relation of the altered cleavage planes to the polar cytoplasm. An additional experiment carried out on *Ascaris* egg by Stevens (1909) showed that UV irradiation of the polar eggs cytoplasm results in failure of any primordial germ cells to develop. Centrifugation experiments on *Ascaris* eggs were performed by Guerrier (1967), with results basically in agreement with Boveri's conception of the organization of this egg. Boveri's studies do not address completely the issue of germ line determination, since they do not test the ultimate function of this lineage, *viz.*, the provision of fertile gametes. However, they demonstrate that the key first step of this process is indeed specifically induced by the polar egg cytoplasm.

The relation between the embryonic germ line stem cells and the definitive germ cells ultimately located in the gonads has been traced clearly in *C. elegans*, by using fluorescent antibodies that react specifically with granular inclusions of the germ cell cytoplasm. These inclusions, which have been called P granules, are dispersed throughout the cytoplasm at the time of fertilization; but, they coalesce in the posterior polar region of the egg during pronuclear migration, and are thence segregated to the P1 cell. This process is probably mediated by cytoskeletal microfilaments, as it can be blocked by cytochalasin D treatment. As cleavage progresses the P granules are segregated successively to P2, P3, and ultimately 94 cells. The P granules revealed by the fluorescent antibody are later present exclusively in the germ line progenitor cells Z2 and Z3, the products of P4 and in differentiated larvae they exist only in the gonadal germ cells. Electron-dense bodies that are probably identical with the P granules have also been visualized in the polar cytoplasm of *C. elegans* eggs, and in the embryonic germ line blastomeres. In the uncleaved fertilized egg these structures are located in the peripheral regions of the polar cytoplasm, and by midcleavage they have assumed a perinuclear location in the primordial germ cells. The immunofluorescent P granules are similarly disposed, and they remain perinuclear throughout oogenesis, when they again disperse into the distal cytoplasm of the egg. It is not clear whether the anterior-posterior axis of the *C. elegans* egg is determined as a result of events following fertilization or is a preformation that is set up during oogenesis.

Maternal mutations are known in *C. elegans* that alter the early cleavage pattern so that two cells form in place of P1. Wood *et al.* (1983) showed that in one such mutant the germ line specific P granules are distributed to the two cells that share the cytoplasm normally destined only for the P2 blastomere, a result that seems exactly to parallel that observed by Boveri in centrifuged and dispermic *Ascaris*.

Distinctive cytoplasmic inclusions that are similar in distribution and ultrastructure to the *C. elegans* P granules and are confined to the germ cell lineage are observed in many forms. They were noticed by classical cytologists in the germ cells of crustaceans, chaetognaths and insects. These structures have now been recognized at the ultrastructural level in germ cells of many other animals, among which are annelids, dipteran insects, ascidians, teleosts, anuran amphibians, and various mammals, in which they are usually referred to as *nuage*. In animals where germ cell determination is precocious, and the germ cell lineage is get aside early in cleavage, the germinal granules are typically found in the region of egg cytoplasm that is to be included in the primordial germ cells, as are the *C. elegans* P granules. In animals such as the mouse, where germ cell determination does not occur until later in embryogenesis, they are not evident until the definitive germ cells can be identified. Classical writers such as Hegner (1914) thought that the germinal granules of the polar egg cytoplasm, often referred to as *polar granules,* might themselves function as localized germ cell determinants. However, as the following review of germ cell lineage specification in insects and amphibians shows, it is not yet possible to distinguish between this view, which attributes causal significance to the presence of polar granules, and an alternative. This is that these inclusions may function (in some unknown way) in later stages of germ cell differentiation, but not in the *determination* of the germ cell lineage. In organisms where germ line segregation is at least initiated by means of localized elements of the egg cytoplasm, as in *Ascaris*, such granules could be performed during oogenesis and distributed to that section of cytoplasm which is to be inherited by the germ line stem cell, much as other maternal components of subsequent use during development appear to be localized (e.g., tubulin, yolk, mitochondria, etc.). At present it indeed seems reasonable to regard germinal granules as cytotypic markers for the germ cell lineage.

Germ Cell Lineage in the Eggs of Drosophila

The differentiation of primordial germ cells was studied in chryosomelid beetles by Hegner (1911, 1914). Germ cell formation can be said to initiate when nuclei arrive at the polar region of the oblong egg, and enter a special region of cytoplasm which appears to function as the germ cell determinant. Hegner succeeded in selectively destroying the polar cytoplasm with a hot needle before the peripheral movement of the nuclei had occurred. The injury induced by the needle is rapidly walled off by the forming blastoderm, and normal development of a differentiated gastrula and ultimately a hatching insect takes place. However, primordial germ cells are absent in embryos descended from the cauterized eggs, and adults developing from them are sterile. No genomic material is directly affected, since no nuclei are in the vicinity of the polar cytoplasm at the time of cauterization. Hegner's experiments show that when the polar cytoplasm is destroyed, the capacity of the embryo to elicit germ cell differentiation is lost.

The primordial germ cells in *Drosophila* eggs arise at the posterior pole of the egg in much the same manner.

The *pole cells* the first cells formed in the embryo and they lie outside of the blastoderm wall. Geigy (1931) showed that UV irradiation of vegetal pole cytoplasm prior to the migration of the cleavage nuclei into this region results in otherwise normal, but agametic animals. Many subsequent experiments demonstrated that the same cleavage nuclei that when surrounded by polar cytoplasm give rise to pole cells, instead become somatic blastoderm nuclei if the polar region of the egg is destroyed, so that they are surrounded by other cytoplasm. Similar observations have been made with a large number of other insect species. For example, Brown and Kalthoff (1983) and von Brunn and Kalthoff (1983) demonstrated that UV irradiation of a sharply defined region of the polar cytoplasm in *Smittia* eggs that had not yet initiated nuclear division interferes with the migration of nuclei into the polar cytoplasm, prevents subsequent pole cell formation, and also affects that survival of those pole cells that do form. This case is of particular interest because the action spectrum for UV inhibition of pole cell formation closely approximates the UV absorption spectrum of RNA, except for a slightly higher relative efficiency at wavelengths > 260 nm. In addition all the effects of UV irradiation on the pole cells are photoreversible, which is usually taken to suggest a nucleic acid target (Jackle and Kalthoff, 1980). Togashi and Okada (1983) reported an action spectrum for UV inhibition of *Drosophila* pole cell formation, with resultant adult sterility, that peaks at 280 nm and is also high at 254 nm. In both this and the *Smittia* study the sensitive target areas were shown to be confined to a thin polar region of the cortical cytoplasm. The UV rays could not have reached the zygote nucleus in the *Smittia* experiment, nor in the case of *Drosophila* any of the nuclei, which at the stage when the irradiation was performed are located deep in the interior of the egg.

The sterilizing effect of UV irradiation offers an experimental opportunity to demonstrate that the polar egg cytoplasm is specifically required for the formation of functional primordial germ cells. It has been shown by many workers that injection of unfractionated posterior-pole cytoplasm, but not anterior pole cytoplasm, rescues eggs that would otherwise have been sterilized by UV irradiation. The injected eggs are able to produce pole cells, and frequently developed into adults that contain fertile gametes, in contrast to the agametic adults obtained from the irradiated controls. It is interesting that the cytoplasmic substances required for rescue of pole cell formation can be physically separated from those necessary to restore adult fertility. Thus while unfractionated posterior pole cytoplasm contains both activities. Ueda and Okada (1982) found that a 28,000 X g centrifugal pellet fraction promotes pole cell formation only. Experiments of Okada and Togashi (1985) indicate that the component required for pole cell formation is an mRNA that can be extracted from unfertilized eggs. However, though injection of UV irradiated eggs with this RNA fraction restores the formation of pole cells of apparently normal structure, fertile flies are never obtained.

Convincing evidence that *causal pole cell determinants* are localized in the posterior region of the *Drosophila* egg is provided by the cytoplasmic injection experiments. In these studies posterior pole cytoplasm was injected into the anterior or ventral regions of recipient eggs. Pole cells displaying cytological structures unique to the germ cell lineage were thereby induced to form at the ectopic locations where the cytoplasm had been injected. Among the ultrastructural features used to identify pole cells are intranuclear organelles that consist of an electron dense cortex and in some species an electron lucid core, called *nuclear bodies*. Illmensee *et al.* (1976) also found that posterior pole cytoplasm from stage 13-14 ovarian oocytes induces the formation of ectopic pole cells when injected into the anterior ends of recipient eggs, but cytoplasm extracted from younger oocytes (stages 10-12) does not display this activity. The polar cytoplasm of these oocytes contains some morphologically identifiable polar granules, but evidently lacks some necessary constituents. The special significance of this demonstration is that the localized determinants for pole cell formation are shown to be present prior to fertilization and ovulation. They are thus in the classical sense a *prelocalization,* of a morphogenetic substance that is synthesized and anchored to the polar region of the egg during ovarian oogenesis. An elegant feature of the cytoplasmic transfer experiments of Illmensee *et al.* was the use of donor and recipient strains bearing different genetic marking. For example, the anterior pole cell induction experiment was carried out in flies of *mwh e (multiple wing hair, ebony)* genotype, and to test the ability of the ectopic pole cells that were formed to serve as germ line progenitors these were transplanted to the posterior regions of *y w sn (yellow, white, singed)* hosts. A very small percent of the resulting adults display germ line mosaicism, as expected if the induced pole cells gave rise to functional gametes. A similar result was reported for pole cells induced to form ectopically in ventral regions of the embryo.

This review briefly summarizes much evidence that the posterior polar cytoplasm can specifically induce *pole cell formation,* and that it contains substances *necessary* for germ cell formation. However, additional demonstrations of the ability of ectopically induced pole cells to give rise to functional gametes is clearly required to secure the proposal that the polar cytoplasm actually *determines the germ cell lineage.* The demonstration of Ueda and Okada (1982) that injection of certain cytoplasmic fractions into UV irradiated eggs restores pole cell formation but not adult fertility shows that this is not simply a semantic issue. A difficulty is that in *Drosophila* not all pole cells give rise to germ cells, though the ultimate fate of those that do not find their way to the larval gonad remains controversial.

The germinal or polar granules of *Drosophila* eggs have been characterized in great detail. At oviposition these granules consist of dense, membrane-free particles surrounded by cluds of ribosomes (Mahowald, 1968). In various species of *Drosophila* these particles fuse or fragment during early embryogenesis, and the ribosomes and polyribosomes initially associated with them are no longer observed after the pole cell stage of embryogenesis. By the time the germ cells are located in the larval gonad, the polar granules have apparently given rise to characteristic fibrillar structures applied to the nuclear membrane. This progression of forms is similar to what is observed in both amphibian and mammalian oogonia. The fibrillar structures remain throughout the oogonial stage, but are absent in oocytes. Polar granules can again be observed in *Drosophila* oocytes during vitellogenesis. Though they are already localized in stage 10-12 oocytes, as noted above the polar cytoplasm of these oocytes is not yet competent to induce pole cell formation. This observation at least distinguishes the cytological entities recognized as ovarian polar granules from the functional properties of the mature posterior pole cytoplasm. The *Drosophila* polar granules have been partially purified from pole cells, and there is evidence from two-dimensional analyses of their proteins for a single major species of about 95 kilodaltons. This protein is synthesized during oogenesis, prior to stage 10.

The effects of several maternal mutations that result in sterility have recently been interpreted as consistent with a mechanistic role for the polar granules in germ cell determination. Among the examples of interest are a *grandchildless (gs)* mutation of *Drosophila subobscura* which in homozygotes prevents pole cell formation and thus results in sterility. Ultrastructural examination shows that among other things this genetic lesion affects the deposition of polar granules during oogenesis. Some *gs* eggs lack any polar granules, while others possess small numbers of polar granules that are often incorrectly localized. Unfortunately, this is not the only cytological lesion in *gs* eggs, as both anterior and posterior poles appear abnormally organized, and during early cleavage, nuclei never descend into the posterior pole cytoplasm. A *gs* mutation of *D. melanogaster*, *gs(1)N26*, also prevents nuclear migration into the posterior cytoplasm but has not effect on polar granule disposition. On the other hand, various alleles of another *D. melanogaster grandchildless* mutant, *tudor*, provide a good correlation between the amount of polar granule material and the ability to produce pole cells. Again, however, this is not the unique effect of the mutation, some alleles of which also disturb embryonic segmentation. Perhaps this mutation interferes with a (cytoskeletal?) process by which several different determinants are localized during early development, and the abnormal polar granule deposition in mutant eggs could be merely a visible index of the disturbance, rather than the immediate cause of sterility. A number of other recessive maternal mutations of the *grandchildless* phenotype are also known, but most of these also produce somatic embryonic defects.

Germ Cell Lineage in Anuran Amphibians: In frogs the germ cell lineage originates early in cleavage from blastomeres forming at the vegetal pole of the egg. In normal development the vegetal pole cytoplasm or "germ plasm" is distributed to all of the first four blastomeres, and each has the capacity to generate a functional germ cell lineage. The primordial germ cells reside in the endoderm during early development, where they undergo several divisions, and after gastrulation they migrate to the germinal ridges where they are located by stage 25 in *Rana pipiens* tadpoles, or by stage 44-47 in *Xenopus laevis* tadpoles. At this point they are in a state of mitotic arrest. Although if undisturbed these cells subsequently give rise to the functional germ cells of the organism, they are not yet irreversibly determined to do so, even by this late stage. Thus Wylie *et al.* (1985) showed that fluorescein-labeled primordial germ cells from stage 45 *Xenopus* tadpoles give rise to a variety of different ectodermal, endodermal, and mesodermal cell types if implanted into the blastocoel of a late blastula stage embryo.

Many workers showed that UV irradiation of the vegetal region of the cytoplasm of frog eggs results in a sharp decrease in the number of primordial germ cells, and ultimately, in some animals, in agametic gonads and sterility. These classical experiments suggested that peripheral components of this cytoplasmic region, which is incorporated in the initial precursors of the germ cell lineage, might be required for specification of this lineage. The significance of vegetal pole cytoplasm for germ line differentiation was confirmed in many subsequent experiments in which this region of the cytoplasm of uncleaved eggs or of 2- or 4-cell embryos was severely damaged by ultraviolet irradiation or by microsurgical means.

The number of primordial germ cells observed in the early tadpole (about 35 in normal stage 47 *Xenopus* tadpoles has traditionally been used to assay the effect of exposing eggs and early embryos to UV irradiation. Thus the germinal ridges of stage 47 tadpoles derived from vegetaly irradiated *Xenopus* eggs, zygotes, or 2-cell embryos were often found to be devoid of germ cells. Irradiation at the 8-cell stage, however, has no specific effect on the number of tadpole primordial germ cells. Similarly results were obtained earlier by Smith (1966) m *Rana pipiens,* and control experiments showed that parallel irradiation of the animal pole cytoplasm produces no visible defects, and certainly none in the germ cell line. Furthermore, Smith (1966) demonstrated that the damage to cytoplasmic targets resulting from ultraviolet irradiation of *Rana* eggs can be compensated by injected into the vegetal pole of cytoplasm from the vegetal pole of an unirradiated egg. In a significant fraction of cases the recipient eggs developed into stage 25 tadpoles in which the germinal ridges bore primordial germ cells, while controls receiving no vegetal cytoplasm or cytoplasm from the animal pole lacked germ cells. Rescue of UV irradiated *Rana chensinensis* and *Xenopus* eggs can also be accomolished by injection of vegetal pole cytoplasm. A crude centrifugal fraction of vegetal but not animal pole cytoplasm from non-irradiated eggs was shown to be effective in restoring germ cells to the adpole germinal ridge. In addition, injection of these fractions into non-irradiated eggs is reported to induce the formation of up to twice the normal number of primordial germ cells.

It has recently been seen that primordial germ cells eventually reappear in the germinal ridges of tadpoles raised from some vegetally irradiated eggs, albeit on a much delayed schedule. Thus while the germinal ridges of stage 25 *Rana pipiens* tadpoles grown from UV irradiated eggs lack primordial germ cell, two weeks later these are present, and they ultimately give rise to differentiated gametes, Ikenishi and Kotani (1979) showed that in *Xenopus* embryos raised from vegetally irradiated eggs, fewer primordial germ cell divisions occur, and after the tailbud stage many of the primordial germ cells initially formed disappear. Furthermore, the surviving germ cells remain in the central endoderm, at tailbud stage, while in control embryos they have already migrated to the lateral and dorsal regions. These observations occasion a somewhat different interpretation than classically proposed of the effect of cytoplasmic UV irradiation. Since surgical removal or disturbance of the vegetal pole cytoplasm does eventually result in sterility or partial sterility there may indeed be a cytoplasmic determinant for the anuran germ cell lineage localized at the vegetal pole of the egg. However, the *UV sensitive component* seems to control only the replicative or migratory behaviour of the primordial germ cells rather than their fundamental specification. We are reminded of the parallel observations in *Drosophila* where, as we have seen, the determination of the germ cell lineage is a process that has several components, and pole cell formation can be separated experimentally from the specification of a functional germ cell lineage.

The role of polar granules in the anuran germ cell lineage specification process is also indistinct. These bodies, also known as the "germinal plasm," are initially found in the outer regions of the vegetal pole cytoplasm, and in ultrastructure they closely resemble the polar granules of insect eggs. They are present in unfertilized eggs, having been localized just beneath the vegetal pole cortex during maturation. They may be anchored to the egg cortex, as if the eggs are inverted, they do not redistribute gravitationally as do the yolk platelets. During cleavage these organelles are incorporated in the primordial germ cell precursors, and they are still present, with associated ribosomes, in these cells at the blastula and gastrula stages. By the time the primordial germ cells are localized in the germinal ridges they no longer contain identifiable polar granules. Instead a fibrous component is found applied to the nuclear membrane.

Though transition stages have not been convincingly described, these could derive from the earlier polar granules, an interpretation that is supported by the analogous progression of form undergone by insect polar granules. In anuran amphibians the fibrillar perinuclear structures persist through most of oogenesis, though they are absent in mature oocytes. Typical polar granules reappear at about the time of germinal vesicle breakdown. Several correlative items of evidence, in addition to the natural history of these germ cell specific organelles, have been adduced in support of the idea that polar granules are involved in germ cell specification. Thus loss of sensitivity to UV irradiation by the 8-cell stage occurs just when the polar granules move inward from the cortex of the egg so that they no longer lie within the shallow penetration range of the ultraviolet irradiation. Furthermore, when *Xenopus* eggs are cenifuged so as to move the polar granules inward precociously, irradiation. Furthermore, when *Xenopus* eggs are centrifuged so as to move the polar granules inward precociously, irradiation at the 2-cell stage fails to affect the number of primordial germ cells in the germinal ridges of stage 47 tadpoles. This result is significant, since only large, dense particles would be likely to have been affected by the low centrifugation forces applied in this experiment, 150 g for 60 seconds. On the other hand, there is on visible effect of UV irradiation on polar granule ultrastructure, and even if these particles were the essential UV targets, their function in germ cell differentiation could be confined to effects on replication and migration.

Polar granules are not evident in the vegetal pole cytoplasm of cleavage stage urodele eggs, and germ cells in these eggs apparently arise from dispressed regions of the marginal zone of the animal ectodermal cap rather than from the vegetal endoderm, as in anurans. Though they may develop from specific cells of the marginal zone, the differentiation of primordial germ cells in urodeles probably requires inductive influences from adjacent caudal mesoderm. The earliest reported appearance of structures resembling polar granules in urodele eggs is at gastrula, though the definitive primordial germ cells of urodele larvae ultimately resemble those of anurans in their ultrastructural characteristics. Thus, the germ cell lineages in urodeles would appear to arise by a different route than in anurans, though in both groups route than in anurans, though in both groups the primordial germ cells ultimately possess the same cytoplasmic organelles. The least artificial view may again be that amphibian polar granules are cytoplasmic structures required later in germ cell differentiation, and that they appear together with whatever localized germ cell determinants may exist in the egg cytoplasm only when the germ cell lineage is set aside at the very beginning of cleavage, as in the anurans.

Cytoplasmic Localization in Molluscan and Annelid Eggs

A characteristic feature of early cleavage in many molluscan and annelid forms is the transient extrusion of a lobe of vegetal pole cytoplasm, which following cytokinesis is resorbed into one of the embryonic marcromeres. In the typical case the first cleavage *polar lobe* is resorbed into the CD macromere, the second cleavage polar lobe into the D marcromere, the third cleavage polar lobe into the ID macromere, and so forth. As will be recalled it is the D quadrant of molluscan and annelid eggs that gives rise to the mesentoblasts and is responsible for the dorsoventral organization of the embryo. The polar lobe is often attached to the remainder of the embryo by only a thin strand of cytoplasm. Its removal is easily accomplished without immediate injury to the embryo, which continues to divide approximately on schedule. However, the ultimate developmental consequences of the removal of this anucleate element of egg cytoplasm are both dramatic and specific. In this section we review the experimental effort that has been made over the last 90 years to establish a causal relation between the polar lobe cytoplasm and the establishment of the dorsoventral axis of the embryo, and to understand the nature of the localized determinants that it apparently contains.

(i) Morphogenetic Significance of the Polar Lobe Cytoplasm: Crampton found that deletion of the first cleavage polar lobe of *Ilyanassa* eggs causes the asymmetric cleavage pattern and the delayed division schedule that normally mark the D quadrant to disappear. Cleavage of embryos from which the polar lobe had been removed is radially rather than bilaterally symmetric, and Crampton was unable to recognize in these embryos the 4d stem cell from which the mesentoblasts derive. Wilson (1904b) followed up these preliminary results with an extraordinary study of the effect of polar lobe removal on embryogenesis in the scaphopod molluse *Dentalium.* These experiments demonstrated that deletion of the first cleavage polar lobe irreversibly blocks the appearance of all the special D quadrant cell lineages. He reported that the mesentoblasts and the mesodermal germ bands never form, and that other larval structures are missing after lobe removal, such as the apical tuft, and the larval shell and foot. Wilson (1904b) compared the capacities of isolated blastomeres with those manifested by lobeless embryos, and showed that the latter develop in exactly the same way (or fail to develop in exactly same way) as do embryos deriving from AB blastomeres or from single A, B and C blastomeres, none of which produce the mesodermal primordium. From these observations it was possible to infer the existence in the polar lobe of cytoplasmic determinants for the D quadrant cell lineages that build the dorsoventral structures of the embryo.

Polar lobe removal is now known to result in the same characteristic complex of developmental defects in a number of molluscan and annelid species. Among the organisms in which this phenomenon has been studied are the gastropod molluscs *Bithynia* and *Ilyanassa* the lamellibranch mollusc *Mytilus* the annelid *Sabellaria* as well as *Dentalium,* the subject of recent investigations Verdonk (1968). Many workers that have confirmed and extended the 1904 conclusions of Wilson. This is not an invariable phenomenon in eggs bearing polar lobes, however. Thus removal of the small polar lobe of the polychaete annelid *Chaetopterus* has only minor developmental effects, and instead vegetal region cytoplasm included in the polar lobe is functionally analogous to the polar lobe cytoplasm in the other species mentioned.

Normal development is shown in which Fig. 8.5, in the D quadrant cells cannot be distinguished. The structures of the advanced larvae descendant from lobeless embryos are compared to those of normal *Ilyanassa* larvae. Lobeless larvae fall to develop heart, intestine, statocyst, operculum, velum, external shell, eyes and foot. On the other hand, lobeless larvae possess active muscle, nerve ganglia and nerve endings, stomach, some velar tissue with cilia, digestive gland, mantle gland tissue, and pigment cells. Removal of the first cleavage polar lobe cytoplasm thus does not block all cell differentiation, only *certain* differentiation, though ultimately it produces a cascade of delayed organizational defects.

The catastrophic though specific effects of first polar lobe removal illustrated in these figures suggests that diverse morphogenetic determinants are located in the vegetal pole cytoplasm of the CD blastomere by the 2-cell stage. The most straightforward presumption would be that during later cleavage these determinants are physically transferred to the stem cells from which lobe dependent lineages arise, e.g., 4d. The latter cell is the direct ancestor of both the primary mesentoblasts (Me^1 and Me^2) and the primary entoblasts (E^1 and E^2). Thus only embryos in which 4d is present produce heart and intestine at the veliger larval stage. In *Ilyanassa* deletion of 1d produces no observable effect, but deletion of 2d causes absence or impairment of the shell, while 3d forms the primordium for the left half of the larval foot (Clement, 1976). The inference that morphogenetic determinants for these tissues are transferred directly into the d quadrant micromeres during cleavage was tested in *Ilyanassa* and in *Dentalium* by determining the effect of ablation of the D marcromere at successive cleavages. Deletion of the D macromere after the 1d micromere has been given off tests whether the morphogenetic factors of the polar lobe still

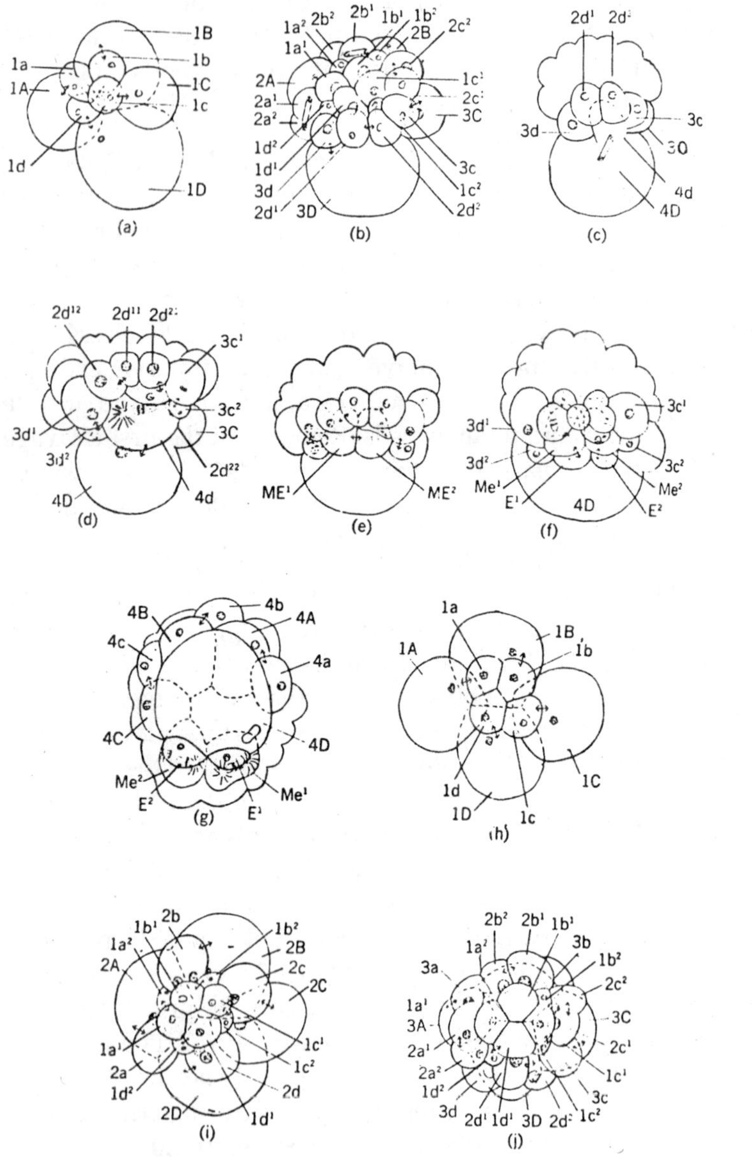

Fig. 8.5. D quadrant cell lineage in Ilyanassa absoleta *embryos.* Upper cases letters refer to macromeres and lower case to micromeres.

Preceding numerals indicate the macromere division at which the micromeres are given off, and superscripts indicate subsequent micromere division products E^1 and E^2. primary entoblasts; Me^1 and Me^2. primary mesentoblasts. (a) and (b) Normal cleavage showing first, second, and third quartet micromeres, from camera lucida drawings of stained whole mount preparations. (c)-(f). Normal cleavage showing the early derivatives of the first mesentoblast progenitor cell. 4d. (g) The egg has been oriented so that the vegetal pole is toward the observer. Further division of Me^1 and Me^2 will produce the primordial mesoderm columns. (h)-(j) Symmetrical cleavage after removal of the polar lobe at the trefoil stage. The D quadrant is named arbitrarily, as it cannot be differentiated by inspection from the B quadrant.

reside in the macromere or have been shunted into the micromere, and similarly for D macromere deletion after 2d, 3d, or 4d have a arisen. Clement (1962) found that in *Ilyanassa* removal of the 4D macromere after the formation of 4d has no effect on later differentiation, and the resulting embryo is normal except for its small size. This experiment also helps to eliminate the possibilities that the effects of polar lobe removal are due to general injuries, starvation for nutrients, or to disruption of an embryo-wide "morphogenetic gradient." On the other hand, removal of the D macromere before the 2d micromere has been formed results in as severe an inhibiton of morphogenesis as removal of the first cleavage polar lobe or of the whole D quadrant at the 4-cell stage. If the 3D macromere is deleted, i.e., after 1d, 2d and 3d have been formed, the resulting larvae display velum, eyes, foot, and some shell, all of which are polar lobe dependent structures, but they still lack the mesodermal primordium. The morphogenetically significant polar lobe contents therefore appear to be shunted into 3d and 4d micromeres. Essentially similar results are reported for *Dentalium*. Here again removal of the D macromere when only 1d has been produced is equivalent in its effects to deletion of the polar lobe, while consequences of decreasing severity occur if the macromere is ablated after 2d, or both 2d and 3d micromeres have appeared.

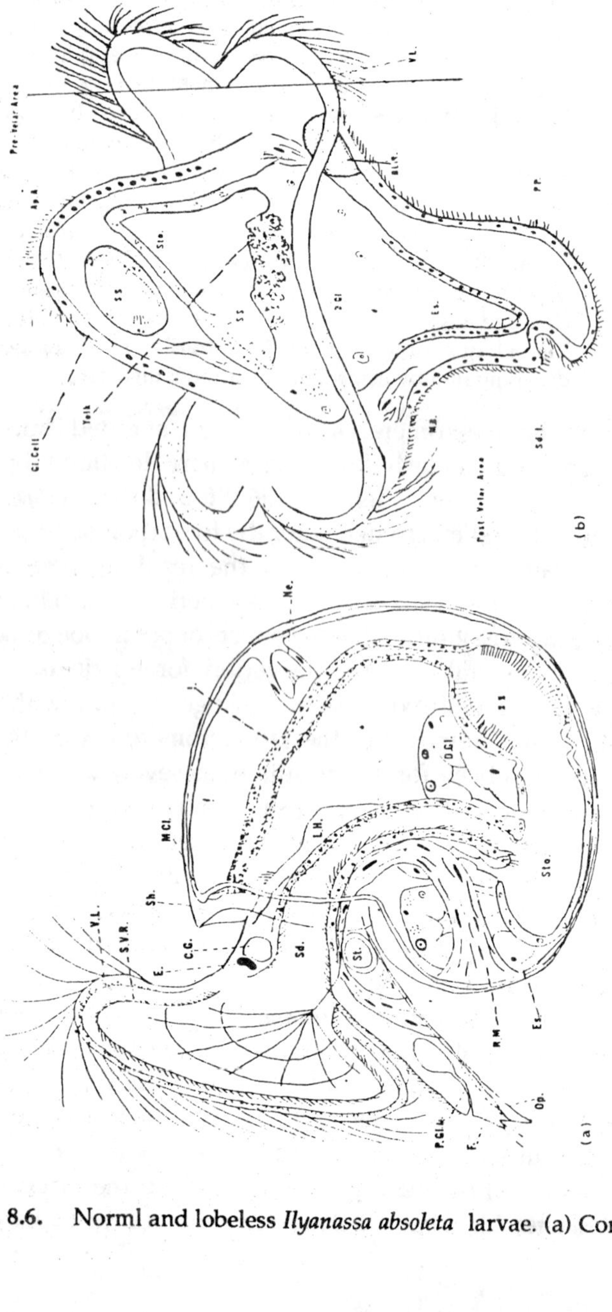

Fig. 8.6. Norml and lobeless *Ilyanassa absoleta* larvae. (a) Composite

reconstruction of normal 8-to-9-day veliger larva of *Ilyanassa*, with left velar lobe omitted, (b) Reconstruction of a lobeless larva based on whole mount and sectional material. Ap. A., apical area: Bi. M., birefringent mass; C.G., cerebral ganglion: D.Gl., digestive gland; E., eye ; Es., esophagus; F., foot; Gl. Cell, ectodermal gland cell; I., intestine; L.H., larval heart; M.B., muscle block; M.C., mantle colar ; Ne., nephridium; Op., operculum; P.P., posterior protusion; P. Tl.D., pedal gland duct; R.M., retractor muscle; S.S., style sac; Sd., stomodaeum; Sd. I., stomodeal-like invagination; Sh., shell; St., statocyst; Sto., stomach; S.V.R., second velar row; V.L., velar lobe.

Dentalium vulgare embryos from which 4d is removed also develop into normal larvae. The general conclusion from both sets of experiments is that the morphogenetic elements localized in the polar lobe are progressively distributed during cleavage to the specific micromeres where they will function.

In annelid and molluscan larvae this organ consists of a plate of differentiated cells, several of which bear long cilia, overlying a plexus of neurosecretory cells that is later associated with the cerebral commissure. *Dentalium* embryos from which the first polar lobe has been removed form no apica tuft, but differentiation of the apical organ proceeds normally in embryos from which the second polar lobe has been deleted. The apical organ forms from progeny of the 1c an 1d micromeres (1974). Microsurgical experiments indicate that the apical organ determinants are localized in the upper region of the first polar lobe and they are later shunted into the upper regions of the C and D blastomeres. Thus, the whole lower end of the lobe and 50-60% of the volume of cytoplasm that it contains can be ablated without loss of apical tuft formation. Determination of the cell lineage that produces these tuft cilia, and probably of the neuronal components of the apical organ as well, occurs after resorption of the lobe into the CD macromere, evidently as a consequence of the inclusion of elements of the upper lobe cortex or cytoplasm in the 1c and 1d micromeres.

In *Ilyanassa* the polar lobe cytoplasm *represses* the ability of micromere derivatives other than those normally destined for the apical plate to produce cilia, and this aspect of apical organ differentiation has also been analyzed in the polychaete annelid *Sabellaria.* In the latter organism the first cleavage polar lobe bears the determinants for the apical tuft and for the posttrochal chaetae. As in *Dentalium, Sabellaria* embryos from which the second cleavage polar lobe has been removed produce apical tufts. Render (1883) found that any combination of isolated macromeres that includes the C macromere. i.e., ABC, BC, or C alone, will form an apical tuft. Cultured D macromeres never give rise to an apical tuft, but, if the *second* cleavage polar lobe is removed and the C and D macromeres are then cultured, *both* are now able to produce apical tufts. It follows that the second cleavage polar lobe contains a repressor of apical tuft formation, while the first contains only the determinants that induce apical organ differentiation.

The elements that induce apical tuft formation are left in the CD macromere after resorption of the first polar lobe, whence they are distributed to both the C and D macromeres, and eventually to the apical precursor micromeres. The elements that repress apical tuft formation are included in the smaller second polar lobe, and thus are confined to the D macromere after resorption, so that only the C quadrant micromeres are able to produce apical tufts. In accord with this interpretation Render (1983) showed that suppression of second polar lobe formation by treatment with a dilute SDS solution prevents apical tuft formation in both C and D isolates, since the repressive cytoplasmic factors are now included in both macromeres. For formation of posttrochal chaetae, on the other hand, the D cell must be included in the isolate. Determinants for the posttrochal chaetae also remain in the second polar lobe, and are returned to the D macromere alone.

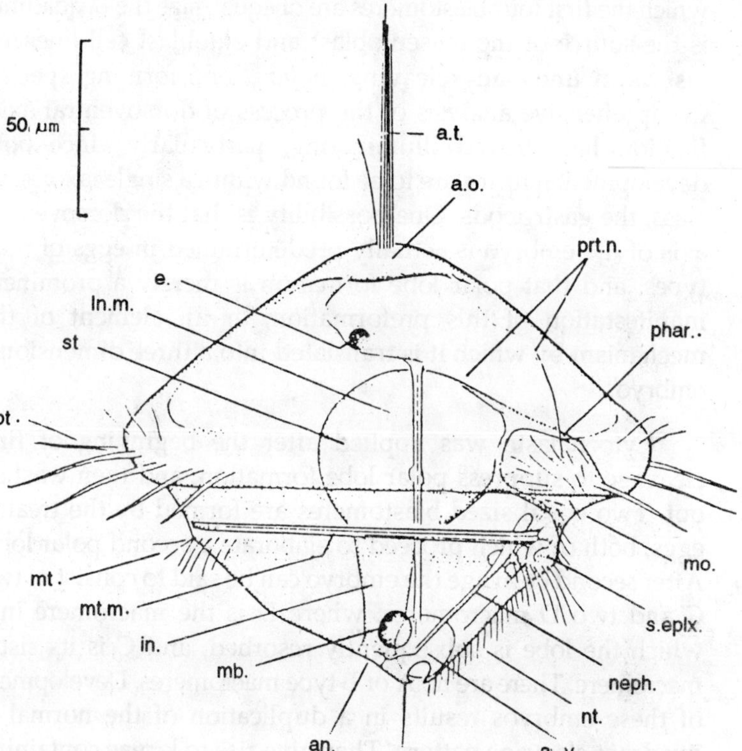

Fig. 8.7. Schematic diagram of the structure of a 2 day old trochophore larva of *Spirobranchus polycerus*. Apical tuft (a.t.), apical organ (a.o.), prototroch (pt.), metatroch (mt.), stomach (st.), pharynx (phar,). The single eye (b), protonephridium (neph.) (only one of pair shown), pretrochal nerves (prt.n.), mouth (mo.), anus (an.), suboral complex (s. cplx.), neurotroch (nt.), anal vesicle (a.v.), mesoblast (mb.), intestine (in.), metarochal muscle (mt.m.), and longitudinal muscle (ln.m.).

Determination of the Donsventral Axis in Equally and Unequally Cleaving Moluscan Eggs

In molluscan eggs the do nor produce polar lobes and in which the first four blastomeres are of equal size the D quadrant is the source of the mesentoblast and entoblast cell lineaces, just as in unequally cleaving, polar lobe forming species. Comprehensive analysis of the process of dorsoventral axial fixation has proved illuminating, particularly since both developmental forms are to be found within a single taxonomic class, the gastropods. One possibility is that the dorsoventral axis of the embryo is actually predetermined in eggs of both types, and that polar lobe formation is merely a prominent manifestation of this preformation, or an element of the mechanism by which it is translated into a three dimensional embryo.

Cytochalasin was applied after the beginning of first cleavage to suppress polar lobe formation, and then washed out. Two equal sized blastomeres are formed by the treated eggs, both of which proceed to elaborate a second polar lobe. After second cleavage the embryo can be said to consist of two C and two D macromeres, where D is the macromere into which the lobe is subsequently resorbed, and C is its sister macromere. There are no A or B type macromeres. Development of these embryos results in a duplication of the normal D quadrant cleavage pattern. They give rise to larvae containing double the normal number of polar lobe dependent structures, e.g., two dorsal shell glands, four statocysts, and four mantle cavities, but lack any A or B quadrant derivatives such as the stomodaeum. This experiment shows that the *disposition of the polar lobe cytoplasm is sufficient to determine the dorsoventral axis,* since two such axes form if this cytoplasm is distributed to both macromeres at first cleavage. Choice of the blastomere into which the first cleavage polar lobe flows in the normal egg may be a matter of chance, i.e., the exact position of the cleavage plane with reference to the thin cytoplasmic stalk by which the lobe is connected at the trefoil stage. The future dorsal end of the egg is in any case clearly *not irreversibly predetermined.* On the other hand, the *vegetal* location of the first polar lobe is a genuine preformation, since the annular constriction by which it arises always occurs in a plane perpendicular to the animal-vegetal axis. Cytological evidence reviewed below shows that the animal-vegetal axis is already clearly evident prior to fertilization.

In equally cleaving gastropod eggs the macromeres remain for sometime both morphologically and morphogenetically equivalent. It has been shown for three different species that either of the first two blastomeres can give rise to complete larvae a result that contrasts directly with the restricted potential of the AB blastomere in unequally cleaving molluscan eggs. In equally cleaving gastropod eggs the macromere that gives rise to the primary mesentoblast, 4d, cannot be distinguished until after 5th cleavage. This macromere, 3d, is then identified by the central position that it assumes in the embryo, its non-synchronous cleavage, and the asymmetric disposition of its progeny. Deletion experiments have shown that prior to 5th cleavage all of the four macromeres retain the capacity to serve as the "D macromere, " and to produce the primary mesentoblasts.

In normal embryos the 3D macromere is usually, but not necessarily. One of the two "vegetal cross furrow" macromeres. The event that specifies which will become the D macromere, and hence that determines the dorsoventral axis, is apparently an inductive interaction with the first quartet micromeres across the blastocoelic cavity, quadrant function is assumed by that macromere first achieving the greatest number of contacts with micromeres during the interval between 4th and 5th cleavage. This has been established for *Lymnaea stagnalis* and *Patella vulgata* embryos by deletion of first quartet micromeres. Removal of a sufficient number of these micromeres precludes the differentiation of any D quadrant macromere, and in partial deletions the macromere that assumes the role of 3D can be determined by the position of the remaining micromeres. If the intercellular interaction between macromeres and micromeres is prevented by cytochalasin treatment, the micromere cap remains completely or partially radially symmetric. These experiments show furthermore that in *Lynaea palustris* all four quadrants of the micromere cap remain undetermined and capable of forming dorsal, lateral, or ventral head structures until after the crucial cell contacts occurs, i.e., during the pause in cleavage following the 24-cell stage. It follows from these and the other results reviewed that the dorsoventral axis of neither the micromere nor macromere blastomere tiers is primordially established, both arising inductively by means of the interactions between them. A variety of other cell contacts form elsewhere in the embryo during the same interval in which these micromere-macromere interactions occur. For example, in *Patella* embryos functional intercellular coupling is also established at this point, as indicated by the diffusion of iontophoretically injected lucifer yellow dye from blastomere to blastomere. However, contacts of this kind appear not to be required for dorsoventral determination, since they are absent from the 3D-micromere junctions, and conversely, they still occur in embryos from which the first quartet micromeres have been deleted.

Establishment of the dorsoventral axis in equally cleaving gastropod eggs is thus accomplished by an *intercellular* interaction, while the same end is achieved in unequally cleaving gastropod eggs through an *intracellular* mechanism. The contrast is nicely illustrated by the effect of first quartet micromere deletion in embryos that form polar lobes. Though severe defects in larval head formation result, the whole first quartet can be removed from *Bithynia tentaculata* or *Ilyanassa* embryos. without affecting determination of the dorsoventral axis and of the specific cell lineages responsible for its construction. The role of the polar lobe is illuminated by the comparison. It is the device by which determinative cytoplasmic agents or structures initially resident in the egg cytoplasm are spatially disposed to the appropriate blastomeres. This can be seen in

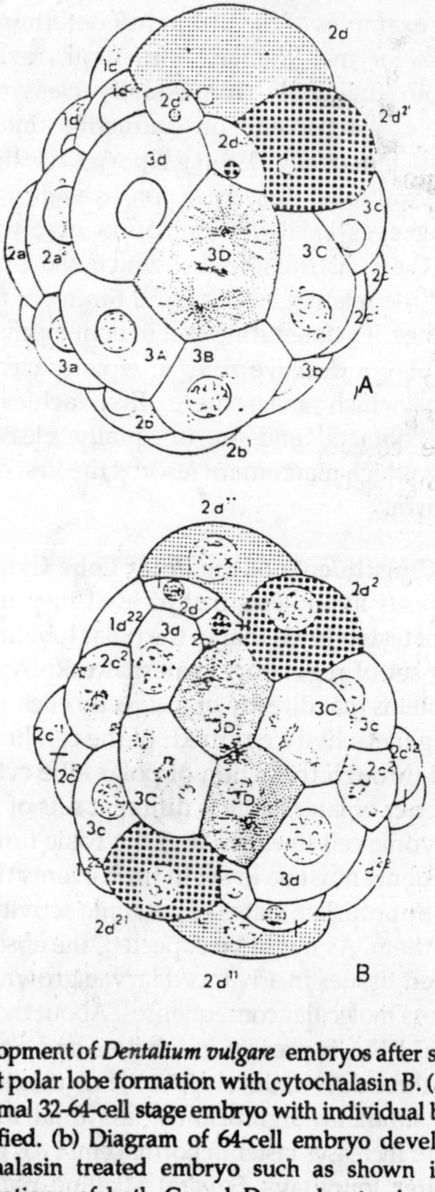

Fig. 8.8. Development of *Dentalium vulgare* embryos after suppression of first polar lobe formation with cytochalasin B. (a) Diagram of normal 32-64-cell stage embryo with individual blastomeres identified. (b) Diagram of 64-cell embryo developed from cytochalasin treated embryo such as shown in (a) note duplication of both C and D quadrant micromeres and absence of A and B quadrant micromeres.

detail in the examples of the apical tuft determinants, and the determinants for mesentoblast formation, reviewed above. Whereas both unequally and equally cleaving gastropod embryos rely extensively on inductive interactions for organogenesis, the equally cleaving embryos use this mechanism for dorsoventral axial determination as well, while in polar lobe forming eggs axial determination involves the inbuilt location of the constructions by which the polar lobes are formed, and the relative location and timing of the successive cleavage planes. It is interesting that in both mechanisms chance appears to play a decisive role, in equally cleaving eggs in determining which macromere first achieves sufficient micromere contacts, and in unequally cleaving eggs in determining which macromere resorbs the first cleavage polar lobe cytoplasm.

Molecular Constituents of the Polar Lobe Cytoplasm?

An obvious hypothesis that at least in principle has the advantage of testability is that in the polar lobe are sequestered a particular set of regulatory maternal mRNAs. These could code for proteins that directly or indirectly elicit the differential patterns of gene activity required of the cell lineages that are determined through the action of polar lobe cytoplasm. Such lineages sooner or later express different sets of genes than do other embryonic cell lineages, and the basic function of polar lobe constituents must be to set in train events that eventually impose the appropriate states of genomic activity in the nuclei exposed to them. As might be expected, the absence of certain differentiated tissues in advanced larvae grown from lobeless eggs has gross molecular consequences. About the same amount of ribosomal RNA is present in lobeless and normal *Ilyanassa* embryos at three days of development, but after this the lobeless embryo accumulates significantly less total RNA. Similarly DNA content increases faster in normal embryos than in lobeless embryos after four days. Several alkaline phosphatase and esterase isozyme forms that normally appear in advanced larvae are also lacking in larvae derived from lobeless embryos.

The presence of maternal mRNA in polar lobes was established initially by Clement and Tyler (1967), who showed that *Ilyanassa* polar lobes continue to synthesize protein for at least 24 hr after isolation at the trefoil stage. The spectrum of proteins synthesized in the isolated polar lobes was analyzed by 2-dimensional gel electrophoresis. Out of the several hundred species visualized there are no newly synthesized polar lobe proteins that are not also synthesized after equivalent incubations in lobeless or normal cleavage stage embryos. Though this result provides no support for the proposal that certain species of maternal mRNA are sequestered exclusively in the polar lobe, nor is it excluded. Proteins whose pI's fall outside of the range of the isoelectric focussing gradient, i.e., any whose pI is >8 or <5, and proteins coded by rare messages, would not be included in these anlyses. The latter class of course includes the large majority of the diverse species of maternal mRNA present, if measurements carried out on other embryonic material can be considered a guide. An interesting observation is that several proteins are synthesized in isolated polar lobes after 25 hr of incubation *in vitro* that are not synthesized at 4hr, and that these same changes occur in normal and in lobeless embryos, and after actinomycin treatment of these embryos. This set of molecular changes thus originates at the translational (or post-translational) level, and the polar lobe evidently includes the cytoplasmic elements necessary for their occurrence.

The prevalence of a few other newly synthesized protein species increases sharply during the development of normal *Ilyanassa* embryos, but not in isolated polar lobes. Their appearance could be the result of embryonic transcriptional activity, as also implied by the effect of actinomycin, which alters the pattern of protein synthesis during the first 24 hr. However, the same changes occur in lobeless embryos as in normal embryos and thus there is no evidence that the regulation of these protein species is dependent on polar lobe constituents. The sole observation that assigns to the polar lobe any direct effect on embryonic transcriptional activity is a quantitative one. In mesentoblast or gastrula stages the absolute rate of total RNA synthesis in lobeless embryos is only about 55-60% of that in normal embryos. The difference may be significant, since cell division in lobesame number of nuclei throughout the period studied. During these stages the newly synthesized RNA measured by labeling is mainly of heterogeneous, non-ribosomal nature, and is probably a combination of nRNA and mRNA.

Cytological Observations

Numerous ultrastructural studies have been carried out on the polar lobe cytoplasm. Unusual membrane-bound organelles are present in the cytoplasm of the large polar lobes of *Ilyanassa* and *Dentalium,* as well as mitochondria and other undefined particulate inclusions. However, none of these structures are required for polar lobe function, and they cannot contain the morphogenetic determinants of the lobe, since they can be redistributed to other parts of the egg by low speed centrifugation without disturbing normal development. On the other hand, a special organelle termed the vegetal body that by the same test is functionally important has been demonstrated in the small polar lobe of the gastropod *Bithynia,* and similar structures are present in other species that produce small polar lobes. The vegetal body consists of a dense aggregation of small vesicles. Though the *Bithynia* polar lobe occupies <1% of the egg volume, it morphogenetic determinants are functionally similar to those of the *Dentalium* and *Ilyanassa* polar lobes. Thus, lobeless *Bithynia* embryos fail to form mesentoblasts and lack eyes, foot, intestine, organized shell, operculum, etc., though they successfully develop digestive gland, ganglia, and some muscle. In *Bithynia,* however, the developmental capacity of the C macromere is the same as that of the D macromere, and lobe-dependent structures form if either the C or D macromeres are present. The progressive distribution of the vegetal body vesicles parallels that of the lobe morphogenetic determinants. The vegetal body is shunted into the CD cell at first cleavage, whereupon the vesicles disperse, and are inherited by both the C and D macromeres. The staining reactions of the vegetal body suggest that its vesicles contain RNA but this remains to be established by biochemical measurement. Centrifugation at speeds of about 1400 x g dislocates the vegetal body away from the polar region of the uncleaved egg in about 50% of cases, and these eggs form polar lobes that lack the vegetal body. Van Dam *et al.* (1982) found that about 50% of centrifuged eggs from which the first polar lobes had been removed developed into *normal embryos,* a result that is never observed after deletion of the lobe from uncentrifuged eggs. Inclusion of the vegetal body in one of the two blastomeres, in the absence of the polar lobe structure, thus suffices for normal development, and it follows that the vegetal body is the site of the lobe determinants. Seen in this light the function of the polar lobe appears essentially to be the distribution of the special package of vesicles to one or the other of the first two blastomeres.

In eggs with large polar lobes organelles similar to the vegetal body are not present, and as we have seen centrifugal forces that totally redistribute the particulate cytoplasmic elements fail to dislodge the morphogenetic determinants. Thus it is likely that these determinants are anchored in the cortical structures of large polar lobes, and that they exist initially as regional differentiations of the lobe plasma membrane and cortex. For example, the apical tuft determinants in the first cleavage *Dentalium* polar lobe are apparently fixed in the upper region of the lobe. The extrusion and resorption of the lobe can be considered as mechanisms for the spatial disposition of the cytoplasmic determinants to the appropriate blastomeres. These movements are accompanied by a complex of cytoskeletal changes.

Scanning electron micrographs have provided dramatic instances of regional cortical differentiations at the polar lobe. The vegetal pole is seen to be clearly marked by distinct surface features prior to the onset of cleavage. Furthermore, at least in

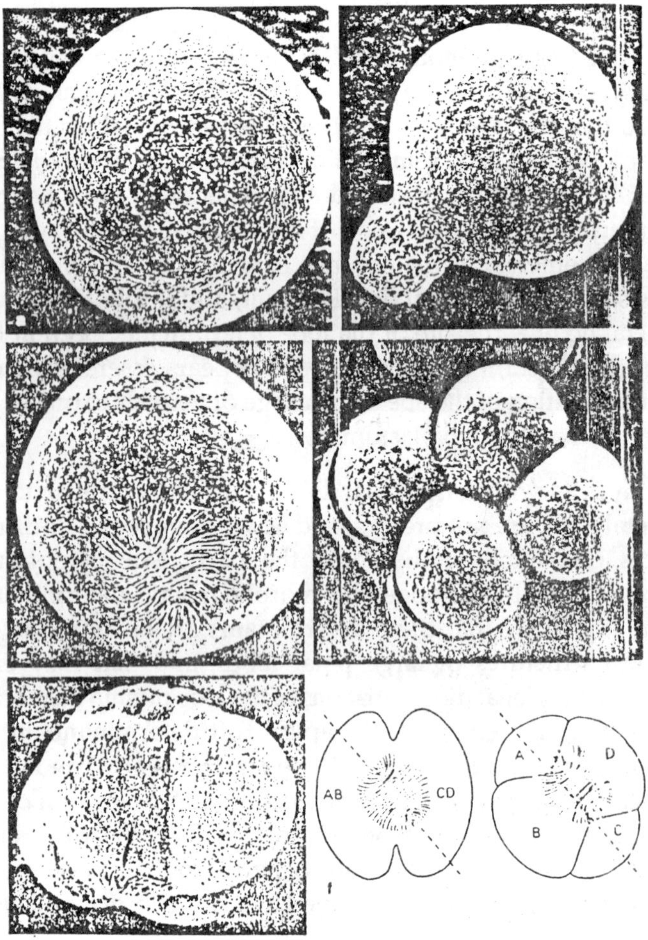

Fig. 8.9. Surface structures of the vegetal pole in gastropod eggs, (a)-(c) Scanning electron inicrographsy (a) Vegetal pole of egg of *Nassarius reticulata*, 1st meiotic division, Vegetal pole is marked

by a circular interruption in the external carbohydrate coat, containing microvilli, and surrounded by a spiral array of folds. The ring of folds marks the location of the constriction that will give rise to the first polar lobe. About X325. (b) Polar lobe of uncleaved *Buccinum undatum* egg. The lobe is covered with branched surface ridges absent from rest of egg. About X 180. (c) Uncleaved egg of *Crepidula fornicata* after first meiotic division. Oriented folds are present at the vegetal pole. About 400X. (d) *Crepidula* embryo at 4-cell stage. Surface folds are confined to D macromere. In region where 2nd polar lobe has been resorbed. About X2 10. (e) *Crepidula* embryo. 24-cell stage. Micromeres are visible at the animal pole. The surface folds marked by arrow are visible on 3D macromere. About X 225. [(a)-(e)]. (f)Diagram showing relation of bilaterally symmetrical pattern of folds at the vegetal pole of *Crepidula* egg to the 1st and 2nd cleavage planes. The 1st cleavage plane is oriented at about 50° from the axis of symmetry marked by the dashed lines

the egg of *Crepidula* the plane of first cleavage is also predetermined, since the longitudinal orientations of the polar mocrovillar patch bears a fixed relation to the future plane of cleavage, which is retained during the subsequent cleavages.

The initial specification of the vegetal pole in polar lobe bearing eggs may occur far back in oogenesis. Animal-vegetal differences in distribution of microvilli and other cytological features can be observed in the uncleaved egg of the gastropod *Nassarius reticulata*, and could derive originally from the polarized morphology of the oocyte-follicle cell complex. Thus during the growth phases of oogenesis the area of contact is confined to the base of the oocyte, and this becomes the future vegetal pole. An interesting inference is that the animal-vegetal polarity of the egg is imprinted by intercellular surface interactions occurring in the ovary. Prelocalization of vegetal pole structures extends as well to internal structures such as the vegetal body, which is already localized at the future vegetal pole in growing *Bithynia* oocytes. Maternal components are also responsible for the asymmetric "spiral" cleavage of molluscan eggs, as shown by genetic studies on the maternal inheritance of left and right handed body forms in the gastropod *Lymnaea*. The orientation of the adult body form in gastropods in determined by the direction in which the micromeres are given off during early cleavage (see the discussion of spiral cleavage in Wilson. Freeman and Lundelius (1982) showed that cytoplasm from right handed or dextral eggs injected into genetically sinistral eggs can transfer the property of dextrality. The direction of cleavage is thus not determined by the orientation of the ovarain oocyte with respect to the follicle cells. The cytoplasmic elements that confer this property are synthesized during oogenesis, and provide an excellent additional example of the performation of components specifying spatial organization in the molluscan embryo..

The studies of polar lobe function reviewed in this section provide an observer of early development with a comprehensive and detailed description of the biological aspects of a cytoplasmic localization process. It seems inescapable that a characteristic set of dorsoventral determinants is carried in the polar lobe cytoplasm, and that these are responsible for specification of many of the early cell lineages of the embryo. The spatial distribution of these determinants is an epigenetic, developmental process mediated by the disposition of the polar lobe constituents. However, the structures of substances that later exert morphogenetic effects evidently include or consist largely of preformed maternal components. Their origins in oogenesis, their nature, and their molecular mode of action, which directly or indirectly ultimately affects the embryo genomes, remain unknown.

The Development of Localization

The distinction was drawn earlier between spatial assignments in the egg that are prelocalizations, and those that are constructed ontogenically following fertilization. Of course, as Driesch and Wilson pointed out, prelocalized features of the mature egg are the product of prior ontogenic processes that have taken place during oogenesis. The animal-vegetal axes of ascidian and some molluscan eggs, for example, are established during oogenesis, while the anterior-posterior or dorsoventral axes that provide the bilateral organization of the embryo are established by the developmental redistribution of maternal cytoplasmic elements. While common in marine eggs, this scheme of development is by no means universal. In some animals both axes are prelocalized in the unfertilized egg and in others neither. The examples considered in the following provide some insight into the possible nature of both prelocalized and ontogenically localized determinants, and the processes by which they may be deposited in given regions of the egg.

Prelocalization in the Cortex of Dipteran Eggs

By the cellular blastoderm stage the cortical region of the *Drosophila* egg is already is mosaic of determined cell types. Thus when allowed to develop in intracoelomic culture, cells from anterior and posterior regions of the blastoderm give rise only to the expected anterior or posterior adult structures. In addition, Schubiger (1976) demonstrated that partial embryos formed by ligation at the blastoderm stage can produce all the anterior, thoracic, or caudal adult parts that would normally derive from them.

Blastoderm fate maps were based initially on histological reconstructions of the development of organ anlagen and then on analyses of genetic mosaics induced during the first few divisions of the zygote nuclei, in which the developmental disposition of clones of cells bearing visible morphological markers can be followed. By the latter method have been determined the approximate location and number of the primordial cells for adult cuticular structures, as well as for some larval organs. A more direct approach to mapping the morphogenetic values of the blastoderm cells is microablation, by pricking the egg, local cautery or UV irradiation. An UV laser microbeam was utilized by Lohs-Schardin *et al.* to map the primordia for the thoracic and abdominal epidermis of the first instar larva and for adult structures. Cephalic larval structures were mapped by microsurgical blastoderm cell deletions by Underwood *et al.* (1980). In the experiments of Lohs-Schardin *et al.* (1979a) the UV microbeam was 20 μm in diameter, and affected a patch of about 15 cells (the embryo is about 500 μm x 175 μm and at the stage of irradiation in these experiments it contains about 5000 cortical cells). Specific defects in the larval epidermis resulted from the localized irradiation, and these experiments demonstrate that about one-third of the blastoderm cells comprise the progenitors of the entire larval epidermis. Extensive fate map data have also been obtained by injection of horseradish peroxidase into individual blastoderm cells and other cytological observations. The overall result of these studies is to confirm that long before the appearance of visible manifestations of differentiation (except for the precociously appearing pole cells) the blastoderm has acquired a complex pattern of specification, in which are encoded both embryonic axes as well as may specific regional assignments. The following evidence shows that in contrast to any of the eggs so far discussed, *both embryonic axes are prelocalized* in the unfertilized dipteran egg cytoplasm, just as are the determinants for pole cell formation.

Origin of the Axial Determinants in Drosophila: The bilateral rather than radial symmetry of the unfertilized dipteran egg is evident from its external morphology. Thus the future dorsal side is less convex; the future anterior pole of the embryo is marked by the micropyle. The opening in the chorion through which the sperm enters: the two chorionic filaments are located bilaterally at the future dorsal posterior region; and the germinal cytoplasm is already localized at the future posterior pole.

A number of maternal mutations are known that disturb primary axial determination in the embryo. There are so far identified 10 different maternally functioning genes, mutations in which alter the prospective fates of ventral and lateral cells of the blastoderm to that of the prospective dorsal epidermis. A well-known mutation of this type is *dorsal (dl)*. There are two *dorsal* phenotypes. Embryos from homozygous *dl/dl* females express the "recessive phenotype," which is characterized by a lack of ventral, mesodermal, and neural structures. Though at the blastoderm stage the embryo appeals normal, invagination of the presumptive mesoderm fails, and the embryo becomes a yolk-filled epidermal tube that lacks any internal organs. A "dominant phenotype" is temperature sensitive. The eggs of *dl*/+ females develop normally at 22° C but arrest at 29°C, due to the absence of muscles and portions of the ventral epidermis, tissues arising normally from the ventral regions of the blastoderm. Nusselein-Volhard *et al.* (1980) concluded that vental cells of these embryos carry out assignments normally relegated to the more lateral cells, in particular the formation of epidermis instead of mesodermal primordia. These and other observations show that embryonic specification of ventral cell types requires the maternal products of the wild-type *dorsal* gene, and also of the other maternally acting genes that when mutated produce *dorsal* phenotypes. The *dorsal* gene has been cloned and found to produce a single 2.8 kb poly (A) RNA species, which is present in eggs up to the blastoderm stage, after which it disappears. Studies with germ line mosaics demonstrate that the activity of the *dorsal* gene is required only in germ line cells and *in situ* hybridization using a *dl* probe shows that the gene is active in the nurse cell nuclei at stages 5-11 of oogenesis. After stage 11 dl^{+}RNAs are transferred to the oocyte, as are the products of many other genes.

Of the ten maternally acting loci known to be required for specification of the normal dorsoventral egg axis, the effects of eight, *viz., dorsal, tube pipe, snake, easter, Toll* ree, *spatzle* , and , *pelle,* have been shown to be at least partially rescued be injection of wild-type egg cytoplasm, and all except *dorsal* can also be rescued to some extent by injection cf poly (A) RNA extracted from very early wild-type embryos. The rescuing activity for *pelle* mutants disappears from the poly (A) RNA fraction at an earlier stage of cleavage then from the cytoplasm, suggesting that translation of the maternal mRNA is by this stage complete while for certain other of these mutants, the rescuing activity is instead preferentially located in the poly (A) RNA fraction of the cytoplasm. Thus translation of *dorsal* mRNA may occur particularly early in development (and /or in oogenesis). The *dorsal* cytoplasmic component required for rescue is by the blastoderm stage spatially localized, since cytoplasm removed from the ventral regions of donor embryos is the most effective in rescue experiments. The mutant effects of the *toll* locus that can be rescued by RNA injection are those caused by recessive mutations, which produce phenotypes similar to those of the other dorsalizing maternal loci. Dominant *Toll* mutations (*Toll*^{D}) have the opposite effects, resulting in the appearance of patches of ventral denticles over the entire dorsoventral circumference, and the absence rather than overextension of dorsal structures (Anderson *et al.*, 1985a). Injection of wild-type cytoplasm into *Toll* mutantys restores their ability to produce dorsal structures, and the site of injection defines the ventral region of the structures formed. Thus, whereas lack of *Toll* function results in *dorsalization*, as does lack of function of the other nine loci (all recessive), these observations are consistent with the observation that the dominant *Toll*^{D} gene results in *ventraalization*, due to production of excess *Toll* product. Double mutants combining *Toll*^{D} and recessive mutant alleles of *gastrulation*, *defective*, *snake*, *easter*, etc., lack *both* ventral and dorsal pattern elements but develop lateral structures, while the *Toll*^{D} and *dorsal* combination is totally dorsalized. In any case the demonstration that dorsoventral polarity can be restored to mutant embryos by injection of wild-type egg cytoplasm of poly (A) RNA shows explicity that *maternal transcripts synthesized in the meroistic egg chamber contain genetic information required for normal axial specification in the embryo.*

A large number of maternally acting genes are also required specifically for establishment of the anterior-posterior morphogenetic organization of the *Drosophila* egg. Expression of mutant alleles of seven such genes in cells of the female germ line carried in mosaic mothers has been shown in each case to suffice for complete production of the mutant embryonic pheno. The developmental abnormalities caused by these seven recessive maternal lethals fall into three morphological classes. Mutations in *tyorso* and *trunk* result in failure to develop structures of the extreme anterior and posterior ends of the embryo, and in addition they cause development of structures normally arising more towards the midregion to occur in the vicinity of the anterior and posterior poles. Thus, for example, *torso* embryos typically terminate at the posterior end with cuticular patterns characteristic of abdominal segment A7. Yet, these mutation to not prevent cellularization of the blastoderm, and their effects apparently result from alterations of blastoderm cell assignments. These effects are large scale, disturbing the global fate map of the egg. Thus maternal mutations at *valois, vasa, staufen,* and *tudor* loci that prevent the formation of polar granules and pole cells also cause deletions of abdominal segments, and in the case of *staufen,* head structures; and mutations at the *exuperantia* locus cause replacement of head structures by an inverted posterior end. It is not clear, however, whether loci such as these are *directly* responsible for the spatial patterns in which maternal morphogens might be distributed. Thus, for example, threr is some evidence that another gene that when mutated may produce a torso-like phenotype, *viz., lethal (I) pole hole,* may be involved primarily in some common cellular function such as proliferation or cell formation and similarly, many *valois* eggs display severe defects in cellularization in the blastoderm.

Among maternally acting mutations that affect establishment of the anterior-posterior axis are genetic defects that interfere with determination of *either* anterior of posterior embryonic structures, which implies (but does not in itself prove) the existence of separate maternal factors necessary for cephalic and for caudal cell specification. A well-known example is provided by *bicaudal (bic),* a recessive maternal point mutation. Two other loci, which are semidominant, *BicC* and *BicD,* produce similar defects. Morphological defects produced by mutations at any of these three loci consist either of deletions of various amounts of anterior structure, with otherwise normal retention of embryo polarity, or of replacement of the deleted anterior structures with duplicated posterior structures in reverse orientation. Possibly this mirror image symmetry is the result of a secondary reorganization along the anterior-posterior axis induced in embryos suffering the most extensive deletions of anterior segments. Determination of pole cells is not affected in *bicaudal* embryos, and these remain situated normally at one end. The *bicaudal* mutations affect both the specification of particular cell lineages, for example earising malpighian tubules to form from cells that in normal embryos would give rise to cephalic structures, and the general pattern, for example, the number of posterior segments in the embryo. The variation in the phenotypes of double abdomen embryos is interesting, in that the most anterior structures formed , suggesting that the *bicaudal* mutations affect the distribution of a morphogen that extends over large regions of the egg, rather than being sharply localized to one area of the cytoplasmic cortex, for example. The temperature sensitivity of the *BicD* lesions have been used to demonstrate that, as expected, function of the gene is required only during oogenesis, and it is interesting that oogenesis in females homozygous for *BicC* mutations is blocked at stage 8.

In seems likely that the distribution of products of maternally acting genes that effect anterior-posterior embryonic axis specification would reflect the polarity of the egg. We have already noted evidence for dorsoventral localization of maternal products of genes affecting that axis. Mlodzik *et al.* (1985) isolated a gene by its sequence homology with a *Ubx* homeobox and though the morphological significance of this gene, named *caudal* is unknown, the distribution of its maternal transcripts provides a possible example. These transcripts are produced in the nurse cells during oogenesis, and at fertilization are present in the egg. *In situ* hybridization shows *caudal* is unknown, the distribution of its maternal transcripts provides a possible example. These transcripts are produced in the nurse cell during oogenesis, and at fertilization are present in the egg. *In situ* hybridization shows *caudal* RNAs to be more or less evenly distributed during cleavage, but at the syncytial blastoderm stage when they are located in the cortex, they have largely disappeared from the anterior end of the egg up to about 20% of total egg length. The concentration of these-transcripts increases monotonically for the next 50% of egg length, and they are present at their maximum level for the posterior 30% of egg length. At cellular blastoderm stage their distribution sharpens, possibly by means of zygotic transcription, and they accumulate in a single sharp band not far from the posterior pole, where are located precursors of the proctodaeum and regions of abdominal segments 9 and 10. Such observations focus attention on the *origins of the primordial polarity of the egg,* which provides an initially oriented matrix for the establishment of differential concentrations of maternal gene products.

An interesting insight into the processes by which polarity might initially be imposed derives from morphological observations on another maternal mutation, *decephalic,* which produces an abnormality converse to that arising from *bicaudal* mutations. As the name implies, *dicephalic* embryos of extreme phenotype develop head furrows at both poles, and their cuticular structures suggest a mirror image duplication of the cephalic segments, while differentiation of posterior structures is suppressed. In contrast to bicaudal and *dorsal,* the *dicephalic* mutation visibly affects the construction of the nurse cell-oocyte complex. In *dicephalic* ovarioles the 15 nurse cells form two clusters positioned at opposite ends of the oocyte, in place of the normal asymmetric arrangement in which the nurse cells are all at the future anterior pole. The chorion of *dicephalic* eggs contains two micropyles, one at each end. The implication is that the anterior determinants are localized near the sites of the nurse cell oocyte junctions, or perhaps are loaded into the oocyte by the nurse cells. Thus, the spatial organization of the ovariole may be translated into the spatial organization of the egg, and the preformation of the anterior-posterior axis may descend from intercellular interaction that took place during oogenesis. We have encountered a mechanism of this nature earlier, *viz.*, the determination of the vegetal pole of the *Nassarius* oocyte at the unique site of its contact with follicle cells.

Characterization of the Prelocalized Anterior Determinants in Smittia: In the unfertilized egg the UV sensitive elements of the anterior determinants are symmetrically distributed within a polar cone of cytoplasm about 20 µm in diameter. Since little of the incident UV energy penetrates beyond 5-10 µm of the surface, *these elements must lie in or near the cortex of the egg.* After 5-7 hr, when in *Smittia* the nuclei begin their migration toward the egg surface, the UV sensitive targets disperse over the entire anterior half of the egg. Double abdomen embryos can then be produced by a low flux of widespread anterior irradiation, a treatment that is not effective earlier. The action spectrum of double abdomen induction by UV light displays a major peak at 285 nm, and a minor peak at about 257 nm. An interpretation consistent with this result is that the anterior determinants include a ribonucleoprotein target. Additional evidence that also suggests a functional RNA component includes the observation of Kandler-Singer and Kalthoff (1976) that anterior determination can be blocked by RNase admitted through a small puncture made in the anterior end of the egg. Up to 40% of eggs so treated display the double abdomen syndrome. Punctures made in other regions of the egg, denatured RNase, or other enzymes do not have this effect. Furthermore, the UV induction of double abdomen is photoreversible. As much as 60% increase in the frequency of normal embryos occurs if the irradiated eggs are subsequently exposed to long wavelength UV or blue visible light. UV irradiation sufficient to induce double abdomen embryos causes extensive formation of pyrimidine dimers in the ribosomal RNA of these embryos, and photoreactivation treatments that rescue UV irradiated embryos also stimulate the repair of most of these dimers. The same would presumably be true of other RNA species within the UV penetration zone. In any case the conclusion that localized maternal RNAs are required for axial specification in *Smittia* is consistent with that drawn earlier for *Drosophila* from the demonstration that maternal mutants such as *snake* can be rescued by injected wild-type RNA.

Potentially useful protein markers for both anterior and posterior determination have recently been identified in *Smittia* eggs. An *anterior indicator protein* of about 35 kd is synthesized in the anterior but not posterior blastoderm cells of normal embryos. This protein in produced in cells from both halves of the blastoderm in double cephalon embryos, and again exclusively in the anterior end after photoreversal of irradiated eggs. A *posterior indicator protein* of about 50 kd displays a complementary distribution. This protein is synthesized in both ends of the blastoderm in UV induced double abdomen embryos, and its appearance is restricted to the posterior end in irradiated embryos rescued by photoreversion. Synthesis of both indicator proteins occurs long in advance of any visible anterior-posterior differentiation. As the regional synthesis of these proteins is predicted perfectly by experimental interventions that affect the prelocalized cytoplasmic determinants, it follows that molecular regulation of these blastoderm cell functions occurs during the earliest stages of the axial specification process.

(ii) Disposition of Maternal Cytoplasmic Determinants During Early Cleavage

Localization Patterns in Crenophore Eggs: At the opposite extreme from the eggs of dipterans and other insects in which both embryonic axes are prelocalized are the eggs of ctenophores and hydrozoans, in which until development begins neither axis is specified. Cytoplasmic determinants are nonetheless present in ctenophore eggs, as was shown convincingly by classical experimentalists. These early experiments as well as more recent blastomere isolation studies were reviewed by Reverberi (1971a). At the 8-cell stage the embryo contains two pairs of external cells denoted "E" cells, located on either side of the four inner cells, which are denoted "M" cells. The derivatives of E cells and M cells have different morphogenetic fates which are expressed accurately in isolated blastomere experiments. In normal embryos the micromeres given off by the E macromeres produce fused rows of large swimming cilia, or comb plates, and the micromeres derived from M macromeres form other structures, including mouth, apical organ, and most spectacularly, photocytes. These cells contain a calcium activated photoprotein which enables them to produce flashes of light. The embryo gives rise to a bilaterally symmetrical lava organize.

Freeman (1977) showed that in three species of ctenophore the embryonic oral-aboral axis is determined by the site of the first cleavage furrow. Though in normal eggs this usually coincides with the location of the polar bodies, which provide a stable orientation marker on the egg surface, that is not which provide a stable orientation marker on the egg surface, that is not invariably the case, and in centrifuged eggs it is often not the case. The cleavage furrow apparently forms near the zygote nucleus, which is easily shifted by low centrifugal forces. Normal development ensues in centrifuged eggs whatever the plane of first cleavage, and the consequent orientation of the oral-aboral axis, with respect to the polar bodies. The future oral pole develops wherever the cleavage initiates, and the plane of this cleavage is the future sagittal plane of the larva. The second cleavage occurs at right angles to the first and its plane becomes the future tentacular plane of the larva. Thus neither axis is prelocalized in the egg, and the first and second cleavages are the causal events in the orientation of both.

The comb plate and photocyte determinants are localized at different times, and in the unfertilized egg are resident in different components of the cytoplasm. The comb plate determinants are associated with the cortex. Segregation of these factors occurs between the 2- an 8-cell stages, when they become localized in the M and E macromeres respectively. Experiments of Freeman (1976 a,b, 1977) imply that localization in this egg occurs as a *consequence* of cleavage Prior to the 2-

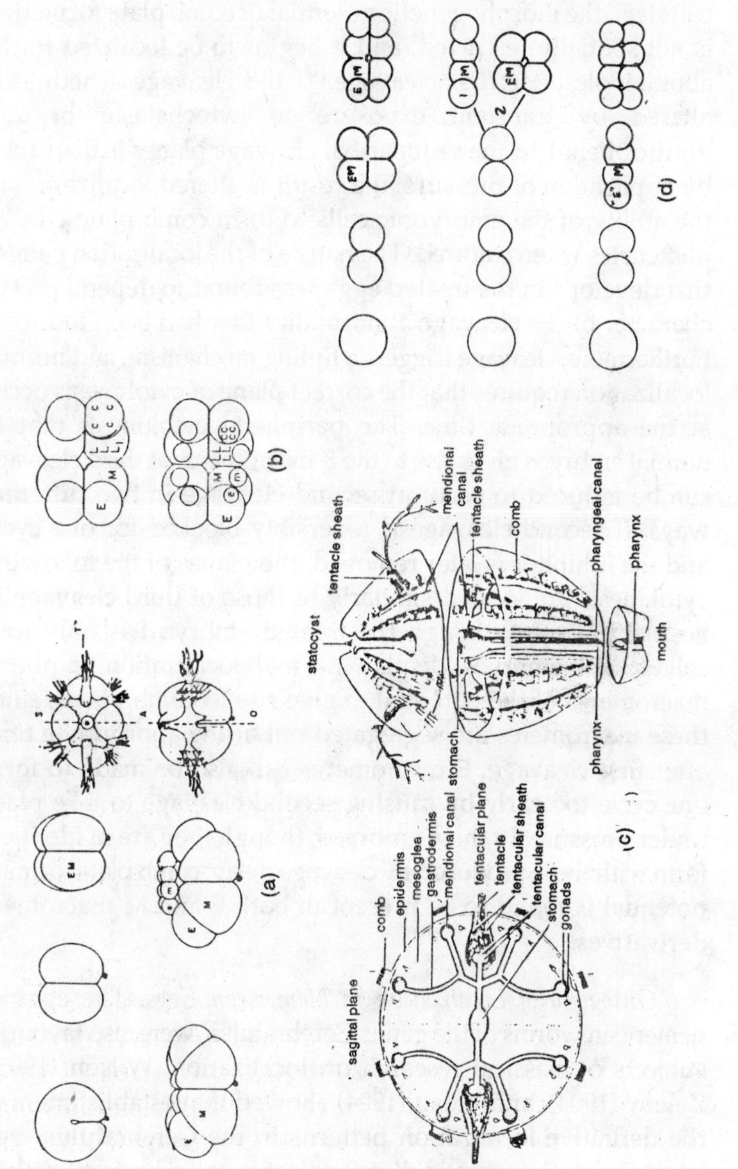

Fig. 8.10. Cleavage and localization in the cteophore egg.

cell stage the morphogenetic potential of comb plate formation is not spatially restricted, and it begins to be localized to the aboral pole after first cleavage. If the cleavage schedule is altered by transient exposure to cytochalasin or 2,4-dinitrophenol, or the sequence of cleavage planes is disturbed by application of pressure, the result is altered localization of the ability of the embryonic cells to form comb plate cilia or photocytes when cultured. The nature of the localization pattern that develops in the treated eggs was found to depend on the character of the cleavage abnormality that had been induced. Furthermore, cleavage triggers a timing mechanism, and normal localization requires that the correct plane of cytokinesis occur at the appropriate time. The peripheral cytokinesis that in normal embryos gives rise to the E macromeres at third cleavage can be induced to occur at second cleavage in two different ways. If second cleavage is reversibly blocked for one cycle and the inhibitor is later removed, the planes of the following cytokinesis are located similarly to those of third cleavage in normal embryos, though the treated embryo has only four cells. Such embryos display normal localization in the E macromeres of the potential to give rise to comb plates, since these macromeres are segregated out at the appropriate time after first cleavage. E macromeres can also be made to form one cycle too early by causing second cleavage to take place under pressure. In these embryos, though they are of identical form with those produced by cleavage delay, comb plate forming potential is found to be present in both E and M macromere derivatives.

Ontogeny of Localization in Nemertean Eggs: The eggs of nemertean worms of the genus Cerebratulus were also favourite subjects of classical research on localization. Wilson (1903). Zeleny (1904), and Yatsu (1904) showed that establishment of the definitive localization patterns in the Cerebratulus egg begins with the meiotic divisions set in train by fertilization, and it is complete until the 8-cell stage. The morphogenetic factors that specify the apical tuft and the gut, among other structures, have by then segregated to their respective progenitor blastomeres. Thereafter, the blastomeres behave in an irreversibly determined fashion. However, in the unfertilized egg the morphogenetic factors are globally distributed, although the animal-vegetal axis is clearly predetermined. Thus, for example, Yatsu (1910) showed that normal larvae may be obtained either from fertilized animal or vegetal half eggs. A developmental map for the apical tuft and gut determinants of the Cerebratulus lacteus egg was constructed by Freeman (1978), by determining the morphogenetic potential of egg fragments isolated at various stages after fertilization. These factors independently move in opposite directions during the meiotic maturation stages. They achieve their final distribution just before the transverse third

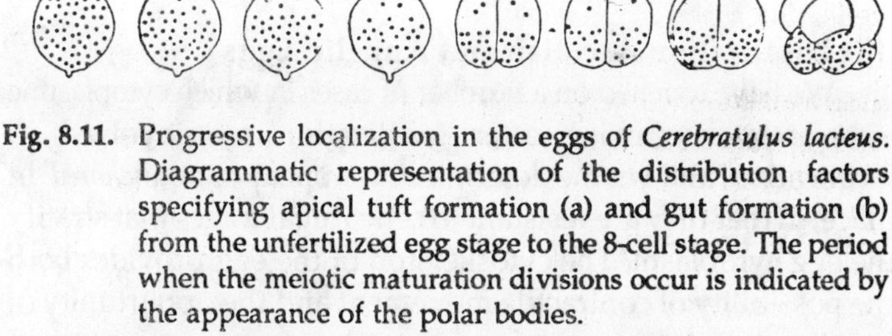

Fig. 8.11. Progressive localization in the eggs of *Cerebratulus lacteus*. Diagrammatic representation of the distribution factors specifying apical tuft formation (a) and gut formation (b) from the unfertilized egg stage to the 8-cell stage. The period when the meiotic maturation divisions occur is indicated by the appearance of the polar bodies.

cleavage. In this egg, in contrast to the molluscan examples examined earlier, cytochalasin treatment does not interfere with the progressive mechanism of localization. Nor does alteration of the cleavage planes by pressure. On the other hand a drug that suppresses aster formation, ethyl carbamate, arrests the localization process for as long as it is present. When the drug is washed out new asters form, and localization resumes. Freeman (1978) also showed that removal of the meiotic asters with a microneedle has the same effect as ethyl carbamate, while precocious induction of asters with hypertonic sea water accelerates the localization of the gut and apical tuft determinants. Though it is clear from these results that asters formation is required for localization in the cerebratulus egg, the function of these organelles in regard to localization remains unknown.

There are clearly many diverse mechanisms by which maternal cytoplasmic determinants are specially distributed. These include intercellular interactions during oogenesis, the disposition of maternal cytoplasm in polar lobes, and the mediation of localization by cytokinesis in early cleavage, or by global cytoplasmic movements triggered at maturation or fertilization. An underlying element of homology in some of these various processes may be the engagement of the cytoskeletal apparatus of the egg, as long ago argued by Conklin (1905a). The crucial role of asters in the localization processes of *Cerebratulus* eggs may be interpretable in these terms. In the following we consider some recent evidence on the participation of cytoskeletal structure in the ontogeny of localization patterns.

(iii) Localization Ascidian and Annelid Eggs

We have touched on a number of cases in which cytoplasmic *movements* occurring after fertilization are involved in localization, and yet the determinants appear to be *anchored* in place, in that they are resistant to centrifugal forces that stratify the egg cytoplasm. The cytoskeleton of the egg provides both the possibility of contractile movement and the opportunity of localized anchorage for particles, organelles, and macromolecules. Current evidence specifically implicates the cortical cytoskeleton of ascidian eggs, both in the initial bipolar contraction of the surface elements of the cytoplasm that occurs immediately after fertilization in the localization of the posterior muscle cell determinants. As will be recalled the yellow myoplasm that marks the positions of these determinants is localized by means of cytoplasmic movements that occur prior to cleavage. Conklin (1931) showed that low speed centrifugation fails to interfere with the localization patterns of ascidian eggs. As described in the legend to that figure the pigmented myoplasm remains in place after detergent extraction that removes three-fourths of the egg protein and almost all of the bulk RNA and lipid. The pigment granules are seen beneath a mesh-like structure termed the plasma membrane lamina, which is composed primarily of membrane proteins, and is supported by action mocrofilaments. Beneath this lies a complex filamentous lattice that may consist largely of the intermediate filament proteins desmin and vimentin, and to which the pigment granules are attached. After fertilization the plasma membrane lamina, which initially covers much of the egg, contracts over the surface carrying its inclusions, and thus forms the localized yellow crescent. The actin microfilaments of the cortical cytoskeletal complex could contribute to the structural integrity of the cortical myoplasm domain, and also provide components of the contractile engine that drives the spatial segregation of the domain. By analogy with what is known of echinoderm eggs cytoskeletal myosin could be present, and might provide the motive force. Jeffery (1982) showed that in eggs of the ascidian *Boltenia villosa* the location of the myoplasm cresent can be specified by presentation of a Ca^{2+} inonophore,which apparently acts as the focal point for the tortical contraction, just as might the penetrating sperm in the untreated egg.

A potentially interesting aspect of cytoskeletal localization in ascidian eggs is that a major fraction of the maternal poly (A) RNA is retained after detergent extraction, though most of the ribosomal RNA is not. The possibility thus exists that within spatially localized cytoskeletal domains could be sequestered *specific* maternal poly (A) RNA sequences. However, a two-dimensional examination of the translation products of mRNA extracted from isolated yellow crescents failed to reveal any such mRNA localization. Among the mRNAs present in the yellow crescent myoplasm are some coding for actin but it is important to note that these maternal actin message code for cytokeletal rather than the muscle actin isoform. Since only the most prevalent mRNAs of the yellow crescent have been examined, perhaps the main significance is the observation that there are some mRNA tightly associated with it. Cytoskeletal components other than the actin microfilaments are responsible for binding these mRNAs since the cortical actin lattice can be disrupted without causing their release. The possibility remains to be excluded that lower prevalence maternal mRNAs could be specifically in the myoplasm.

Maternal poly(A) RNA is also anchored to large cytoplasmic structures in the egg of *Chaetopterus pergamentaceus.* A typical annelid localization pattern is established in this egg by means of a dramatic series of cytoplasmic rearrangements occurring during the meiotic maturation stages that immediately precede first cleavage. Lillie (1906) showed that regional differentiation occurrs in *Chaetopterus* eggs even if cytokinesis is repressed by treatment with KCl, just as it does in cytochalasin treated ascidian zygotes. Cytoplasmic reorganization in both KCl activated and fertilized *Chaetopterus* eggs is blocked by colchicine, which affects microttubule organization, but not by cytochalasin. Stratification of the internal cytoplasm by centrifugation also fails to disturb the localization patterns of *chaetopterus* eggs, and centrifuged eggs develop normally. *In situ* hybridization with H-poly(U) demonstrates that the egg cortex, where the key morphogenetic movements of the cytoplasm take place, is also the site of maternal poly (A) RNA accumulation. The cortical localization of the maternal poly(A) RNA persists in first cleavage, and extends to the polar lobe formed at that time. After low speed centrifugation most of the maternal poly(A) RNA is sedimented to the centrifugal pole of the egg. This observation shows that correct localization does not require the normal cortical disposition of the bulk of the maternal poly(A) RNA. Nonetheless, the result is interesting because it indicates that in this egg heterogeneous RNAs are associated with particulate elements, which for certain sequences could as well be anchored in structural components of the cytoskeleton, just as are the yellow pigment granules and the maternal mRNAs of the ascidian myoplasm.

We have encountered in this review several kinds of evidence suggesting that among the morphogenetic determinants of the egg cytoplasm are maternal RNAs, which may be prelocalized in the egg, or progressively localized after fertilization. Examples for which this has been proposed include the UV-sensitive targets involved in germ cell determination in anuran amphibians and dipteran insects; the anterior determinants of the *Smittia* egg; the early embryo poly (A) RNAs that on injection may result in rescue of maternal axial specification mutants in *Drosophila;* and the polar lobe determinants of the *Bithynia* vegetal body. A theory of localization based on regionally confined maternal regulatory RNAs was originally proposed by Davidson and Britten (1971). The attractiveness of the proposal that the egg stores regulatory RNAs that are or will be localized is based on two features. First, there is the argument in principle that regulatory information transcribed during oogenesis is ultimately required to induce new, differential patterns of genomic function in embryonic cell lineages, information that could reside either in maternal mRNAs coding for regulatory polypeptides, or in regulatory RNA sequences functioning in some other manner. Second, there are observations reviewed in this section that at least in some eggs maternal poly (A) RNAs are associated with regionally localized cytoskeletal elements. This provides in principle the outline of a mechanism for the regional distribution of morphogens. To proceed further will be a difficult task, as it would require experimental demonstration that specific maternal RNA or protein determinants are both necessary and sufficient for cell lineage specification in the embryo.

Regulation, and Localization in the Eggs of Sea Urchins, Amphibians and Mammals

Cytoplasmic factors of maternal origin that ultimately affect the specification of embryonic cells appear to function in the early development of all the creatures so far considered, though the means by which such factors are spatially localized are many and various. The comparative arguments for localization as a general and basic mechanism for early development are thus strong. Nonetheless, as for our predecessors, the problem of explaining regulative behaviour remains difficult. The localization theory has seemed to some observers incomplete, in that it does not obviously predict that subelements of the embryo, when isolated or placed in combination with other subelements, will in some cases produce structures other than those they are normally destined to form, including complete, normally constructed larvae. Regulative behaviour has been regarded as a major challenge ever since the first reproducible observations of this kind were described by Driesch (1891). Several different factors are now understood to contribute to regulative developmental processes in experimentally perturbed embryos, none of which in any way exclude specification, in the *normal* course of events, by means of cytoplasmic localization. Thus, Wilson pointed out that even on the strict assumption of cell lineage determination by localized morphogenetic agents, certain blastomeres would be expected to behave in a *totipotent* manner if isolated and cultured, while others should give rise to partial embryos, according only to the geometry of the localization patterns with respect to the planes of cleavage. We have also seen that localization patterns are in some organisms established ontogenically, at a relatively leisurely pace, so that the morphogenetic potential displayed by a given subembryonic fragment depends on the stage at which it is tested. In addition, a strong relation exists between the extent to which cell lineages are determined through inductive intercellular interactions during early embryogenesis, and the capacity for regulative behaviour that subelements of the embryo will manifest. As an example already considered, in equally cleaving gastropod eggs in which the dorsoventral axis is determined *inductively*, any of the four macromeres can give rise to this axis, while in unequally cleaving gastropod eggs only the D macromere retains the necessary maternal cytoplasmic factors. Another variable factor is that the structural elements responsible for the spatial disposition of localized determinants are sometimes apparently very delicate, so that in early stages localization is *physically labile,* and is easily reversed by experimental manipulation. The eggs of certain deuterostome groups, including sea urchins, amphibians and mammals, provide illuminating examples both of lability and of early inductive interactions between blastomeres.

Regulation in the Sea Urchin Egg

Localization of determinants along the animal-vegetal axis of the unfertilized sea urchin egg was demonstrated by Boveri (1901 a,b), in a study carried out on eggs of *Paracentrotus lividis.* In this species a fortuitous subequatorial band of pigment granules arranged in a plane orthogonal to the animal-vegetal axis indicates the future orientation of the embryo. These granules also indicate the location of cytoplasm that is later included in the cells of the archenteron. By shaking the eggs into pieces and fertilizing the fragments. Boveri showed that only fragments containing the pigmented cytoplasm are able to gastrulate and to carry out gut formation. Horstadius (1928) confirmed that the pigment layer indicates the eventual axial orientation of the embryo, and showed that animal half egg fragments develop in a manner similar to the animal halves of 16-cell cleavage-stage embryos, while the partial morphogenesis carried out by vegetal egg fragments corresponds to that observed in 8- or 16-cell vegetal half embryos. Horstadius (1937b) extended these findings to *Arbacia punctulata,* orienting the eggs individually soon after fertilization, and cutting them in half at the pronuclear fusion stage. The animal halves cleaved equally, producing no micromeres, and gave rise to spherical structures with enlarged apical tufts. The vegetal halves, however, formed micromeres and gastrulated, with archenteron and skeleton formation ensuing. It follows that cytoplasmic factors required for micromere, skeleton, and archenteron formation are localized in the vegetal region of the egg cytoplasm even before fertilization. The orientation and distribution of these materials remain unchanged as the egg cytoplasm begins to be divided up among the blastomeres. Like lobeless *Dentalium* and *Ilyanassa* eggs, the animal egg fragments in these experiments possess complete genomes, but fail to differentiate several important cell lineages.

Boveri (1901a) noticed two additional external features that identify the prelocalized animal pole in banded *Paracentrotus* eggs. These are the minute channel known as the *jelly canal,* and the position of the polar bodies. Schroeder (1980a,b) showed that the jelly canal, visualized by immersion of unfertilized eggs or oocytes in ink, can be used to define the location of the animal pole in unfertilized sea urchin eggs that lack an oriented band of pigment granules. On elevation of the fertilization membrane the ink mark left at the site of the canal remains visible, this mark occurs *opposite* the micromeres at the 16-cell stage. The animal-vegetal prelocalization is thus a general characteristic of sea urchin eggs, and is not confined to those with banded pigment layers. Contrary to Boveri and other authors, both classical and modern, the jelly canal is not a micropyle. Sperm can enter and fertilize the sea urchin egg at any point on its surface with respect to the animal pole. Irrespective of the point of sperm entry, the orthogonal first and second planes of cleavage intersect along the preformed animal-vegetal axis. Figure demonstrates that the polar orientation of the egg is a prelocalization that extends back into oogenesis, since the jelly canal is present even in immature oocytes. A subequatorial band of pigment granules similar to the present in *unfertilized Paracentrotus* eggs appears *during early cleavage* in *Arbacia* eggs. Observations of Schroeder (1980a) suggest that the pigment band forms by means of an upward contraction of the cortical cytoplasm that carries the pigment granules away from the vegetal pole. The implication is that a similar event has already occurred in *Paracentrotus* eggs by the time they become available for fertilization, and in this species the pigment granules have been shown to be structurally associated with the cortical cytoskeleton in both unfertilized and fertilized eggs. A prelocalized animal-vegetal polarity is also manifested in various cytoskeletal features of starfish oocytes, including the distribution of actin-filled external spikes and of certain acidic vacuoles, both emeshed in the contractile cortex: the distribution of pigment granules: and the position of the premeiotic aster (Schroeder, 1985).

Cytological studies of meiotic activation in starfish oocytes and of fertilization in sea urchin eggs that lie outside the scope of this discussion have revealed a complex cytoskeletal architecture in the echinoderm egg. Networks of microtubules and microfilaments are present in the cortex, and cortical myosin, actin, and fascin have all been identified in the starfish oocyte. The imperviousness of the axial organization to stratifying centrifugal force exerted in any direction with respect to the egg axis, imply that the oriented elements important for animal-vegetal prelocalization reside within, or are anchored to, a regionally differentiated, preformed, cortical cytoskeleton. Mabuchi and Okuno (1977) showed that in the starfish egg cytokinesis, which involves a cytoplasmic contraction that is normal rather than tangential to the surface, is inhibited by microinjected anti-myosin antibodies. A reasonable interpretation is that the tangential cortical contractile processes occurring during localization in echinoderm, ascidian, amphibian, and many other eggs are also driven by contractile actin-myosin complexes of the cytoskeleton, and that these structures are associated with the cortex as a whole by actin microfilaments that in many eggs can be dissociated by cytochalasin treatment.

The oral-aboral axis of the sea urchin embryo is not evident at a gross level until the gastrula stage, when the ectodermal wall on that side thickens and shortens, the archenteron develops an asymmetric tilt to the oral side, the oral rods of the skeleton appear, and the secondary mesenchyme cells contract the inner wall of the blastocoel. In undisturbed embryos the oral-aboral axis is apparently specified as early as the 8-cell stage of cleavage, however, although even at late cleavage stages meridional half embryos and embryos in which the animal half has been rotationally reoriented with respect to the vegetal half are all able to reproduce new oral-aboral axes and thence to develop normally.

Localization in Amphibian Eggs

Amphibian eggs have been studied since ancient times, and from the 19th century on this familiar material has been the subject of experimental attempts to understand the relation between the visible structure of the egg and the axial co-ordinants of the embryo. The animal-vegetal axis is a preformed feature, which evidently originates far back in oogenesis. Thus, in immature amphibian oocytes the germinal vesicle is always located toward the future animal pole, and during vitellogenesis the large yolk platelets are concentrated mainly toward the vegetal pole. In species whose eggs are coloured the pigment granules are usually confined to the animal hemisphere, probably an adaptation that provides protection from both above and below, since due to the dense vegetal yolk platelets the eggs normally float animal pole upward. Certain macromolecular constituents also seem to be distributed along the animal-vegetal axis even in oocytes and in uncleaved zygotes. For instance, Capco and Jeffery (1982) found that the poly (A) RNA in advanced vitellogenic *Xenopus* oocytes is concentrated along the vegetal pole cortex and in a perinuclear region on the vegetal side of the germinal vesicle. Shortly after fertilization poly(A) RNA concentrations are higher toward the animal pole. Out of several hundred protein species resolvable on two-dimensional gels a few are particularly localized in the vegetal cytoplasm of matured *Ambystoma* oocytes and a variety of protein species have been reported to be distributed in a gradient along the animal-vegetal axis in newly fertilized *Xenopus* eggs, though most are not. A detailed study was reported by King and Barklis (1985), in which poly(A) RNA was isolated from animal, middle and vegetal thirds of eggs sectioned perpendicular to the animal-vegetal axis and translated in a cell-free system. The proteins synthesized were displayed by two-dimensional electrophoresis, and by this means five mRNA species of the approximately 600 whose products were resolved are found to be confined to the vegetal region of the unfertilized egg, while others are preferentially (i.e., 10 to 100-fold) concentrated there. One species was identified that is significantly more prevalent in the animal region. Several clearly localized maternal mRNAs have also been recovered from a cDNA clone library constructed from unfertilized *Xenopus* egg poly(A) RNA, including one clone representing a transcript localized at the vegetal pole, and three representing animal pole messages. All of these transcripts belong to the low abundance class of maternal messages of which there are $\geq 10^4$ diverse species stored in the egg. Though primordially imposed during oogenesi, the animal-vegetal axis of amphibian eggs can be at least partially reversed by causing the eggs to develop in an inverted orientation. Eggs treated in this manner arrest after gastrulation, but transplantation experiments reveal that reversal of some axial specifications has taken place, so that, for example, neural structures now derive from explanted cells of the original vegetal hemisphere.

Unlike the animal-vegetal axis, the dorsoventral axis of the amphibian egg is an ontogenic construction, that is specified only after fertilization. In appropriately pigmented eggs the location of the future dorsal pole is forecast by the appearance of a zone of cortical cytoplasm that is depleted of pigment granules, traditionally known as the *grey crescent*. Classical observers, beginning with Newport (1854), noted that the plane of first cleavage usually becomes the sagittal plane of the embryo, and that in most (though not all) cases this plane bisects the grey crescent. The latter normally forms *opposite* to the point of sperm entrance. Since the sperm can enter, or will fertilize if experimentally deposited at any spot on the animal hemisphere, the egg is initially radially symmetric, and before fertilization all equatorial loci thus possess the capacity to participate in the formation of either ventral or dorsal structures.

At gastrulation the lower edge of the grey crescent becomes the dorsal lip of the blastopore, which is the site of formation of the dorsal structures of the neurula. The inductive process by which these structures arise was revealed in a series of studies begun by Spemann and associates. Dorsal structures such as notochord, somites and neural tube form as a result of a sequential series of inductive interactions. Attention was initially focused on the induction of dorsal axial structures during gastrular invagination, carried out by cells of the "Spemann organizer", which is located above the dorsal lip of the blastopore. So powerful is effect of the organizer and so plastic the condition of the cells prior to invagination that organizer cells transplanted to the prospective ventral side of the early gastrula induce there a secondary dorsal axis.

A consistent and until recently widely accepted interpretation was that the cells of the Spemann organizer are initially specified by maternal cytoplasmic factors localized after fertilization in the grey crescent area of the egg. This theory found direct support in experiments of Curtis (1962), who reported that prior to the 8-cell stage transplantation of a small piece of cortical grey-crescent egg cytoplasm to the prospective ventral side of the embryo suffices to produce a second dorsal axis, just as does ventral transplantation of cells of the dorsal organizer region at the gastrula stage. Unfortunately these results appear to have been the consequence of experimental artefacts and a wholly different understanding of the role of localization in the specification of dorsoventral axis of amphibian eggs has emerged from recent cell lineage, transplantation and cytological observations.

Origin of the Dorsoventral Axis

The fate map of the pregastrular amphibian embryo shows that the cells located in the dorsal marginal zone ultimately give rise to mesodermal structures of the dorsal axis. It is some of these cells which in the Spemann experiments display the capacity to induce dorsal structures. The problem we now address is whether the initial determination of these cells as mesoderm progenitors with dorsal inductive capabilities in fact occurs as a self-differentiation, such as might follow from their inheritance of cytoplasmic factors localized on the dorsal side. Explantation experiments carried out in *Triturus* embryos and in *Xenopus* embryos suggest that the dorsal marginal cells of the animal blastomere cap have become fully determined by the early blastula stage. Thus, if these cells are excised after this stage and cultured, they produce various mesodermal structures, including myotomes, notochord, mesothelium, etc., but if isolated earlier they give rise only to ciliated epithelium. From these and similar studies, it can be concluded that in *Xenopus* determination of dorsal mesoderm precursors takes place between the 64-cell and the 512-cell stage. In *Rana* and *Ambystoma* as well, *the capacity to induce a second axis,* i.e., behave as a Spemann organizer, appears to have been fixed in dorsal marginal cells in the course of the blastula stage. However, as these tests challenge the state of determination of the dorsal organizer precursor cells, they do not reveal directly the mode of their initial specification.

It is now evident that both the self-differentiation of early dorsal equatorial blastomeres, and the inductive differentiation of cells in this region by vegetal pole blastomeres on the dorsal side lying *below* the original gray crescent are involved, inductive determination of mesoderm in animal cap cells by vegetal region cells was demonstrated in the urodele *Ambystoma* and in *Xenopus,* in experiments in which the normal axial mesodermal precursors were excised at the blastula stage, and the remaining animal hemisphere ectoderm, when combined with the vegetal hemisphere, was shown to be able to recreate the axial mesoderm anew. The dorsoventral orientation of the induced mesoderm is controlled by that of the vegetal hemisphere in these recombinants, and the induced dorsal mesoderm is then capable of behaving as a Spemann organizer, Gimlich and Gerhart (1984) demonstrated explicitly that the ability to induce dorsal mesoderm is originally resident in a few *vegetal* blastomeres at the 64-cell stage. In their experiments the original dorsoventral axial orientation of the egg was destroyed by UV irradiation of the vegetal hemisphere, and the irradiated embryos were shown to be rescued by transplantation of one to three blastomeres from the quadrant underlying the prospective dorsal marginal region. Injection of fluorescent label demonstrates that the transplanted vegetal cells are themselves ancestral to no mesodermal derivatives, and their progeny are later found incorporated only in the gut. On the other hand, similar experiments, in which two dorsal equatorial blastomeres drawn from the third cell tier of a normal 32-cell embryo are transplanted into equivalent positions in UV-irradiated hosts, show that these blastomeres possess the capacity to rescue dorsal axis formation by *self-differentiation.* Thus injection of fluorescent lineage tracers demonstrates that the progeny of these transplanted cells contribute directly to the notochordal and other mesodermal structures of the organizer region. The inductive activity of the vegetal blastomeres is shown by the experiments of Gimlich and Gerhart (1984) to be sufficient to account for the specification of the future organizer region. Nonetheless, the evidence reviewed clearly implies that there exist maternal cytoplasmic factors that specify the cells inheriting them to give rise to dorsal mesoderm, including that of the Spemann organizer, and that these factors are localized in the dorsal equatorial blastomeres of the early cleavage state embryo. The inductive mechanisms also operating no doubt assist in imposing the appropriate spatial patterns of dorsal specification. We have seen that the vegetal pole cytoplasm is also the location of determinants for the germ cells, at least in anuran amphibians. Though, the animal hemisphere blastomeres display innate tendencies to differentiate as epidermis, they seem in general to remain plastic and undertermined into the early blastula stage, pending their reception of appropriate signals from self differentiating blastomeres below. In the sea urchin embryo as well the study vegetal pole blastomeres appear to be more highly determined *ab initio,* and the fates of the animal cap blastomeres to be inductively determined, particularly with respect to the second axis. In considering the mechanism under-lying this mode of development it is important to note that in the amphibian

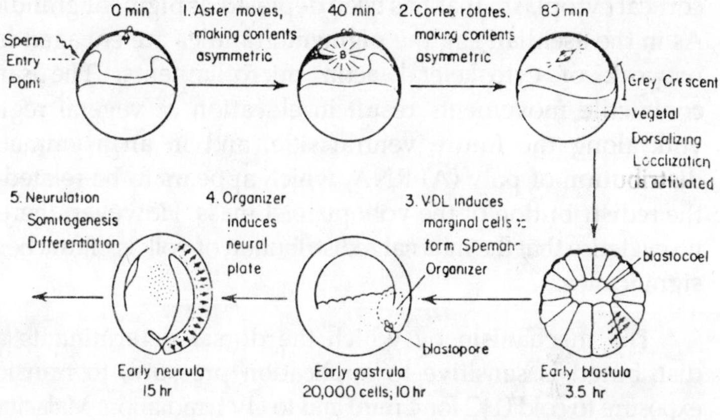

Fig. 8.12. Schematic diagram of a possible sequence of events in inductive dorsal-ventral axis formation in anurans.

embryo the initial inductive interactions required for the determination of the dorsal marginal cells of the animal hemisphere takes place *in the complete absence of genomic transcription.* As reviewed earlier, transcription does not resume in amphibian eggs until the midblastula stage.

The Mechanism of Localization : We now return to the initial process by which the dorsal determinants are localized on the grey crescent side of the egg. The primary dorsoventral asymmetry is caused by the movement of the sperm nucleus, and a displacement of internal cytoplasmic constituents to one side that may be due to formation of the large sperm aster. This is followed, in normal development, by a strong cortical contraction toward the future ventral side, occurring in the plane of the future dorsoventral axis of symmetry, that results in the rotation of the cortex with respect to the deeper cytoplasmic contents. Cortical microtubules and microfilaments are evidently involved in these cytoplasmic movements though the contractile process itself is not sensitive to injected cytochalasin. The consequence of the rotational contraction is the appearance of the grey crescent at the future dorsal side. This is a zone of cortical cytoplasm that has been depleted of pigment granules. As in the ascidian egg the pigment granules are engaged in a network of cytoskeletal actin microfilaments. The same contractile movements result in elevation of vegetal region yolk along the future ventral side, and in an asymmetric distribution of poly (A) RNA, which appears to be related to the redistribution of the yolk platelet mass. However, there is no evidence that the internal redistribution of yolk is of functional significance.

The mechanism by which the dorsal determinants are distributed is sensitive to application pressure, to transient exposure to cold (1°C for 4 min) and to UV irradiation. Malacinski *et al.* (1974) and Chun and Malacinski (1975) found the UV-sensitive neuralizing factors localized mainly in the future dorsal region of the vegetal egg cortex in both urodele and anuran eggs. UV irradiation has a purely cytoplasmic effect, since nuclei from irradiated eggs are able to support development when injected into enucleated eggs as well as do control nuclei, and the irradiation penetrates only 5-10μm or less into the egg cortex. However, ultimately UV irradiation clearly affects subsequent differential gene function in the embryonic cells, and thus alters the distribution of protein products that are confined to dorsal or ventral tissues. Cold, hydrostatic pressure, and UV irradiation all prevent axial determination if applied to the zygote between about 0.4 and 0.8 of the interval between fertilization and first cleavage. Thus the subsequent differentiation of axial dorsal structures is suppressed, to a variable extent, by these treatment, and when most effective they result in embryos that consist only of an ectodermal tube containing ondoderm and some mesenchyme cells. All three treatments probably affect the cytoskeleton of the egg, and interfere with the redisposition of cytoplasmic constituents occurring during the period of sensitivity that precedes first cleavage. Microtubules are stabilized by D_2O, and Scharf and Gerhart (1983) found that if immersed in D_2O, cold treated eggs display much better axial development. The effect of UV irradiation on the cytoskeleton is not known. However, to some extent D_2O also rescues axial formation in irradiated eggs. Unlike some examples considered earlier, the action spectrum for the UV inhibition of axial differentiation in amphibian eggs peaks at 280 nm. indicating a protein rather than a nucleic acid target. An additional argument that the UV effect could be due to damage to processes mediated by microtubules is that UV irradiation also blocks grey crescent formation as well and this is known to be sensitive to treatment with colchicine as well.

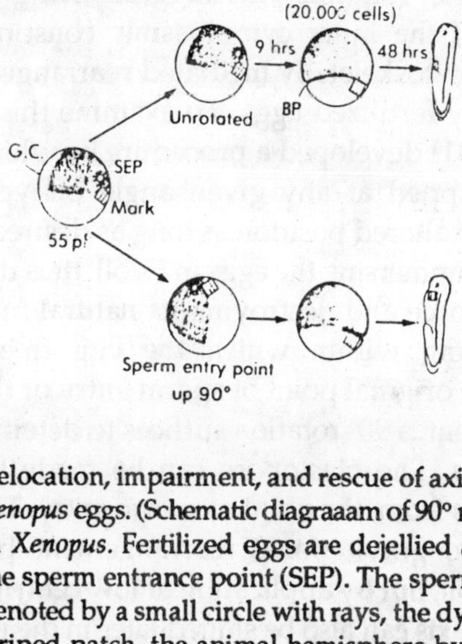

Fig. 8.13. Relocation, impairment, and rescue of axial determination in *Xenopus* eggs. (Schematic diagraaam of 90° rotation experiments in *Xenopus*. Fertilized eggs are dejellied and marked below the sperm entrance point (SEP). The sperm entrance mark is denoted by a small circle with rays, the dye mark by a square striped patch, the animal hemisphere by stippling, and the grey crescent by lighter stippling. In the control series (top line) the unrotated egg initially forms its blastopore lip, 180° opposite the mark, which denotes the site of sperm entrance. In eggs rotated sperm entrance side up (grey crescent down) for the same period, the opposite result is obtained: the blastopore forms on the same side as the mark (sperm entrance side).

The localization pattern established in the sensitive interval before first cleavage is remarkably labile. It has been known for a century that the dorsoventral axis can be shifted merely by rotating the egg and holding it in a turned position after grey crescent formation Schultze (1894) showed that double axes from in anuran eggs that are inverted just before and during first cleavage. This observation was extended to the the 2-cell stage by Penners and Schleip (1928a, b), and was analyzed as a consequence of the gravitationally driven movements of the heavy internal yolk mass with reference to fixed cortical components. Ancel and Vintemberger (1984) showed that in parthenogenically activated *Rana* eggs, which lack the aster normally induced by the sperm, the location of the grey crescent and the future dorsal axis can be perfectly controlled by a brief rotation. This suggests that gravitational rearrangement of the inner cytoploasmic constituents may *substitute* for the cytoskeletally mediated rearrangements that occur in normally fertilized eggs. To examine this inference, Gerhart *et al.* (1981) developed a procedure by which *Xenopus* eggs could be tipped at any given angle off vertical, and maintained in the altered position as long as desired. This was accomplished by immersing the eggs in Ficoll, thus dehydrating the privitelline space and destroying its natural function as a lubricating bearing within which the egg freely rotates. Regardless of the original point of sperm entry or the position of the grey prescent, a 90° rotation suffices to determine a new dorsoventral axis. The blastopore can be made to form at, rather than across from, the point of sperm entry. The eggs are most sensitive to gravitational rearrangement before grey crescent formation, but by application of low centrifugal force, the dorsoventral axis can also be shifted later in the precleavage interval. Centrifugation at two successive times in different orientations, or centrifugation followed by gravitational reorganization, induces the formation of doubly axiated larvae. This result is consistent with the view mentioned above, that potential dorsal determinants are initially distributed radially around the egg. and that in normal development they become active on the future dorsal side following contact with, or additions form, the relocated deep cytoplasm. The major translocation occurring in normal eggs is a rotation in the plane of the future dorsoventral axis, that results in a 30° displacement of the subcortical cytoplasm relative to the egg surface.

The destructive effects of cold or pressure treatment, and of UV irradiation can be overcome merely by tipping the egg. Egg irradiated so as to completely inhibit dorsal development are rescued by being held for 30 min 90° off axis. It follows that in tipped eggs the gravitationally driven reorganization substitutes for the cytoskeletally driven reorganization that occurs in unirradiated fertilized eggs. In both cases a necessary interaction along the future dorsoventral axis must occur between the deep cytoplasm and the cortical components. There is no evidence that such interactions are involved in localization of germ cell determinants at the vegetal pole of the anuran egg, and rotation does not rescue primordial germ cell formation from the effects of UV. irradiation as it does axial determination. Nor are germ cell determinants redistributed in inverted eggs.

As for the eggs of several other species that we have considered, there is evidence that amphibian eggs also contain localized maternal morphogens that appear to promote certain pathways of self-differentiation in the blastomeres that inherit them. Thus, for example, blastomeres of the subequatorial zone tend to produce differentiated muscle cells: blastomeres of the vegetal pole give rise to endodermal cell types; The progeny of cultured blastomeres of the animal pole region differentiate as epidermis. The studies considered here show that dorsoventral axial determination depends on *structural* rearrangements, mediated in normally developing eggs by cytoskeletal cortical movements. There is no evidence that the cortical rearrangements which specify axial orientation are to be equated with the direct localization of specific maternal morphogens. Thus a series of ecentrifugation experiments of Black and *Gerhart* (1984) show the exertion of gravitational force at various angles with respect to the animal-vegetal axis may provoke the appearance of an embryonic dorsal pole at unexpected locations, the immediate cytoskeletal structure of which should not have been perturbed asymmetrically in the treated eggs. For example, eggs centrifuged down along the animal-vegetal axis produce dorsal poles in equatorial regions located randomly with respect to the sperm entry point. Rather than directly redisposing a specific maternal "dorsal morphogen," treatments that affect the cortical cytoskeleton of the egg should perhaps be thought of as altering a spatially oriented *matrix*, in which such morphogens may at some later time regionally associate, Ultimately there indeed is localized in the vegetal blastomeres of the 32-cell embryo the capacity to induce the overlying cells to give rise to the "organizer" tissue on which later axial development depends.

Cell Lineage Determination in Mammalian Eggs

Mammalian embryos differ from all others considered in this review in both their slow absolute rate of early development and in the relative temporal order in which various developmental events occur. Thus first cleavage in the mouse embryo occurs only toward the end of the second day after ovulation; there are only 10-30 cells on day 3; and 100 cells are not attained until day 4. The rabbit egg, which is about 8 times the volume of the mouse egg, cleaves somewhat more rapidly, and at two days it contains 16 cells; at 3 days 128 cells, and at 4 days over 1000 cells. A major distinction of mammalian embryos is the early developmental stage at which they switch from the use of maternal mRNA to mRNA synthesized in the embryo nuclei. In the mouse embryo, which has been studied the most intensively, mental mRNA has largely disappeared as early as the two-cell stage. The synthesis of new proteins from embryo transcripts becomes a prominent feature at this time. Furthermore, the early mammalian embryo develops from the beginning in a nutrient environment, so that relative to other embryos, net growth, including rRNA synthesis, starts very early. The storage of relatively huge amounts of maternal cytoplasmic constituents such as ribosomes and yolk is for the same reason unnecessary.

Most of the differentiated cell lineages giving rise to the tissues and organs of the fetus appear after implantation of the blastocyst in the uterine wall, which in the mouse occurs at day 6. The first overt differentiation of cell types in the embryo can be distinguished at blastocyst formation late in the 5th cleavage stage, with the separation of trophectoderm cells from the inner cell mass (ICM). The trophectoderm is a functionally differentiated cell lineage, as discussed in detail below, which after implantation gives rise only to extraembryonic membranes. The ICM includes the progenitors of all cell lineages participating in the construction of the embryo proper, as well as of some additional extraembryonic structures. The first specific cell layer to derive from the ICM appears with the delamination of the primitive endoderm. This takes place late on the fourth day of development in the mouse. The slow space of determination within the ICM has been demonstrated by transplanting genetically marked primitive endoderm and other ICM cells into dissimilar host blastocysts. Gardner (1982, 1984) demonstrated by this means that primitive endoderm cells are already definitively committed since single cells of this layer from 5 day embryos participate only in the development of endodermal tissues in the postimplantation fetus. However, they give rise both to the visceral yolk sac endoderm, and to the parietal endoderm sheath. Even in postimplantation (7 day) embryos the endoderm cells remain plastic with respect to their ability to be included in either parietal or visceral structures. This decision appearently depends on contacts with other cell layers. Individual cells other than those of the primitive endoderm of the 5th day ICM (usually described by the misnomer "primitive ectoderm") may give rise to all other types of fetal tissue, including the definitive embryonic gut, and mesodermal and neural structures. We have already seen that germ cell determination in the mouse embryo also occurs in the "primitive ectoderm," and not until after the 7th day of development. Probably many of the primitive ectoderm cells have the capacity to give rise to germ cells. It is most unlikely that the late determination of cell lineages within the ICM is in any way influenced by localized maternal cytoplasmic components. Though it is difficult to exclude this categorically, in view of the extreme plasticity observed in chimeric recombinations of mammalian blastomeres. Determination in the post-implantation mammalian embryo instead is probably mediated by intracellular interactions, but little is yet known of the detailed mechanisms.

Discussion of the role of cytoplasmic localization in mammalian embryogenesis has thus focused on the separation of the trophectoderm from ICM cell at the early blastocyst stage, a process which has been studied in unusually great detail. The caveat should be kept in mind. however, that the separation of a determined layer of *extraembryonic* cells from the totally uncommitted embryonic progenitor cells of the ICM is not necessarily a phenomenon that is analogous to the determination of muscle or germ line or skeleton-forming cell lineages *within* the embryo.

Blastomere Totipotency

A fundamental observation is that the cells of the early mammalian embryo appear completely totipotent when tested by culture of isolated blastomeres or of partial embryos, and in chimeric blastomere recombination experiments. Individual blastomeres from 2-, 4-, and 8-cell rodent and rabbit embryos can develop into complete blastocysts. Blastomeres from 8-cell mouse embryos fused with late morulae are able to participate in normal development and fused embryos up to the late morula stage also develop normally. Furthermore, although cells from the inside of the morula normally give rise to ICM and those from the outside normally give rise to the trophectoderm, inside cells isolated from 32-cells morulae or from the ICM's of blastocysts remain totipotent. Not until the late blastocyst stage do the ICM cell lose the ability to regenerate trophectoderm. The outside cells of 16-cell morulae also retain the capacity to contribute to the ICM. along the same lines Ziomek *et al.* (1982a) showed that experimentally produced aggregates of pure inside cells, pure outside cells, of mixed blastomere populations from 16-cell morulae will all give rise to blastocysts that implant and develop normally. In normal embryos cells ancestral to both trophectoderm and ICM are present at late as the 32-cell stage. Marked clones of cells descended from single outer blastomeres that had been injected with horseradish peroxidase at the 16-cell morula stage. These clones can be seen to include progeny in both trophectoderm and ICM. After the 32-cell stage in normal embryos there is no further allocation of new cells to the ICM, and this process seems mainly to be complete by the 16-cell stage. Soon after its formation the mouse blastocyst contains about 14 prospective ICM cells, and there are about 20 prospective trophectoderm cells.

The totipotency of the blastomeres of rodent embryos at least through cleavage and morula stages implies a large reliance on inductive intercellular processes during normal early development. The plasticity manifested by these cells under the artificial conditions of the experiments reviewed also indicates the difficulty of determining whether cytoplasmic localization might be involved in the differentiation of the trophectoderm in undisturbed embryos. Thus even if particular spatial relations predicting cell fate were to exist in the normal embryo, these assignments might be very easily reversed, and hence not easily detected.

Origin of the Trophectoderm Cell Lineage in Normal and Experimentally Perturbed Embryos : By the 6th cleavage there are about 45 trophectoderm cells, and these can be functionally distinguished from the ICM cells by a number of cytological, physiological, and molecular criteria. The trophectoderm cells are linked to one another by characteristic junctional complexes, and they are distinguished in ultrastructure from ICM cells by a number of both surface and integral cytoplasmic features. They perform the special function of pumping fluid into the interior of the blastocoel. On presentation to the uterine wall they elicit an implantation response, and in their absence implantation cannot occur. Ultimately they give rise to the extraembryonic ectoplacentral cone, to a population of endopolyploid giant cells that divide no further, and probably to chorionic. Two-dimensional analyses of the proteins synthesized by the trophectoderm cells of preimplantation embryos have revealed a number of species not included in the set of proteins synthesized by ICM cells at the same stages. After implantation some of these proteins continue to be synthesized in the extraembryonic layers descended from the trophectoderm. A monoclonal antibody against intermediate filament proteins of tropphectoderm cells has been described and a cDNA clone representing the message for this trophoblast-specific protein has also been recovered.

Since cells on the outside of the morula usually give rise to the trophectoderm, it was proposed that *outside position* is the causal factor in the determination of this cell lineage, and conversely, that *inside position* leads to the altrnative developmental pathway, participation in the ICM. This idea was supported by experiments of Hillman *et al.* (1972), in which the fate of marked cells inserted in chemeric blastomere combinations was shown to depend in the expected way on their positions. Though as we have seen each blastomere of the 4-cell embryo is totipotent, Hillman *et al.* (1972), showed that if placed on the outside of a 4-to 16-cell embryo the progeny of the test blastomere is often found only in the trophectoderm, and after implantation, in the yolk sac. If surrounded by other blastomeres, on the other hand, its progeny is usually recovered in the ICM and in the post-implantation fetus. Blastomeres from the inner regions of the 16-cell embryo can also be steered toward either trophectoderm of ICM differentiation by being placed on the outside or inside; respectively, of arrays of 15 other blastomeres of the same age. In these last experiments the test blastomere and its clonal progeny were labeled with a fluorescent stain. However, Ziomek and Johnson observed that by this stage test outer blastomeres display a predisposition to give rise to trophectoderm, though not an irreversible one. Ultimately the ICM is covered by a trophectoderm layer, and this may be the event that represses further delamination of trophectoblasts, by permanently eliminating an "outside" environment. Several other kinds of experiments contribute to the impression that external or internal position exerts a powerful influence. Thus, for example, alkaline phosphatase, which is an inner cell mass maker, is reported to appear in chimeric aggregates only when these include blastomeres totally surrounded by other blastomeres. Stern-flattened mouse embryos under a glass plate and showed that in these embryos trophoblast cells appear even in internal regions where the blastomeres about the glass interface rather than other cells.

The observations may be interpreted in two distinct ways. The external cells could become committed to the trophectoderm pathway as a direct result of their position on the outside of the embryo, as proposed earlier, or alternatively, in undisturbed embryos their natural process of commitment could be mediated by their *inheritance* of cytoplasm localized radially at the periphery of the embryo. The latter interpretation is allowed because observations on intact embryos have shown that in the course of cleavage, morula and blastocyst formation, relatively little radial or migratory displacement of cell occurs. Thus cells marked with fluorescent or enzymatic stains, or by injection of oil droplets, give rise to progeny that remain clonally clustered detailed study of the cell lineage of the cleavage stage mouse embryo and of the geometry of early intercellular contacts has been carried out by Graham and his colleagues. At third cleavage one of each pair of daughter cells normally lies deep within the embryo while the other occupies a peripheral position. By injecting oil drops into inside and outside blastomeres (Graham and Deussen (1978) showed that as expected the inner cells of these pairs contribute preferentially to the ICM, while oil drops injected into the outside cells are normally recovered in trophectoderm cells. The important point is that this radial correlation with blastomere fate reflects an earlier intracellular cytoplasmic distribution. Thus oil droplets injected in the peripheral cytoplasm of 2-cell stage blastomeres also tend to be distributed to later trophectoderm cells, while those injected centrally ultimately appear in both trophectoderm and ICM. Thus in normal development there appears to be a relatively fixed radial relation between the peripheral cytoplasm at early cleavage, the peripheral cell layer of the 8-cell embryo, and the outer trophectoderm progenitors of the 16-and 32-cell morula. There is unfortunately insufficient evidence to indicate whether relevant regional localization preexists in the uncleaved mammalian egg or the fertilized ovum.

A closer focus on the events occurring at the 8-cell stage reveals a radial intracellular *polarization* of the blastomere cytology, which is mediated by intecellular contact. At 4th cleavage the peripheral cytoplasmic regions are physically segregated to the outer cells of the 16-cell morula. The early 8-cell stage blastomere is spherical, and both surface and internal features are arranged in a homogeneous way. By the end of the 8-cell stage a major cytoplasmic reorganization has occurred. The nuclei move toward the basal, inner surface of the cell, where contact with an adjoining cell is established. At the apical or external pole endocytotic vesicles accumulate. These have been displayed in cytological preparations by enzymatic staining of ingested horse-radish peroxidase. Cytoplasmic actin also accumulates at the polar regions, as visualized with fluorescent antibodies. The most extensively studied changes occur near the surfaces. At the basal side gap junctions and tight junctions form; arrays of microfilaments and microtubules are erected parallel to the lateral walls of the blastomere; and the apical surfaces are distinguished by localized arrays of short microvilli. These can be easily detected at light microscope magnifications by the density of binding sites for various fluorescent ligands that they provide. The overall consequence of this polarization is a change in form of the embryo from a loose aggregate of spherical cells to a tightly compacted, radially organized embryo, in which intercell contacts have been maximized within, while the outer surface has become covered with microvilli. When 16-cell morulae are disaggregated into couplets of sister blastomeres and examined, the *outer cells* display the same polar cytological features as do the *outer regions* of the 8-cell blastomeres. The inner cells, which are derived from the basal and lateral potions of the 8-cell blastomeres, remain non-polar. Outer cell features, such as the external microvillar surface, are also retained by the outside cells of the 32-cell morula and the trophectoderm cells of the early blastocyst. The fate of the progeny of individual polar cells at the 8-cells stage depends on the plane of cleavage. If this bisects the pole, two new polar cells are formed, both of which become trophectoderm, while if it is normal to the polar axis a trophectoderm and an ICM progenitor cell are formed.

These observations imply directly that the peripheral cytoplasm of the polarized 8-cell stage blastomere acts as a determinant for trophectoderm differentiation, while the cells inheriting the internal cytoplasm are destined to become ICM cells, i.e., unless they are placed in circumstances where they are induced to polarize anew. The early determination of trophectoderm and ICM precursors at once affects their biosynthetic activities. Thus Handyside and Johnson (1978) showed that marker polypeptides synthesized only in the trophectoderm cells of the definitive blastocyst are not synthesized in inner cells of the 25-30 cell morula, while polypeptides characteristic of the later ICM are.

Polarization is an inductive process which requires intercellular contact. Furthermore, continued cell contact is needed to maintain the states of polarization initially imposed. When cultured *in vitro*, polar 16-cell blastomeres display trophectoderm-like behaviour, and they wrap arround non-polar cells, while apolar cells cultured together tend to polarize, and to produce aggregates including both external trophectoderm-like cells, and inner, ICM-like cells. Polarization occurs *in vitro* when a blastomere from an 8-cell embryo that has not yet undergone compaction is brought into asymmetric contact with other blastomeres. Polarization is induced in the 8-cell stage blastomere by contact with the surface of a 2-cell, 4-cell, 8-cell or 16-cell stage blastomere but not with an uncleaved zygote or an unfertilized egg. As we have seen, the 2-cell stage, when this inducing ability first appears, is also the point at which the biosynthetic processes of the embryo become dependent on its own transcription. Induction of polarity does not require the formation of intercellular junctional complexes, since in normal embryos it precedes the formation of these structures; since 8-cell blastomeres polarize when placed in contact with 2-cell blastomeres, but form no junctional complexes; and since ionic media, antesera, and drugs that block junction formation do not prevent polarization. Perhaps the most convincing demonstration that contacts with competent inducing cells account for polarization is that the axis of polarity can be controlled by the geometry of the contact. The apical pole always forms at a point opposite the contact, and in the case of multiple contacts, at a point equidistant from them. Within the intact 8-cell embryo, this property would naturally result in a radially organized embryo in which the apical faces of the blastomeres are oriented outwards, since the greatest number of intercellular contacts is in the inside.

The ability of the early cells of the rodent embryo to undergo polarization as a result of cell contact lies at the root of their totipotent behaviour in isolated blastomere as chimeric recombination experiments. When isolated, the blastomeres divide, and then induce polarization in each other, with the eventual segregation of cells that inherit the apical cytoplasm of the new external poles and give rise to trophectoderm precursors. In chimeric aggregates cells placed on the inside do not polarize, because they are symmetrically surrounded by contacts on all sides. However, on the outside they polarize anew, whatever their original orientation, and thus give rise to trophectoderm cells. Since outside cells tend to form extended contacts with underlying cells and to wrap around them, their dimensions expand in the plane normal to the axis of polarity.

The plane of cytokinesis in these cells this is usually radial, and two new polarized outside cells are generally formed at each division in the undisturbed morula or in complex aggregates. However, isolated outside cells may appear totipotent since the plane of cleavage now frequently bisects the axis of polarity, producing both new outside and a new, nonpolar inside cell. In short, the experiments that display the totipotency of mouse embryo blastomeres perturb these cells in two fundamental ways. They cause repolarization, and they result in reorientation of cleavage planes. Neither effect is inconsistent with the view that in *undisturbed embryos* the determination of the trophectoderm cell lineage begins with the segregation of radially localized cytoplasm into the outer cells, though their specification clearly remains reversible at least through the 32-cell stage.

The ability fo the early cells of the rodent embryo to undergo polarization as a result of cell contact lies at the root of their totipotent behaviour in isolated blastomere as chimeric recombination experiments. When isolated, the blastomeres divide, and then induce polarization in each other, with the eventual segregation of cells that inherit the apical cytoplasm of the new external poles and give rise to trophectoderm precursors. In chimeric aggregates cells placed on the inside do nt polarize, because they are symmetrically surrounded by contacts on all sides. However, on the outside they polarize anew, whatever their original orientation, and thus give rise to trophectoderm cells. Since outside cells tend to form extended contacts with underlying cells and to wrap around them, their dimensions expand in the plane normal to the axis of polarity. The plane of cytokinesis in these cells are generally formed at each division in the undisturbed morula or in complex aggregates. However, isolated outside cells may appear totipotent since the plane of cleavage now frequently bisects the axis of polarity, producing both new outside and a new, nonpolar inside cell. In short, the experiments that display the totipotency of mouse embryo blastomeres perturb these cells in two fundamental ways. They cause repolarization, and they result in reorientation of cleavage planes. Neither effect is inconsistent with the view that in *undisturbed embryos* the determination of the trophectoderm cell lineage begins with the segregation of radially localized cytoplasm into the outer cells, though thir specification clearly remains reversible at least through the 32-cell stage.

9

The Formation of Different Cell Types in Animal Embryos

The molecules which control development must determine which cells should synthesize particular proteins by acting either on the transcriptional or post-transcriptional controls of gene expression. At present, the identity of these molecules is unknown (see previous section) and here we will discuss certain experiments which animal embryos which provide information about the properties which these molecules must possess.

The success of nuclear transfer in amphibia makes it certain that the membranes and cytoplasm of the egg can induce nuclei to participate in all the interactions which are required for the diffrentiation of the cell types of the embryo. We will therefore be concerned with experiments on the one-cell egg before its organization is altered by many cleavage divisions; much of later development consists of cell interactions during morphogenesis.

There are at least three kinds of theory which seek to explain how the membranes and cytoplasm of the egg may organize genetic expression in the multiplying nuclei of the cleaving embryo. These theories may be called 'mosaic', 'reference points' and 'non-instructive;' and extreme versions of these theories will be described.

THEORIES OF DEVELOPMENT

(a) Mosaic Theory

This theory proposes that within the membranes and cytoplasm of the egg there are groups of different instructions. These are thought to be precisely localized in the cytoplasm so that, as cleavage proceeds, these instructions are packaged around the nuclei of different cells and control their behaviour. The instructions are considered to be chemicals which determine what a cell should synthesize and the ways in which the cell should move and interact with other cells. The implication is that there are many different types of instructions, each grouped in different parts of the egg.

(b) Reference Points Theory

A reference points theory also proposes that the cytoplasm and the membranes of the egg contain instructions. In this case, it is thought that the instructions issue from reference points and that their influence varies with distance from the reference points; as cells are formed so they find themselves in different positions relative to the reference points and as a consequence they become different from each other.

Traditionally embryologists have thought of reference points as areas of the egg which set up embryological fields of influence. They were thought to do this by being sources of gradients and cells were believed to be able to detect the concentrations of substances in these gradients and thus determine their position. In practice, if the reference points provide any signal whose effect changes with distance then they can serve their function in the embryo and there is no compelling reason to believe that they consist of gradients. The reference point theory contrasts with a mosaic theory in that there are thought to be rather few kinds of instructions involved (say four or five different kinds). Both the mosaic and the reference points theories require that the egg cytoplasm and its membranes should contain localized regions which either contain or issue instructions about development. A 'non-instructive' theory does not depend on the localization of instructions in the egg.

(c) Non-instructive Theory

This theory of development proposes that the membranes and cytoplasm of the egg contain no instructions for the unclei which are formed during cleavage. It is proposed that as the egg is cleaved, so unclei come to lie in different positions relative to each other and as a consequence become different. For instance, as a consequence of post-fertilization metabolism the nuclei on the outside of the embryo may be in cells which contain more nutrients than those on the inside and conversely the inside cells may contain more metabolic waste products than those on the outside. It is proposed that a particular environment is created around each nucleus as a consequence of cleavage. It is also thought that these environments cause the sells to develop different properties so that they may become reference points as development proceeds.

TESTING THE THEORIES

It is important to discover which of these theories best accounts for the development of embryos, because the phenomena of development cannot be explained in molecular terms until we know the class of molecules which may be important in the process. For instance, if a strict mosaic mechanism is operating in the embryo, then one would expect to find the instructions for development contained in large macromolecules which in themselves can carry sufficient information to specify many cell types. Examples of such molecules are nucleic acids, proteins and glycosoaminoglycans; if the egg is a mosaic then the discovery of the distribution of these molecules becomes very important. In contrast, the action of a reference point mechanism is likely to depend on small molecules which can readily move within a cell and across cell membranes, and in this case it would be necessary to study to distribution of such molecules in the egg.

The following tests may be used to find out whether or not an egg does contain localized instructions, and if it does so then it is possible to find out whether or not these instructions are distributed as a mosaic or issue from reference points.

(a) Test of Isolation

For this test, the egg is divided into fragments and the development of each piece is studied in isolation. If the parts of the egg continue to develop as if they were still in place in the egg, then they must contain their own instructions. It is likely that these instructions have a mosaic distribution if the isolated fragments are a small fraction of the total egg volume. In practice, the parts of the egg may develop in two other ways and these are more difficult to interpret.

Parts of the egg often develop fewer structures than they would in the intact egg. The question of whether or not they contain their own instructions cannot always be answered as the expression of these instructions may have been altered by the fragmentation process itself or by the small size of the isolate. In contrast, if parts of the egg form a complete adult then it is possible to conclude that the egg did not contain a fixed mosaic of instructions. It is also tempting to conclude that the egg did not contain reference points either. However, this conclusion does depend on the relationship between the fragments and the distribution of reference points in the egg, and if the egg is divided so that each fragment contains a complete set of reference points then each fragment might still be able to form a complete adult. The development of adults from parts of the egg therefore does not exclude the possibility that the egg contains reference points.

(b) Test of Transplantation

In this test, a part of the egg is transplanted to a new site in another egg. If the transplanted part develops in the same way as it would before transplantation then clearly it must have contained its own instructions. It is also possible to discover if the transplanted part can act as a reference point by observing if the recipient embryo changes its pattern of development in response to the graft does not develop as it would without transplantation, then it is impossible to be certain that it lacks instructions for the expression of these instructions may have been altered either by the operation or by mixture with instructions already present in its new site. Once again, this test can demonstrate the presence by not the absence of instructions.

(c) Test of Destruction and Replacement

This test can be used to distinguish between a mechanism of development based on a mosaic of instructions and one based on reference points. A way of illustrating the application of this test is to compare the egg with a globe of the world. Consider London on the globe. It might influence tourists for two reasons. On a mosaic theory, it would be the kinds of things which are in London which would have the greatest influence on the tourists (Houses of Parliament, Buckingham Palace, etc), while on a reference points theory the tourists would be influenced by the particular longitude and latitude of London i.e., on the distance from the reference points of longitude the latitude, Greenwich and the Equator). If London were scooped from the globe with a large spoon then both the Houses of Parliament, etc., and the locality (longitude and latitude) would disappear and it would be impossible to decide what had previously in fluenced tourists in London. From this discussion it can be seen that if an egg develops abnormally after part has been removed, then either theory could account for the disturbance.

Its only possible to distinguish between the two theories if a parts removed and then another part is put into its place. For instance, if a reference point theory is operating, then the locality of London (and its influence on tourists) can be restored by transplanting Washington in its place because the particular longitude and latitude of London is restored by this operation. In practice the molecules or influence of the reference point would have to move into the Washington graft. While if a fixed mosaic mechanism was operating then Washington would not be able to substitute for London because the Senate has a different influence to the Houses of Parliament.

(d) Iest of Maternal Effect Genes

The mosaic and reference points theories of development suppose that instructions are present in particular regions of the eggs cytoplasm and membranes. The instruction and their position may be determined by nuclear gents and if this is the case, then the mothers genotype should affect development by altering the instructions and down during cogenesis. Take the case where the egg contained a mosaic of instructions; it should be possible to discover genes which in a mother caused the absence or malposition of tissues and organs in her offspring. If such a gene was found then it would be important to show that the same gene did not have a similar effect on development when it was contributed to the embryo by the father and thus it would be possible to show that the gene was primarily affecting the organization of instructions in the egg. Such games are called maternal effect genes to emphasize that the effect on development stems from their effect on the egg.

The test consists of searching for maternal effect mutations which appear to determine the instructions for development inside the egg. If such are found then the test can provide clear evidence for the localization of instructions in the egg.

However, there are many reasons why a failure to detect such mutants does not provide that localized instructions are absent from the egg. First, it is difficult to detect and discover the effects of genes which act early in development. Second the instructions and their localization in the egg may primarily depend on cytoplasmic inheritance and processes of self assembly, with the nuclear genes providing no more then the components of the instructions. For instance the pattern on the cell surface of *paramecium* can be inherited independently of the nuclear genes, and it is not yet clear that egg instructions are not inherited in this manner. If they were inherited in this way then the chances of finding tissue specific maternal effect mutations are reduced.

EXPERIMENTAL USE OF THE TESTS

It is now useful to consider what can be learned about the organization of animal eggs by subjecting then to these tests. We have only included examples in which several tests have been performed on the same kind of egg because a single test does not by itself provide conclusive evidence about egg organization, and may be misleading.

Germ Plasm

The eggs of amphibia and insects often contain distinct germ plasm which is localized in one region of the cytoplasm. This cytoplasm can be recognized by its staining properties and it is included into a group of cells during early cleavage; it can be shown that these cells are the primordial germ cells which will eventually form the eggs and sperm (reviewed for amphibia, for insects. This is then a case where a visible structure in the egg may have a direct effect on the cells which inherit it.

EXPERIMENTS ON INSECT GERM PLASM

A germ plasm is found in many insect eggs and its effect on development has been thoroughly studied in *Drosophila* embryos. The egg of *Drosophila*, like that of most insects, does not cleave in the usual sense. After fertilization, the nuclei divide seven times (about once every 10 min), and migrate out to a position beneath the egg membrane. At this time a small number of nuclei in the posterior pole are surrounded by germ plasm and are cut off into cells; the rest of the nuclei of the embryo divide four more times before cell membranes are formed around them (blastoderm stage, for *Drosophile*. The late formation of cell walls in insect embryos is a useful property, for the organization of the embryo up to the blastoderm stage cannot be caused by cell to cell interactions and must directly depend on egg structure. The germ plasm which is incorporated into the primordial germ cells at the posterior pole consists of characteristic polar granules round which are mitochondria and RNA staining material [reviewed 36]. It now appears that the instruction 'form primordial germ cells' is located in this region.

The following experiments demonstrate the special properties of the germ plasm of the egg. Following other workers, Okada has shown that if the posterior pole is UV-irradiated while nuclei are still at the centre of the egg then no primordial germ cells are formed and the flies which develop into adults are sterile. The irradiation could either destroy a substance or destroy a locality and test (c) was next performed. To distinguish between these two explanations of the results. Okada irradiated eggs at the posterior pole and then injected anterior pole cytoplasm and posterior pole cytoplasm back into this region of the egg. If locality is important, then the injection of any cytoplasm into the irradiated region should restore fertility. However, if a particular instructive substance is destroyed in the irradiated region, then only the reintroduction of this substance into the egg should restore fertility. It is found that only the reintroduction of posterior pole cytoplasm which polar granules can cure the egg's sterility and this result strongly suggests that this cytoplasm containsm the instruction 'form primordial germ cells'.

Illmensee then showed that this instruction is antonomous. The posterior egg plasm, which contain the germ plasm, was transferred into the anterior half of a second egg. The nuclei of the second egg moved into this cytoplasm and formed histologically normal primordial germ cells in the anterior part of the embryo. It was possible to show that these were also functional primordial germ cells. To do this, these cells were transferred to the posterior pole of a third egg. This egg was allowed to develop into an adult and it was found that the transferred cells could develop into viable gametes which transmitted their genetic markers to the next generation. This transplantation *tour de force* is the most perfect demonstration of the localization of an autonomous group of instructions in the egg. It should be noted that the destruction of the germ plasm by UV-irradiation does not disturb the development of the rest of the fly and therefore germ plasm does not act as a reference point.

There are maternal mutations of fertility in *Drosophila,* and so it is possible to apply test(d). One maternal effect mutation is *grandchildless (gs),* which is found in *D. subobscura.* In the homozygous state this mutation causes females to produce sterile sons and daughters. It seems that the mutation has its principle effect on the gonads and does not cause sterility indirectly by effecting other organs in the body larval gonads from the offspring of gs/gs females remain sterile when transplanted to with type hosts, wild type hosts, while recipronily transferred gonads are fertile. This is then a situation in which one would hope to find an obvious morphological disturance such as the absence of polar granules) at the posterior cole of the egg. Unfortunately, all structures in this region appear normal when viewed under the electron microscope: the difference between the embryos of gs/gs mothers and these of wid type mothers seems to be that the nuclei in the affected embryos do not eater the germ plasm directly from the yolk but rather react it indirectly after remaining beneath the cell membrane strileparts of the egg. Since the late migrating nuclei in the strile embryos are eventually surrounded by germ cilsm but never form primordial germ cells, it seems that the cytoplasm can only impose its instructions at a very short period in embryogenesis. If this analysis is correct, them the effect of the *gs* gene may be on the time that nuclei receive an instruction rather man on the instruction itself. Of course, it remain possible that the instruction is altered and that this change cannot be detected by microscopy.

Experiments on Amphibian Germ Plasm

The germ plasm can also be discermed in rections of some amphibian eggs and this account of its behaviour is based on studies with the anurans *Rana temporaria art Xemopus Inevis.* The plasm is first observed after fertilization in pattaches beneath the egg membrane in the vegetal raif of the egg; it migrates towards the vegetal pole and concentrates here during the first cleavage division. The patches adheres to the inside of cell membranes and they are found in two fourteen cell of the early blastulas in the late blastula they move and surround the nuclear membrane, and the cells containing this material migrate into the genital rudiments and differentiate into the germ cells. The association between this material and the primordial germ cells could be fortuitous or it could be that it carries the instruction 'be a primordial germ cell'.

TABLE 9.1

Irradiation and Replacement of Parts of Eiptitiac Egg Cytoplasm

A. Irradiation at either pole

Pole irradiated	*Number of eggs*	*Number developing into tadmoles*	*Percentage with primardial germ cells*
Vegetal	39	39	0
Animal	20	20	100

Results of experiments with 7752 ergs. mm² of aimaviote irradiation.

B. Injection of cytoplasm into eggs irradiated at the *** pole

Source of cytoplasm	*Number of eggs normal after operation*	*Number developing into normal todmoles*	*Percentage of normal taxcles with primordial germ cells*
Vegetal pole from 2-cell	64	51	29
Animal pole from 2-cell	31	28	0

There are now good reasons for believing that such an instruction is carried either in the germ plasm or in material closely associated with it. Test (c) was applied to this material by Smith several years before the similar studies on insect germ plasm described in the previous section. Smith, exposed eggs of *Rana pipiens* to UV irradiation, and showed that if the egg were irradiated at the vegetal pole then tadpoles were formed which lacked germ cells, while if it were irradiated at the animal pole then there was no effect on development. Clearly the irradiation could either destroy a substance or destroy a locality. Employing test (c), it was shown that only the injection of vegetal cytoplasm can cure the sterility of eggs irradiated at that pole and the conclusion must be that this cytoplasm contains instructions for forming primordial germ cells.

The biochemical properties of this instruction are not known in either amphibians or insects. However, the germ plasm of amphibians, like that of insects, contains mitochondria and material which stains with RNA reagents. In the amphibian case it has also been shown that the action spectrum of UV sterilization is not too dissimilar from the UV absorption spectrum of nucleic acids.

Taken together, these experiments on germ plasm provide good evidence for the localization of a particular instruction or group of instructions in a small region of the egg. These instructions do not appear to act as reference points because following their inactivation by UV the rest of the embryo appears to develop normally. We will next consider two cases where a region of the egg appears to issue instructions to other regions and thus appears to be acting as a reference point.

Grey Crescent Region of the Amphibian Egg

Most of the animal half of the amphibian egg is covered by a black layer of pigment granules which lie beneath the membrane. After fertilization or parthenogenetic activation, these granules migrate upwards on one side of the egg leaving a grey crescent slightly darkened by the granules which have been left behind. The cytoplasm and membranes of the grey crescent are eventually incorporated into the dorsal lip of the blastopore and it has been known for sometime that this region of the gastrula organizes the longitudinal axis of the embryo

(Spemann's experiments). The special properties of this region are thought to be the consequence of interactions between the yolky cytoplasm and the material bound to the cell membrane on the animal side of the egg's equator; it is certainly the case that by shifting the relationship between these two structures the position of the blastopore and the number of blastopores in the embryo can be altered.

In a series of elegant experiments with *Xenopus laevis,* Curtis has shown that the organizing ability of the dorsal lip of the blastopore derives from instructions close to or in the membrane of the grey crescent region of the egg. The first experiment shows that normal gastrulation can be blocked by cutting out the cell membrane and adhering cytoplasm in the region of the grey crescent of a fertilized one-cell egg. The embryo divides normally and forms a blastula: cell division continues for sometime in this mass of cells but usually gastrulation does not occur, no axis is formed and the cells do not obviously differentiate. This observation suggests that this region of the egg might issue instructions to other parts.

The next experiment provides evidence that particular substances in the grey crescentregion are important. Employing test (b), the grey crescent region may be transplanted to the vegetal margin of another egg and in this case two dorsal blastopore lips are formed and the embryo develops a double axis; this result cannot be due to the creation of a novel locality by the operation because animal pole material grafted to the same site does not organize a new axis. A further reason for believing that a unique locality is not involved is that if the grey crescent locality is disturbed by removing 5-100 μm^3 of cytoplasm from beneath the membrane, then the grey crescent region can still exert its normal effect and the embryo gastrulates.

It seems most likely that the instructions of this reference point are at first concentrated near the membrane in the grey crescent region and that as development proceeds, so they move away into the cytoplasm. Evidence that this is the case

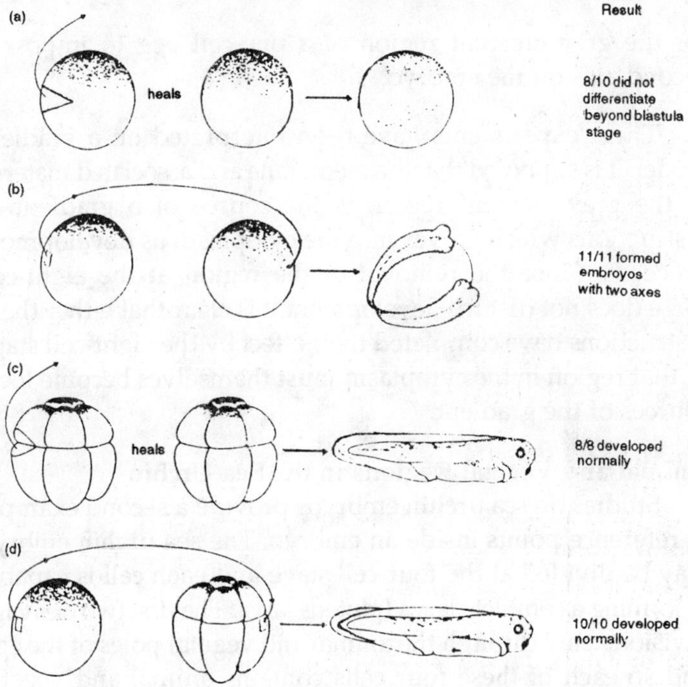

Fig. 9.1. Removal and transplantation of the grey crescent region of *Xenopus laevis*. If the membrane and adhering cytoplasm is removed from the grey crescent region at the one-cell stage, then gastrulation does not occur (A). If this material is removed at the eight-cell stage then development continues normally and this suggests that an instruction has passed from the periphery of the egg towards the centre between these twos stages (c). The *grey* crescent region can be transplanted to a second egg where it causes the formation of a double axis (b). It is unable to impose this effect on an eight-cell embryo (d).

comes from experiments in which the membrane and adhering cytoplasm are removed from the grey crescent region of an eight-cell embryo; in contrast to their removal at the one-cell stage, this operation does not disturb normal development, and it is likely that by this time they are widely distributed in the cytoplasm or have already exerted their effect. It is also the case that once this distribution has occurred, it is impossible for the grey crescent region of a one-cell egg to impose a second axis on the embryo.

These experiments have been interpreted on a gradient model. It is supposed that the membrane and associated material in the grey crescent region is the source of a gradient of instructions which moves into the cytoplasm as development proceeds. Since the removal of the region at the eight-cell stage does not disturb development it is clear that either these instructions have completed their effect by the eight-cell stage or that region in the cytoplasm must themselves become local sources of the gradient.

Animal and Vegetal Regions in the Sea Urchin

Studies on sea urchin embryos provide a second example of reference points inside an embryo. The sea urchin embryo may be divided at the four-cell stage and each cell is capable of forming a complete larva (pluteus larva. The first two cleavage divisions pass through the animal and vegetal poles of the egg and so each of these four cells contains animal and vegetal pole material. This material is now known to contain reference points and the presence of localized regions in these eggs can only be detected if the egg is divided along the equator lying between the animal and vegetal poles.

A famous series of manipulation studies have been conducted on the eggs of the sea urchin *Paracentrotus (Strongylocentrotus) lividus* by Horstadius and others. The fertilized eggs have a subequatorial pigment band and it is possible to observe the segregation of the pigmented cytoplasm during cleavage and to follow the segregation of other parts of the egg by staining the cell membrane and cytoplasm with vital dyes. Using these techniques, it can be shown that particular parts of the egg regularly form particular parts of the pluteus larva during normal development. The cell lineage of normal development.

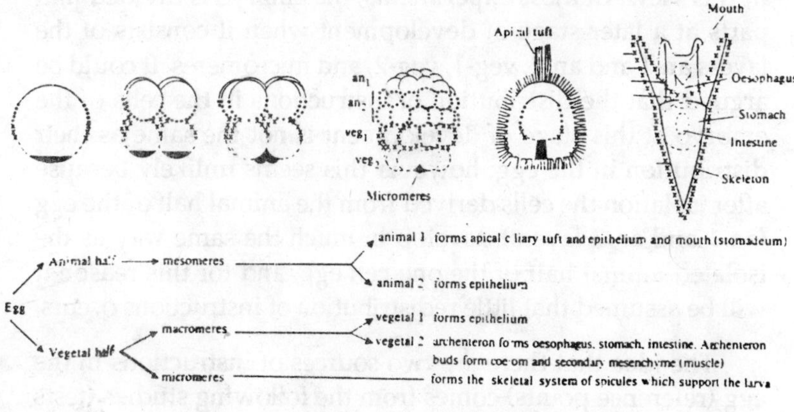

Fig. 9.2. Normal development of *Paracentrotus lividus*, diagram of cytoclasmic segregation and cell fate.

The distribution of instructions can be studied by observing the development of isolated parts of the egg (test a). Ideally one would wish to break the egg up into 20 or so fragments, but in practice it is only possible to study the development of isolated animal and vegetal halves. These are broken from the unfertilized egg and subsequently fertilized in isolation: these parts which contained a female nucleus at isolation are diploid, while those that larked this nucleus at isolation are hoploid, but the pioidy does not influence their development. The development of these isolates is illustrated [solared animal halves form apical tufts and ciliated blastulae; in were cases they contain a mouth but they never form a complete larva. Isoiated vegetal halves usually form larvae which laci in apical tuft and mouth; these usually have a reduced epitheium which restricts the development of the skeleton, and the stomach tends to be large and many exogastrulate.

These observations on half eggs, which show an apparent autonomy of development of some of the structures of the larva, would be consistent with a loose mosaic pattern of instructions in the egg. However, further experiments do not fit this view. In these experiments, the embryo is divided into parts at a later stage of development when it consists of the layers an-1 and an-2, veg-1, veg-2, and micromeres. It could be argued that the distribution of instructions in the cells of the embryo at this stage of development is not the same as their distribution in the egg: however this seems unlikely because after isolation the cells derived from the animal half of the egg (an-1 and an-2 layers) develop in much the same way as the isolated animal half of the one-cell egg, and for this reason it will be assumed that little redistribution of instructions occurs.

The idea that there are two sources of instructions in the egg (reference points) comes from the following studies (tests a and b). First there is evidence which supports the theory that there is the instruction 'form the normal derivatives of the vegetal half of the egg' which has its most powerful source at the vegetal pole. The animal half of the egg (an-1 and an-2 layers) is isolated and combined in turn with the layers veg-1, veg-2, and micromeres. The larvae which are formed from these combinations become progressively more normal as cells are added from regions closer to the vegetal pole. It should be noticed that cell lineage studies have shown that in these combinations the veg-1 cells and micromeres transmit instructions to the animal cells; for instance, in isolation the animal half forms a huge apical tuft which covers most of the surface of the early embryo, while in combination with veg-1 cells the tuft is restricted to its normal apical position. Similarly, in combinations of animal half cells with micromeres, the an-1 and an-2 cells which normally form only the epithelium and mouth of the larva now forms its oesophagus, stomach, and coelom as well. The development range of veg-2 cells is also extended in these combination. Normally they never form the skeleton while here they regularly do so.

In addition to this vegetal instruction, there appears to be another instruction with its most powerful source at the animal pole: its effect opposes that of the vegetal instruction. The evidence in favour of this view again comes from studies on isolated cells and again I will assume that the distribution of instructions in these cells is the same as that in the egg. It is possible to assess the strength of the instruction form the normal derivatives of the animal half of the egg by titrating the development of the layers an-1 and an-2 against the addition of different numbers of micromeres. The an-2 layer requires the addition of only two micromeres to form a normal larva while the an-1 layer must be combined with four micromeres to achieve this result and this observation suggests that the most powerful source of this instruction is at the animal pole. It should be noticed that as in the previous series of experiments, it is the an-1 and an-2 cells which form the oesophagus, stomach, and coelom of the larvae and the range of structures which they form is extended in comparison with normal development.

These results indicate that the egg does not contain specific instructions for every organ (mosaic theory) but rather it appears as if the egg contains two types of instructions whose effects are set up in opposing chemical gradients. In fact, the results need not be explained by a chemical model, any two signals whose effective power changes with distance from the animal and vegetal poles could account for these observations. If a chemical gradient model is applied, it remains possible that in the unfertilized egg, the potential sources of each gradient or signal are located at the animal and vegetal poles (reference points) and that they are not fully expressed until the cell layers are formed: it is only at the multicellular stages that the distribution of these instructions has been studied. Also it is clear from the cell combination experiments that cells away from the animal and vegetal poles are themselves capable of issuing animal and vegetal instructions: in other words they are not dependent on a continuous supply of instructions from these poles in order to exert their effect on other cells of the embryo.

The chemical basis of these two gradients of instructions

is uncertain. However a gradient of reducing potential can be demonstrated by vital dyes and this gradient does coincide with the instruction gradient in both normal development and in a variety of experimental situations. Further discussion on the possible nature of these instructions may be found in reference.

Eggs of Eutherian Mammals

Some of these tests have also been applied to mammalian eggs to discover the features of their organization. These tests have so far failed to demonstrate the presence of localized instructions.

It has recently become possible to study the development of isolated egg fragments (as in sea urchin experiments), but all published work has been concerned with the use of isolated cells from early cleavage stages (test a). In this case, it has been shown that at least one cell of an eight-cell stage rabbit egg can form an adult rabbit and that each cell of a four-cell stage mouse egg can form a complete blastocyst (this is a hollow ball which implants on the ulterus. However, isolated cells do not develop very successfully by themselves and it is necessary to combine them with other embryos to follow their full development. In this way the developmental potential of mouse blastomeres has recently been tested in detail.

Kelly dissociated four-cells stage embryos and surrounded each isolated cell with cells from another embryo; the isolated cell could be distinguished from those of the carrier embryo both by the coat colour genes and the enzyme marker which it contained. Four such composites could be obtained from each dissociated four-cell stage embryo, and they were transferred back to the uterus of a foster mother to continue their subsequent development.

First, it was shown that each isolated cell could form parts of the foetus, yolk sac, and placenta of each of the four composites which developed halfway through pregnancy (at this time the embryo has just started to form muscle blocks, the somites). In several cases, three of the four composites were born alive and developed into adult mice: the isolated cell had formed much of the coat colour and germ cell population of these animals. These results are particularly impressive because, unlike sea urchin experiments, the isolated cells form adult and not just larval structures. Similar experiments have been performed by Gardner on later embryos. In this case, it could be shown that cells isolated from the inside of the blastocyst were able to colonize a carrier blastocyst to which they had been transferred and that the isolated cell could form part of many of the organs of the mice which were born. Taken together all these results suggest that there are no localized instructions in the mouse egg, but it should be noticed that the isolation studies on these eggs are not yet as detailed as those performed on sea urchin eggs, and it is still possible that there are localized instructions along the radius of the egg.

The development of different cell types in the mouse embryo appears to depend on the position which the cells occupy during early cleavage. The evidence for this view comes from experiments in which isolated cells were placed inside or outside another embryo. In an outside position the cells tend to form part of the placenta while if they are placed in an inside position then they tend to form part of the foetus and the adult which develops from it. The conclusion must be that if the cells of these early cleavage stages contain any instruction at all then the effect of these instructions is obscured by the effect of cell position. It is interesting to see if it is possible to envisage a mechanism of development which does not depend on any instructions in the egg.

The following is a model which has been applied to the mouse egg. This model proposes that an instructionless egg is divided by cleavage so some cells lie on the outside and others on the inside of a cellular lump. As a consequence of the environment on the outside, the exposed cells are caused to stick to each other and to pump inwards and these cells therefore form a liquid filled vesicle. In contrast, the environment inside the lump causes the enclosed cells to stick to each other and not to pump, and also it causes the small ball of cells which is formed to stick to the sides of the liquid filled vesicle at some point. This simple model accounts for the fomation of a blastocyst containing two cell types and an inner cell mass at some point; in other words, it can account for the differentiation of cells and the formation of an axis in this cell population, and a series of events of this kind can be used to build up an embryo.

The point of describing such a model is to show that it is possible to conceive of development occurring in the absence of instructions laid down in the egg. It is probable that once the initial differentiation has occurred in such embryos, then the cell groups can act as reference points.

HOW GENERAL IS THE USE OF MOSAIC INSTRUCTIONS AND REFERENCE POINTS?

Despite the wealth of embryological experiments, there are rather few eggs which have been tested as thoroughly as those in the previous section. In addition to the experiments mentioned, there are some beautiful studies on the polar lobes of mollucs. Work with these embryos has clearly shown that there are localized instructions in mollusc eggs, but it is not yet clear whether they are distributed as a mosaie or whether they issue from reference points. Because the correct interpretation of these experiments is still uncertain they will not be discussed and for similar reasons many other experiments have been excluded (for review of other experiments).

The experiments described provide expamples of parts of eggs which contain but do not issue instructions (germ plasm), and of parts of eggs which issue instructions (grey crescent of amphibia and the animal and vegetal halves of sea urchin eggs). In contrast to these, the mammalian eggs appears to contain no instructions and reference points must be created during cleavage. It would next be useful to discover which of these mechanisms usually operates in eggs.

In this section it will be argued that few, if any, eggs contain a detailed mosaic of instructions laid down during oogenesis. The evidence for this view mainly comes from further studies on the eggs of insects and amphibia.

Mutant and UV Studies on Insect Eggs

Several observations have led some workers to believe that the egg of *Drosophila* contains groups of localized developmental instructions in addition to those for primordial germ cells. One experiment which led to this view involved the UV irradiation of parts of the egg. If the mid-lateral cytoplasm is UV-irradiated during early cleavage, then the flies which develop have missing or crippled legs, distorted body segments, missing head bristles, and bent wings. Unfortunately (without the complete use of test c) it is impossible to decide whether the defects are caused by damaging instructions in this part of the cytoplasm or by some effect on locality (e.g., by changing the time at which nuclei move to the locality, cf. observations on the eggs of gs/gs mothers, experiments on insect germ plasm). Another approach has been to look for maternal effect mutants. So far several mutants which effect the development of a particular region of the body have been found. For instance, homozygous *almondex* females (amx/amx) produce a few heterozygous daughters when mated to amx/+ males; 80% of these daughters have a distorted abdomen or thorax while the rest of the body appears normal, and the abnormal ties appear similar to those produced by UV irradiation of midlateral cytoplasm. However, as in the case of *grandchildless*, it is not clear that a particular set of instructions is missing from the egg; it is still possible that the time at which the nuclei encounter a particular locality is altered in these eggs and it is also possible that the movement of instructions from a reference point to this region of the egg is confused.

The evidence from studies on insect eggs is more consistent with a reference point explanation of the differentiation of structures other than primordial germ cells. One recent experiment is particularly revealing. Kalthoff irradiated the anterior one eight of the egg of *Smittla parthenogenetica* before nuclear migration to this region had occurred. The larva which developed consisted of two abdomen, which faced each other and no head structures were visible. These bicasudal larvae could be produced either by altering the effect of some reference point which tissues the instructions "form lead and trurax structures, or by the mosaic of instructions for abdomen development moving around to the anterior aid of the egg and redistributing themselves in the correct arrangement for normal abdomen development. The latter expuanatun seems unlikely, because the germ lasm and primordial germ cells remain in their couse location when similar doune abdomen are produced by UV-irradiation of the amterion pole of another insect *Chironomus dorsalis*. In Kaalthoff experiment the effect of UV irradiation is photo-reversible and thus suggests that the UV-induced effect is mediated by rucleic acids. It is also worth nothing that the bicaudal comditior can be produced by a material effect gene in *Drosomhila*. In the eggs produced by mutant mothers, the polar granues are located at only one pole of the egg and this observation suggests that while the reference points for arranging the structures of the doume abdomen are set up during oogenesis, these cannot in addition determine the position of the polar granules.

10

Influence of Environment on Reproduction

Various factors in the environment affect the maturation, game to genesis, fertilization and all developmental processes. The environmental factors studied so far one moesture and sainity humidity, temperature and photoperiaderm.

Moisture, Salinity

In a large number of animals rainfall and its effects influence reproduction, especially in those which live in dry regions. One can examine the problem of salinity in relation to the rainfall when rain or overflowing streams alter the salinity of brackish water bodies into which they drain, or of the sea. Panikkar and Aiyar (1939) have demonstrated that some marine brackish forms reproduce only when the salinity gets reduced by inshore water draining into the water body. On the other hand, lower salinity inhibits reproduction in several forms which have established there, their population being enriched by periodical migrations from saline waters. The shrimp *Palaemon* which lives in freshwater has one species *(P. carcinus)* which migrates to saline, brackish lake for breeding purpose. In the brackish water form *Gammarus* reproduction is reduced by higher salinities *Arbacta* shows that gonad development gets inhibited at higher salinities *Arbacia* shows that gonad development gets inhibited at higher salinities (75-110% at 20ºC). But Schleiper (1957) found that in the Baltic *Asterias* the gonad of the brackish water forms developed slower than in those inhabiting the North Sea. Kinne (1971) summarizes that where marked salinity changes occur, reproductive rates get modified and in lowered salinities larval release gets affected and brood protection in those forms possessing it is an advantage. Time and duration of breeding season also get affected. Asexual reproduction is more easily resorted to in decreased salimity as in *Cordylophora*. The antractic isopod *Glyptonotus* which is benthic, breeds throughout the year as also *Sphaeroma*, but becomes seasonal in other examples when there is a phytoplankton bloom in summer releasing young to feed on the appropriate food, as in the amphipod *Borallia*. In the insect *Schizomyia* the midge emerges and reproduces soon after rainfall, every month thus affording tender leaves that sprout after the rains, for the new larvae that energe to produce galls. The swarming of the tropical termites usually coincides with rains. Also the emergence of several moths (*Eupterote*) is linked up with the onset of monsoon. Mating occurs in earthworms (*Megascolex, Drawida*) also at this time.

Humidity

Humidity is linked up with temperature, as the relative humidity varies with temperature. Low humidity depletes water and is, therefore, generally antagonistic to egg production. Humidity has to be examined from the general nutritional point, as the diet will yield water to the animal. Insect examples are known where the total weight of water in the animals is 90-92% and the food provides 72% of the same (*in Telea*) or 46% of the total weight of water in the animal contains to the extent of 10% in *Sitophilus* which feeds on stored wheat. 80-90% of RH is usually conducive to reproduction in the majority of insects. There is a decrease in reproduction as this falls. This is so in locusts in *Thermobia* in *Apanteles* and in species of the beetle *Cryptolestes*. Insects pests of stored material, which habitually live on dry grain, lay no eggs if the moisture content of the grain is 8-10% as in *Calandra* and *Rhizopertha*. Though generally laying of eggs stops when RH falls below 50%, in *Trogoderma* it has been recorded that some eggs were laid even at 0% RH.

Temperature

The influence of temperature on the reproductive cycle has been examined in a larger number of cases. In some cases, the developmental history is influenced by temperature and related factors like humidity. It may also in certain cases influence the triggering off of reproduction and subsequent processes like maturation, arrest of reproduction, mating, and oviposition or spawning.

Temperature is the intensity factor of heat. It is variable in different latitudes and special regions of latitudes,—rivers, lakes etc.

In *Hydra,* experimental alteration of temperature for 2-3 weeks will promote sexuality which occurs also in early winter or early summer, the former due to fall and the latter due to rise in temperature; these are suggestive of a major role for temperature than for photoperiod.

For each species the season for reproduction occurs at a fairly constant temperature over its distribution range and this usually commences and lasts longer in warmer regions. This is illustrated by Ostrea which will start breeding when the water temperature is 15-15°C and the breeding season lasts until this temperature falls. In Naples the breeding starts earlier than in England and lasts longer Nelson (1928) has found that spawning occurs at different temperature in different molluscan species, being 10-12°C for *Mytilus, Mya;* 15-16°C for *Ostrea, Pecten;* 20°C for *Gryphaea* and 24-25°C for *Venus,* while temperature lower than 12°C will enable *Hippopus* to breed. It is generally believed that the tropical waters show comparatively little variation in temperature in thge year, though breeding may be restricted to seasons like mid-summer as in *Hippopus.* Often this may be related to other phenomena also like the lunar influence as in *Ostrea* (Orton, 1926), or periods following spring tides *Ostrea* and *Patella* breed at the time when the wind speed is over 20 knots. The presence of sperms near the siphon or a special substance called dianthin, induce breeding in *Gryphea.* While in the examples mentioned above reproduction occurs at a favourable temperature, along with other possible stimulants, there are examples where gonadal activity has been low and the reproduction period is of extended duration in warm water, while cold will initiate more intense gonadic activity and shorter reproductive period. This has been seen in the sea urchin *Strongylocentrotus,* which lives side by side but exposed to different temperature. This animal shows a breeding period of five months in a year, starting from about the beginning of summer. The gonadic activity is stimulated by summer, and by five months after this, the animal spontaneously spawns. Moreover, it is also recorded that *Arbacia* shows two breeding seasons, one in summer when temperature is 17.2 to 21.5°C and the other in winter when it is 1 to 2.7°C. *Arbacia* shows however only one breeding season in the North Carolina waters. The forms which exhibit the twin peaks of spawning, when reared in identical environmental conditions, showed identical patterns. But similar experiments showed that in the scallop *Aequipecten* were the North Carolina and Woods Hole forms were reared simultaneously under similar conditions, their spawning periods did not coincide. It is possible that there may be genetic differences between the populations. It is also to be mentioned that the typical Indian echinoid *Salmacis* breeds throughout the year. In the Adriatic Sea, of the 97 species studied 64 spawned in spring, 49 in summer, 26 in autumn and 18 in winter, this being so because the high and low-temperature preferred forms lived in an overlapping assemblage.

The reproductive activity is optimal in several insects at definitive temperatures: in *Cocinella* it is 16°-20°C in *Leptinotarsa* is is 22°C in *Calandra* it is 25-29°C and in *Thermobia* it is 37°C. The effects also are apparent along different pathways: higher temperature interferes with mating in *Panolis* or sperm transfer in *Ephestia,* while lower temperatures affect egg laying in locusts or fail to mature eggs in *Nauphoeta*. The range of temperature where effective reproduction occurs is also known in some examples; in *Bruchus* it could be 8 to 40°C and 32 to 41°C in *Thermobia*.

Hot springs with a temperature around 55°C has its own characteristic fauna like larvae of *Tendipes,* the brine fly *Ephydra,* crustaceans *Exosphaeroma* and *Thermobathynella.* It said that the brine fly *Ephydra* can live and breed in hot water.

Studying the Antarctic fauna it has been found that breeding occurs in the copepod *Rhincalanus* in April or May. The benthic *Odontaster* and *Orchomenella* interbreeding forms releasing young by later winter or early summer. Arctic invertebrates also have been recognized as winter breeders. All the year round even in these cold waters breeding is shown by *Glyptonotus* (Dearborn, 1967) and *Idotea.* At 0°C the stone fly *Nemoura* mates and the eggs develop in ice caverns.

In terrestrial forms variations occur and this to a large extent is controlled by the adaptations exhibited by the animals to conserve water and avoid excessive dehydration. In *Armadillidium* reproduction occurs in April to November, and daylength perhaps here has a more decisive influence. This is also related to the rainfall which helps in preservation of desiccation. Indiplopods egg maturation occurs within a few days. The rate of production of eggs depends on temperature in species of Iulids. In *Cylinthdroiulus,* it has been found that vertical migration in the soil occurs, the millipedes mating and living during the first three instars under logs; and by about winter the forms move into deeper layers to reemerge by the end of spring. Reproduction occurs throughout the year as in several arachnids *Heterometrus* or during summer as in *Damaens* or in autumn as in *Hermannia.*

The factor that triggers the release of gametes has been studied in some spawning marine animals. In the scallop *Aequipecten,* it has been suggested that because the gamete shedding is during the cooler months than in the warmer months, it might be the seasonal changes that stimulate the shedding. Earlier, Belding (1910) suggested that the northern forms start spawning when temperature is rising up, while the opposite is true in the southern forms. Many forms however do not shed gametes when the temperature falls considerably to interfere with development, reduction in such cases may lead to regression of gonads. Peak spawning of the limpet *Cellena* occurs in the Bay of Bengal during June to August and December to February and under optimum conditions of salinity and temperature, high wind speed appears to hasten the process.

In the social insects, we find very good examples of thermoregulation in the colony. Even social air-conditioning mechanisms are met with here. When temperature falls considerably such mechanisms are of value, but a number of forms exhibit as a last resort migration to more favourable micro-environments, notably in ants. In the bee *Trigona* the queen is left behind to die as she could not fly, and queen rearing is initiated in the new colonies. Additionally, the social insects curtail oviposition by the workers and prevent rearing of new queens: these will be achieved by slaughter or by producing primer inhibitory pheromones (see chapter on Pheromones). In *Coptotermes* the extra virgin females are eaten. A method of birth control seems to be employed by the queen of *Bombus*. The queen constructs the nest, and in oviposition, new eggs are laid in cells above those containing 3 days old pupae. The new eggs laid number from 4 to 16.

The number of eggs will be at the mercy of the workers that would emerge from the cocoons, and workers thus available to nurse the eggs are reduced in number.

The ability to lay eggs is very remarkable when the annual output is taken into account in social forms. However, there is a variability in the egg production in relation to the season, as for example in *Apis* there is a peak of laying in summer when he will produce 1000-2000 eggs a day. During peak period *Bellicositermes* lays 36,000 eggs per day.

Photoperiod

The period of one day consists of alternating illuminated and dark phases and this recurring rhythm forms the photoperiod. The photoperiod influences the life of all animals. Very often the lighted period is known as photophase and the relatively lightless period is known as scotophase. Reproduction occurs at the end of short photoperiods of winter, generally in the spring. The relation between photoperiod and reproduction has been extensively studied in the northern hemisphere. In *Lumbricus* gonad maturation lasts till the end of early summer when mating takes place and oviposition occurs. The period is practically the same in the equatorial-region for earthworms like *Megascolex*, but this time has the added advantage of heavy monsoon rains. Photic control or reproduction has been found in *Piaster* and in *Strogylocentrotus* the long day stimulates gonadic activity, as also perhaps in other sea urchins. However, Holland (1965) has shown in *Strongylocentrotus* that feeding continuously in light and darkness, promotes earlier maturity of the gonads, and it may be a nutritional effect only. However, when food is abundant, no maturity of any precocious type has been found.

In the colder seasons with reduced periods of daylight, all organisms are somewhat adversely affected. The fall in the plant populations will tell upon the animal populations that depend on them for sustenance; there is a marked fall in the animal populations also. Thus nutritional setbacks affect reproduction in all animals. A concomitant fall in the ambient temperature, gradually to extreme cold conditions, also occurs during the period of shorter daylength. Thus photoperiod and temperature are two connected factors in ecology. When there is a fall in temperature, photoperiod and nutritional elements, reproduction also gets modified. This period will mark the period of oviposition, laying a special eggs (as in rotifers, Cladocen. Terrestrial animals, freshwater animals) and sometimes resistant stages to tide over conditions (like statoblasts, encysted stages, emippium; gemmules etc.) Gonadic development in relation to light has been studied in the amphipod *Pontoporeia*. Here constant illumination, in the absence of a dark period (akin to autumn) will not bring about gonadic maturation; this however is not absolute, because as Barnes (1963) has earlier shown in *Balanus* continuous light does not completely inhibit maturatun.

The lengthening day of the spring will promote sexuality in *Hydra pseudoligactis*; also increased photoperiods can promote spavning in *Platyneries* with 12 : 12 darkness and light with extra constant light for 6 days simulating lunor conditions. Winter influences the terrstrial animals more drastically. The grasshopper *Orthacanthacris* and *Anncridum* do not breed even if higher temperatures are provided. But if *Anacridium* females are exposed to longer photoperiods during autumn, eggs begin to grow within three weeks. This indicates that the shortfall in daylength arrests reproduction in this insect. But Norris (1957) has shown that *Schistocerca* requires short daylength of 7-8 hours while longer periods arrest growth of the gonads.

Photoperiodically decided behavioural rhythms have been demonstrated in several arthropods including reproductive ones: Oviposition of the mite *Tetranychus* sex responses in noctuid moths and in oviposition in *Crambus Pectinophora* and *Aedes*. In the insects cited above, the phenomena occur in scotophase but in the mite it occurs at photophase. In the Humenoptera, mating flight occurs during early scotophase in *Parmaporera* late scotophase in *Veromessor* and in photophase in *Solenopsis*. The swarming after emergence, mating and egg laying of the midge *Clumu* is controlled by photoperiod. The short lived female performs all these functions in a period of about two hours, which coincide with the low tide which will expose the sea weeds on to which eggs will be deposited. The eclosion and the reproduction are linked up with the 15 days tidal cycle, involving both the photoperiod and the cycle of the moon.

One phenomenon that has to be discussed in relation to

photoperiod is the associated rhythms, known as circadian rhythms. Broadly speaking, circadian rhythm is unaffected by temperature. Extreme low (near freezing) or high temperatures will make the animal inactive, but restoration of general normal conditions will bring out the rhythms associated with the diel or 24 hours. Locomotor, feeding, mitotic, colour change and a host of other mechanisms occur in a rhythmic pattern spread over a 24 hour (diel) period. This phenomenon is light-sensitive but is not dependent on it, because the cycle will continue irrespective of light or dark conditions but is set in tune with the accustomed patterns. This cyclical rhythmic activity is conditioned by what could be called a physiological clock, its phase is changeable by alterations in photoperiod in nature or low natural temperatures, but the clock will run with a constant rhythm.

Many marine and littoral invertebrates show a circadian rhythm, which is associated with other factors like tidal or lunar rhythms.

A minimum of six rhythms has been recorded in the fiddler crab *Uca*. While one could present evidences of various activities controlled by the physiological clock, the aspect of reproduction has been studied in only a few cases. Studying the bollworm *Pectinophora* it has been recorded that oviposition occurs as a rhythm in DD phase (constant darkness). This circadian phenomenon involves the exact measurement of night (DD phase). The response to pheromones in insects also exhibits a circadian periodicity. In *Trichoplusia* the response is better defined in the dark-period, while light as low as 0.3 lux will inhibit this. The dark cycle phase also promotes singing in crickets and in *Ephippiger*. Continuous darkness promotes prolonged mating songs and in *Grylhus* light inhibits courting. Rhythmic activity is also seen to be associated with the moonlight. This biological rhythm is mostly manifested in the reproduction of some invertebrates. Among echinoderms *Comanthus* and *Strongylocentrotus* show this phenomenon. *Commanthus* spawns annually in October at a time which coincides with the maximum declaration of the moon and is a brief process lasting for only an hour after 3 p.m. A study extending over seventeen years has confirmed this periodicity. Fox (1924) studying *Diadema* living in the Red Sea, saw that the full moon of the breeding season initiates spawning activities. It is at this time that the new germ cells for the next fall, get initiated into activity also, because a single individual will breed several times. Yoshida (1952) observed that in sea urchins the testicular and ovarian histology reveals increased sexual activity during full moon periods, though all of them did not spawn at the full moon. A lunar-temperature cycle has been explained to work in the breeding activity of *Lytechinus*. They suggested that in *Tripneustes* the temperature is a trigger and the gonad size is significant here to be of value in responding. The famous palolo worm breeds in the Pacific, all individuals becoming mature about the same day every year, and their epitokous parts are liberated which swarm on the full moon in November-December. It is suggested that in the Atlantic palolo *Leodice* and the Pacific *Eunice* the quality of mocalight and its duration are critical factors in releasing gametes. In the Malayan waters *Lysidice* breeds in March-April, two or three nights after the full moon. The mediterranean *Platynereis* breeds in May to September, between sunset and midnight during the first and third quarters of the moon. *Diopatra* lays eggs when the tide is high. Among molluscs *Littorina* shows egg laying activity at spring tides. Colour change and feeding rhythms have been elucidated in Uca and it would have been an added information if reproduction had also been studied. Lunar periodicity in insects is seen in *Clunio marinus* which emerges and mates at low tides; in Europe this happens around full or new moon. Another species *C. Takahashi* sticks to tidal periodicity without responding to moon at all. The populations of *Clunis* species show variability which links up with tides, or with moon or may rarely, as the baltic submerge variety, with no relation to tides or the moon. Thus, we see that in invertebrates the animals can register time, of 24 hours of a solar day, or a lunar month or a solar year. However, the precise nature and working of the biological chronometer is little understood.

Diapause

Diapause is arrest of development or growth or activity during the unfavourable changes in the environment. This helps the species to survive the undesirable effects due to climatic conditions. Often this is associated with the building up of reserves in the tissues. Sometimes quiescence is distinguished from diapause, because the arrest relating only to growth and development will be diapause; while quiescence is a torpor phase of inactive condition which gets removed immediately when the factor imposing this is resupplied (like water). Both are sometimes treated as dormancy or the two are treated separately considering diapause as different from a comparatively insignificant phenomenon of quiescence. As a matter of fact several types of phenomena have been found and a segregation of the two becomes arbitrary. Diapause is well-known in insects, while in other examples aestivation and hibernation in summer and winter respectively have been recognized. A number of reviews of the phenomenon are available. Mansingh (1971) classifies dormancy into three categories, viz., hibernation (caused by lower temperature than normal), aestivation (due to higher than optimum temperatures) and athermopause (due to factors like nutritional hygric etc.). Athermopause may itself comprise three categories viz., quiescence (slight and short term adverse weather), oligopause (due to mild and long term climatic changes) and diapause (extent and long term adversity of season and climate). In diapause occur two conditions: teleodiapause which is a high intensity diapause and ateleodiapause which is less intense. There are well-defined characteristics and well marked adverse conditions which bring in the varieties of dormancy.

One well-known example of diapause exercising its influence on reproduction is seen in *Leptinotarsa*. Towards autumn the beetle becomes a burrowing form and hibernates in the soil, shows reduction in the size of corpora allata, and failure of formation of mature eggs. A large number of starving forms go into this phase more readily at this time. It is photoperiodically regulated also, because normal beetles do not show diapause if kept in a 15 hour period of light, which need be only of low intensity as 0.1 lux. It can then lay eggs even if the temperature is as low as 17°C. The diapause syndrome of *Leptinotarsa* in addition has other features also like reduced oxygen consumption, and in the works of De Wilde and his colleagues the chief operative centre in this phenomenon has been detected to lie in the neuroendrocrine system. Other insect examples are also known though not as extensively studied: failure of egg laying in winter even under high temperature occurs in *Sitona Eurygaster, Nysius* and *Limothrips*. Short day treatments can induce diapause in forms like *Pyrrhocoris*. Usually in mosquitoes feeding and maturation of eggs go hand in hand, but in *Anopheles* feeding females will not oviposit in winter which has been described to be due to gonotrophic dissociation probably due to a photoperiodic influence. In *Chrysopa* seasonal reproduction occurs, and diapause characterized by gonotrophic dissociation could be avoided or induced by manipulation of diet or photoperiod. It reproduces from April-June when photoperiod is satisfactory and prey abundant. Then a diet-induced diapause follows in summer. October-December shows short day diapause (LD 14: 10). In *culex,* feeding and egg growth go together as in all mosquitoes, but in winter the tendency for feeding on blood falls and many females do not grow eggs. Towards late winter feeding in blood commences and the animals start growing eggs which perhaps the lengthening day stimulates. Here, however, is a break in dormancy in mid-or late winter but not when the season is over. Aestival (summer) diapause in reproduction occurs in *Hypera*. Emerging in early summer, the beetle goes on feeding but will not lay eggs, and a reproductive delay occurs till autumn, found to be dependant on a 12 hours photoperiod. As in winter diapause, accumulation of reserves occurs before entering the aestival dormancy. Likewise in the laboratory longer daylength produces diapause in *Limnophilus*. Long days induce diapause by aestivation in *Otiorhynchus*, *Agrotis*, *Euplagia* etc., 15 hour daylight produces diapause in reproduction in *Galeruca* even if food is abundant.

Oxygen

How oxygen influence reproduction is not understood. In the polycheat *Capitella* low oxygen tension always inhibits reproduction. When very little oxygen is available the animal is inactive thought it can live for a long time, when oxygen content is 2.9 ppm feeding is regular but there is no reproduction. When it is 3.5 ppm reproduction takes place promptly. Oxygen exercises effects on gametes also. In the absence of oxygen egg of *Arbacia* will live for about 8 hours but there is a tendency for stimulation of parthenogenesis. Oxygen lack damages sperms of *Arbacia*.

Availability of Suitable Hosts

In a number of invertebrates, prolific egg laying is the rule, and the eggs are left to fend for themselves during the course of development. This will be applicable to the cases of even parasites like the cestodes. Under conditions where the parasite and the host have not speciated together, it is suggested that the parasites have a slower evolution. In such instances the parasite can infect multiple hosts, and can pass from one host to another readily. Among helminths digenetic trematodes are not strictly confined to the host species as in Monogenea. This condition holds good in tape worms, and in nematodes, like *Ascaris*, *Hymenolepis* etc., however, avian cestodes show specificity of hosts. Among insects fleas, lice etc., also show a loose relationship to the host—like the human louse *Pediculus* living on monkeys (*Ateles*) and the saki (*Pshecia*) as reported by Ewing (1938). Among parasitic insects oviposition is also influenced by the availability of the host, lesser number of eggs being laid only if the proper host is not available. The non-feeding adult *Diadromus* laid the maximum number of eggs where the weevil host has been available; the number fell if the host was not available or was present only for shorter periods. The weevil forming the host is *Acanthos celides* and this itself laid more eggs if the bean was present. Cadeilhan (1965) observed that if onion is available throughout the period, an average 233 eggs were laid by *Acrolepia* and only 78 were laid if the plant was available only once in three days.

A well-defined host specificity is exhibited by gall making insects. Here the plant and the animal have a more intimate relationship and the gall itself is characteristic of the species of plant and the species of insects that makes it. *Austrothips* produces axillary gall of a specific, large, unilocular type in the plant *Calycopteris* Echinate galls on *Macaranga* will be produced by *Schizomyia* which emerges at a time when the leaves are in the appropriate stage only. The parasitic mite *Proctolaelaps* lives on the cockroach *Nauphoeta* and *Lucophaea*. An attractant to the mite has been recently extracted from the host and purified, found to be a polypeptide with MW 2000 and containing seven amino acids. Other cues also play their part in host discrimination. The body segment of the host has to cover the dorsum of the mite to have a hold on to the cockroach.

A phenomenon of alternation is also known which occurs between bisexual and parthenogenetic reproductions—this is known as Heterogony. The parthenogenetic generation may be single or multiple in that several such generations intervene between two bisexual reproductions. This phenomenon is known in rotifers, aphids, cynipids etc. Among aphids, several morphs are produced chief among aphids, several morphs are produced chief among which are:

Fundatrix — A parthenogenetic female and emerging from a zygote.

Virginopara — Female that is viviparous and parthenogenetic which produces other virginoparae.

Ovipara — Female that lays overwintering eggs, winged.

Gynopara — Vivipara that produces only ovipara (are winged).

Sexupara — Vivipara that produces both oviparae and males (are winged).

Andropara — Vivipara producing only males (are wingless).

Heimalis — Vivipara which are wingless that overwinters.

Sexuals — Females and distinct males, oviparous.

This classification brings out the morphs and their reproductive patterns and points out the influence of the winter. Primary host plant is that on which the animal lives through winter in the egg stage. Often a change in the host plants is visible, the primary and the secondary ones. For example, *Sacchiphantes* has the fundatrix living on spruce and then goes out to the larch on which the over-wintering stages are passed. The virginopara that leaves the primary host and goes over to the secondary host is called Emigrant, while the vivipara that moves over to the primary host plant from the secondary, is called Immigrants. Both wingless and winged viviparae are born and live on the secondary host plants, these are called Exules.

The basis of the production of the various morphs have been to a considerable extent attributed to environmental factors. Crowding has been found to induce vivipara to produce alate virginoparae or sexuals. In *Brevicoryne* crowding helps production of sexuals, and exposure to short photoperiods under crowded conditions produces sexuals. Virgins will thrive on primary host plant and can overwinter even without crowding in forms like *Pemphigus;* short photopriod and low temperature promote sexual production. Nutrition influences heterogony, and this has been found to be so in the soil dwelling *Eriosoma*. Photoperiod can directly act on the forms which live exposed like *Megoura* though it has also been claimed that the photoperiod affects the aphid through the food plant. When the daylength falls below 14 hours and 55 minutes, there will be reproduction leading on to the development of sexuals in *Megoura* at 15°C with a reduction by 15 minutes, if there is 5° rise in temperature. But 23°C is the maximal limit which would be conductive to ovipara production. Lees (1966, 1967) has afforded evidence in *Megoura* that photoperiod operates through the midbrain to produce virgins, and the eyes do not participate in this registration of daylength. He believes that the pars intercerebralis of the brain is the chief region involved. It may also be mentioned that the exposure of the grandparent *Megoura* to long photoperiods before the birth of the mother, followed by exposure to short photoperiod after birth will make her first produce virgins and then oviparae. When the grandmother was exposed to short days and then the mother exposed to long photoperiod, virgins only will be produced. Such a mechanism where long day induces virgin production is not seen in *Dasyaphis* where the virgins exposed to 12 hours darkness respond by producing sexuals, long periods of continuous daylight will induce virgin production and sexuals will arise only after intermittent exposure to 12 hour scotoperiods. Apparently the mechanism is not so simple and general in aphoid reproduction.

It was found that fundatrices of *Aphis* could respond to photoperiod only after some period of sexual generations. This has been found to be so in number of aphids *Aphis chloreus, Brevicoryne, Acrythosiphum* etc. This has been described as the "time factor" by Wilson (1938), "facteur fondatrice" and the "interval timer." The fundatrix produces a substance in a quantity which decreases or gets diluted as generations are passed and has a role to prevent sexuale productions. This view of Bonnemaison presumes that short photoperiods of spring and early summer do not affect virgin production. But when a number of generations have passed the ability of virgins to produce sexuals reaches a level of exploitation. Whether the factor is to be diluted to reach this stage or whether a certain time has to be passed after the first generation, have been studied by Lees (1966) in *Megoura* who maintain that the latter is the condition found in them. Rearing the first and the last born young of a female, he established two lines of aphids, kept, at 15°C and 12 hours photoperiod. This photoperiod and temperature promotes production of sexupara. It was observed that 66 to 76 days after the first young had been born the offspring in both lines could respond to decrease in daylength. In other words, the first born young had one generation more than the last born, with which the experiment started. He, therefore, has suggested that it is a "time interval" and not a "factor" that is behind this activity in *Megoura*.

In three genera : *Megoura, Antherea* and *Pieris* the photoperiodic action is known to be mediated by the direct action of light on the brain. What wave length is active in this respect, is not known, but in *Megoura* an attempt towards its analysis has been outlined; a blue sensitive photoperiod response has been suggested in some insects. Hayes (1971) has suggested a photosensitive pigmented area in the brain of *Antherea* a point which has been criticised. The point requires further elucidation.

11

Aging and Developmental Biology

Membrance Damage and Aging

Many hypotheses have been developed to explain how cells become damaged over the years, leading to aging. One group of models suggests that aging is the result of a leakage of hydrolytic enzymes from lysosomes. Lysosomes are membrane-bound bodies that contain hydrolytic enzymes. There normal function is to digest particles inside of cell vesicles. Clearly, it enzymes leak from lysosome, the cell may begin to digest itself. The membrane surrounding lysosomes might be disrupted by a variety of agents including free-radical compounds that are generated in metabolism or by radiation. They are unstable and contain unpaired electrons. Free radicals can react with the lipid of lysosomal membrance, damaging the membranes and allowing lysosomal enzymes to leak out. Future work should tell whether such enzyme leakage is important in the aging process.

Reduction of Aging Rate by Modification of Diet

Several interesting experiments suggest that some aspects of aging are related to environmental factors. The rate of aging in rats has been significantly reduced by a long-term restriction of dietary caloric intake immediately after weaning, the animals were placed either on a normal diet or on a diet that was nutritionally complete put included just enough calories for show growth. The animals on the latter diet remained stunted in size and juvenile in physiological functions, but resumed normal growth when returned to a normal diet. More importantly, animals on a restricted diet lived as much as 33% longer than animals on a normal diet. The incidence of both cancer and respiratory diseases normally associated with aging was clearly lower in the calorie-restricted individuals.

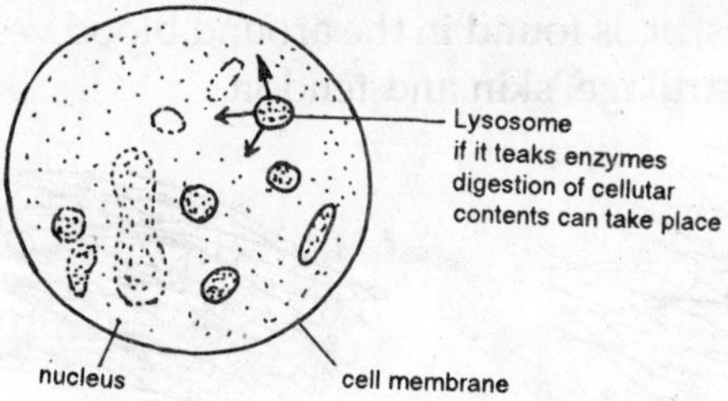

Fig. 11.1. Leakage of hydroloytic enzymes from damaged lysosome.

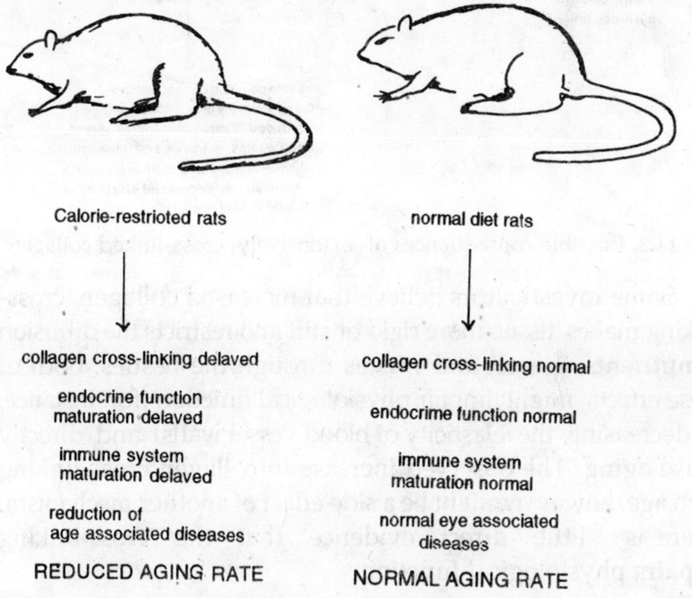

Fig. 11.2. Some consequences of restricted calorie intake in rats

Collagen Cross-linking and Aging

The experiments just described suggest that external factors are involved in aging. The results, however, are subject to various interpretations. The calorie-restricted individuals displayed delayed cross-linking of the protein collagen. Increased collagen cross-linking is typically associated with aging. In fact, tissue age is often determined by measuring the degree of collagen cross-linking. Collagen is present in just about all body tissues; it is found in the around blood vessel walls and in bone, cartilage, skin and tendon.

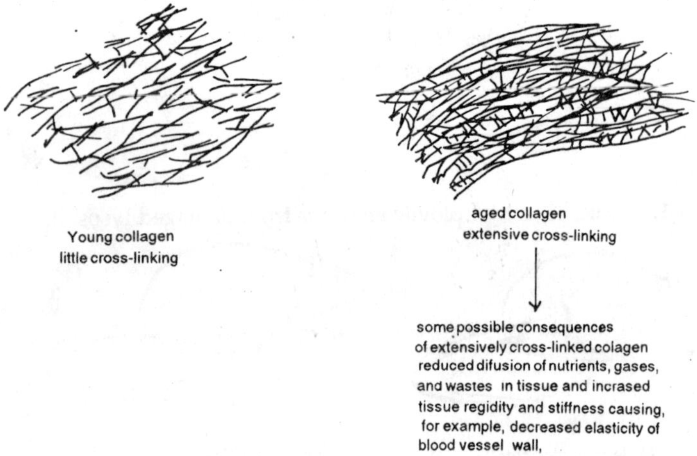

Fig. 11.3. Possible consequences of extensively cross-linked collagen.

Some investigators believe that increased collagen cross-linking makes tissue more rigid or stiff and restricts the diffusion of nutrients, gases, and wastes through the tissues. Both of these effects might impair physiological function (for instance, by decreasing the elasticity of blood vessel walls) and directly cause aging. The observed increase in collagen cross-linking with age, however, might be a side effect of another mechanism. There is little direct evidence that the cross-linking impairs physiological function.

Aging and the Immune and Endocrine Systems

In addition to delayed collagen cross-linking, calorie-restricted animals exhibit delayed maturation of hormonal functions and delayed maturation of the immune system. Several theories of aging consider the immune system and the endocrine system to be key regulators of the aging process.

Reduced functioning of the immune system has been proposed as an explanation for aging. Presumably, individuals with diminished immune function would have increased incidence of age-related diseases such as cancer. Patients whose immune systems have been suppressed for organ transplant they have a much higher incidence of cancer than is normal for their age. The immune system functions best in young individuals and declines with increasing age.

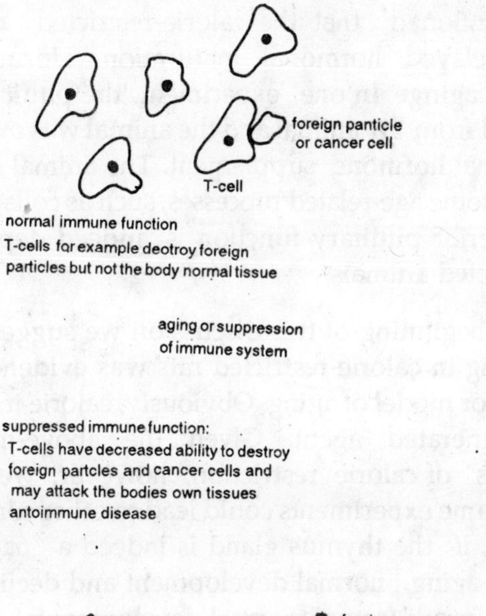

Fig. 11.4. Functioning of normal and suppressed immune systems.

The thymus gland may play a "pacemaker" role in aging. This immune system organ reaches its maximum size early in life and becomes smaller with age. In one experiment, the thymus gland was removed from young mice. The mice acquired "runting disease," a disease in which the immune system attacks normal tissue (*autoimmune* response), causing tissue degeneration. Autoimmune diseases are related to aging. Rheumatoid arthritis, for example, is caused by such a mechanism. Decreased immune function is also related to the increased susceptibility to infection and cancer that appears in old age. Delay in the maturation of the immune system, therefore, may be part of the explanation for the extended life of calorie-restricted animals.

We mentioned that the calorie-restricted rats also displayed delayed hormonal maturation. Hormones are important in aging. In one experiment, the pituitary gland was removed from an animal and the animal was maintained with a minimal hormone supplement. The animal showed a reduction in some age-related processes, such as collagen cross-linking. Anterior pituitary function is indeed depressed in calorie-restricted animals.

At the beginning of this discussion we suggested that reduced aging in calorie-restricted rats was evidence for the extrinsic-factor model of aging. Obviously, calorie intake is an externally generated agent. Given the above-mentioned consequences of calorie restriction, however, we can see that these same experiments could lead to other conclusions. For example, if the thymus gland is indeed a "pacemaker" that controls aging, normal development and decline of this gland would result from a normal developmental program. This program, however, could be modified by externally generated factors such as calorie intake. This suggests that aging is controlled not by internal "clocks" alone or by extrinsic factors alone, but by an interaction between internal "clocks" and extrinsic factors. Extrinsic factors might turn the internal "clocks" on or off, just as a variety of agents turn genes on and off during differentiation.

Internal Aging Clocks

What other evidence favors the intrinsic, or internal "clock," mechanism of aging? Many previously inactive genes become activated late in life. For example, genes that specify traits such as adult-onset polycystic kidney disease, hereditary glaucoma, hairy ears, and Huntington's chorea usually become active only in the later years of life. This suggests that the conditions associated with aging are programmed into our genetic information, but they are repressed in early life. Aging, therefore, might represent the later part of a developmental timetable that includes embryonic development, maturation and aging. Again, externally generated agents probably play a role in turning such "aging genes" on or off.

Adult onset diabetes	Hairy ear rims
Adult onset polycystic kidneys	Huntington's chorea
Pereoneal atrophy (dominant)	Hereditary glaucoma
Dominant muscular distrophy	Dystrophia myotonica

Fig. 11.5. Some genetic conditions that appear to become active during later life.

Additional information regarding the built-in cellular "clocks" has been obtained through studies of isolated cells in culture. Mammalian cells in artificial culture generally go through three growth phase:

1. Establishment, or adeptation to the culture conditions
2. Exponential growth of the cells, which divide rapidly
3. Cessation of growth and eventual degeneration

Although cells form old doners can exist for quit a long time in culture—longer than the life span of the donor would suggest—cells from old donors usually go through fewer doublings than do cells from fetal tissue. Such experiments suggest that normal cells contain some sort of "clock" that controls (at least in part) growth potential and growth cessation. Again, however, environmental conditions can modify the number of doublings that cells undergo in culture. For example, hydrocortisone increases the number of doublings, but seldom by more than 25%. This increase is in the same general range as the life extension caused by calorie restriction in rats (although cell doubling time should not be equated with the overall aging of organisms). These results suggest the existence of a "clock" that can be modified, to some extent, by external agents.

Genetics of Aging

Aging may be caused, in part, by the buildup of cell products that form as a result of accumulated mutations in the genetic material. Defects in the enzymes that control DNA copying would result in the formation of abnormal proteins.

Some diseases caused by gene mutations can accelerate aging. These diseases include Werner's syndrome and the Hutchinson-Gilford progeria syndrome. In Werner's syndrome, the hair may turn gray in the teens, and cataracts, adult-onset diabetes, skin lesions, and sarcomas are common. In the Hutchincon-Gilford syndrome, the individual usually lives only until age 11 and always dies before age 30. Symptoms of the disease include the early loss of hair, retarded growth, enlarge stiff joints, and hardening of the arteries. Such diseases suggest that genetic mutations can indeed cause age-related declines in physiological function.

The life span of a given species is well-defined within a certain range of years or days. The life spans of parents and children in a given family are quite similar. Identical twins usually show even more similar life spans. These observations indicate that the genetic makeup of an individual controls has or her general life expectancy. The digree to which life span can be modified by altering extrinsic factors (such as calorie intake) remains one of the most intriguing questions in the ares of aging today. Perhaps one's genetically controlled life expectancy can be substantially altered. The experiments on altering calorie intake in rodents suggest that such modification may be feasible for human beings.

12

Cancer and Developmental Biology

Cancer is the uncontrolled cell division whereas in the development cell division is a well controlled process that is why both cancer cells and certain embryonic cells are similar in many ways. A study of cancer and embryology will tie together many of the concepts developed in this text. We shall show how our understanding of embryonic cells leads to important insights into a major disease, cancer.

A complete discussion of cancer must include the causes of cancer, the types of tumors, the diagnosis and treatment, and the cellular aspects of cancer. The relation of cancer cells to embryonic cells falls into the last category., We shall begin with the causes of cancer and then unravel the entire story, ending with cellular aspects. In this way, we shall develop an understanding of the cancer problem as a whole. Our discussion of the clinical aspects of cancer will be brief.

CAUSES OF CANCER

Cancer is a disease in which cells grow in an unregulated manner. These cells may eventually kill the individual by growing into blood vessels and causing hemorrhage or by interfering with some vital function. Many things can cause cancer, including radiation, viruses chemicals and genes. Many of the agents that cause cancer may act or the chromosomes of the cells. Certain genes code for proteins that are able to transform cells into a cancerous condition; these genes may be activated by cancer-causing factors listed above. These factors may also act by damaging DNA, thus altering normal genes enough to transform the cell into a cancerous state. Cancer may also be caused by the direct insertion of a viral genome into the host cell genome. As the speculative nature of this discussion indicates, the mechanism by which specific agents cause cancer are uncertain at best. There is little doubt, however, about which agents cause cancer.

Radiation

Both X rays and ultraviolet radiation can transform normal cells into tumor cells. When cells in culture are X-rayed, they may begin to grow in an unregulated manner. That is, they do not stop growing or stop moving when they come into contact with other cells. The inability of contact to inhibit growth and movement, a property easily observed in culture, is one of the criteria used to determine whether cells have become cancerous.

Substantial evidence has accumulated that radiation causes cancer in humans. For example, the survivors of the atom bomb blasts in Japan have a high incidence of leukemia, a cancer of the blood-forming tissues. The frequency of leukemia is proportional to the exposure to radiation during the blasts. Also, a high incidence of leukemia has been observed in people who have received large doses of radiation as treatment for conditions other than leukemia. Radiologists who used some of the early X-ray equipment, which allowed substantial radiation leakage, also experienced a high incidence of leukemia. A variety of studies have shown that ultraviolet radiation from the sun is implicated in human skin cancer. Radiation damage of genes is probably a direct cause of such cancers.

Virus

Although virus particles are associated with some human tumors, it is difficult to prove that viruses are a major cause of human cancer because of the limited availability of human tissue of experimental studies. There is little doubt, however, that viruses can cause cancer in animals. This suggests that viral genes are also a causative factor in human cancer. When we talk about viruses, we must consider two major types:

DNA viruses and RNA viruses. DNA viruses, consist of a core of DNA surrounded by a protein coat; RNA viruses, on the other hand, consist of a core of RNA surrounded by a protein coat. Lipid is associated with the coats of some virus particles. Let us briefly consider first the DNA tumor viruses, then the RNA tumor viruses.

Two extensively studied DNA tumor viruses are polyoma virus and simian virus 40 (SV 40). These viruses typically attach to the cell surface and inject their DNA core into the cell. Once inside the cell, the viral DNA either begins to produce many additional viral particles, killing the cell, or incorporates itself in the host cell's genome. When the latter event occurs, the cell may become transformed into a cancer cell. Viral genes in the host genome may induce cancer by coding for new proteins that alter the behaviour of the cells, or the insertion itself may damage the host DNA so that processes such as growth regulation are impaired.

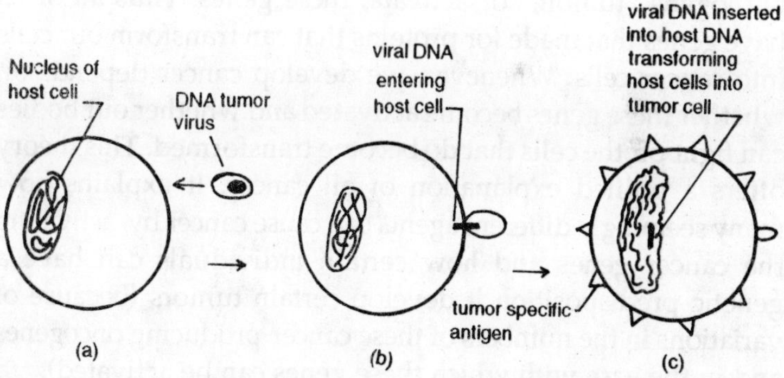

Fig. 12.1. Viral transformation.

RNA viruses have also been implicated in cancer. A well-studied example is Rous sarcima vius (REV), which induces tumors when extract from chicken cumers and inoculated into tumor-free birds. When normal chick cells in culture are infected with RSV, they enter the cancerous state. The transformation as signaled by a cessation of contact inhibition of growth and movement in the cells.

The means by which RNA tumor viruses transform cells into cancer cells has been the select of intensive investigation. The RNA virus apparently attactives to the cell then injects its RNA into the cell. Viral DNA is synthesized or the viral RNA template by the enzyme reverse transcriptase (RNA dependent DNA polyerase). (This enzyme upsets the once acepted dogma that DNA can be synthesized only on a DNA template. This viral DNA then irsert itself in the host DNA, just as the DNA from DNA viras does. It can induce cancer by coding for proteins that a bar cell remavior (such as new cell-surface proteins) or by altering the control of cell division.

A current theory cotmigene theory is that all cancer is caused by viral genes that were incorporated in the genome of the host species long ago and have been passed in from generation to generation. Chemicals; radiation and other cater narnogens "tumon," or activate, these genes. Thus all of us have genes that made for proteins that can transform our cells into cancer cells. Whenever, we develop cancer depends on whether these genes become activated and whether our bodies can fight off the cells that do become transformed. This theory offers a unified explanation of all cancer. It explains how many seemingly different agents can cause cancer by activating the cancer genes and how certain individuals can have a genetic predisposition it develop certain tumors (because of variations in the numbers of these cancer-producing oncogenes and in the ease with which these genes can be activated).

Is there any evidence to support the oncogene theory? Here we shall examine evidence from a few years age. See the end of this chapter for the newest developments relating to human oncogenes and their activation and human anti-cancer genes, or antioncogenes. Many agents that cause cancer, such as methylcholanthrene and radiation, not only transform normal cells into cancer cells but also induce the appearance of type C

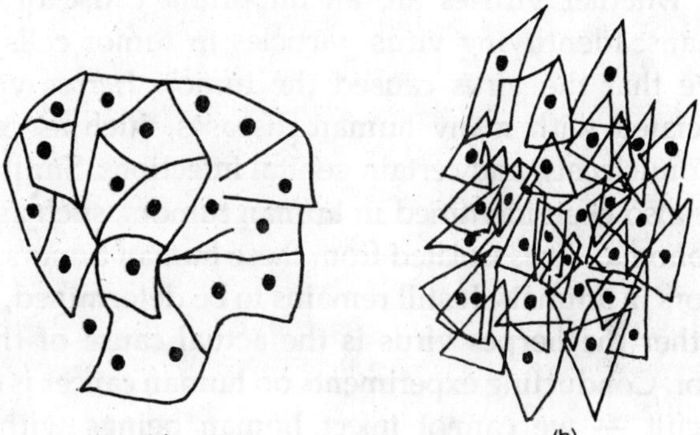

Fig. 12.2. Cultures of normal and transformed

RNA virus particles (type C is a classification of virus structural type) in some of these cells. A virus-specific protein antigen also appears in these cells. These results suggest that cancer-causing agents can promote the expression of viral genes in cells that contained no visible virus. Such experiments, however, do not prove that the viral genome is the cause of cancer. The appearance of viral particles in treated cells could be a secondary effect of the X-ray or chemical treatment.

It is not clear whether all cells contain cancer-causing viral genes in their genomes. Some studies with identical twins suggest that cells in all individuals do not necessarily contain oncogenes. Type C DNA synthesized from type C RNA virus was hybridized with DNa from identical twins, one of whom had leukemia. Only the DNA from the twin who had leukemia hybridized with the viral DNA. This result suggests that the leukemia patient had viral DNA genes, whereas the healthy twin did not. Thus, the viral genes were not present in the fertilized egg, but were picked up or formed by only one twin at some later time.

Although certain viruses cause cancer, it remains to be seen whether viruses are an important cause of cancer in humans. Identifying virus particles in tumor cells does not prove that the virus caused the tumor. *Herpes* viruses are associated with many human diseases, such as cold sores, mononucleosis, and certain genital infections. Similar viruses have also been identified in human tumors, such as Burkitt's lymphoma. Virus isolated from these human tumors can cause tumors in animals. It still remains to be determined, however, whether the herpes virus is the actual cause of the human tumor. Conducting experiments on human cancer is obviously difficult — we cannot inject human beings with virus to determine whether the virus causes cancer. Rather good evidence, however, is being obtained from carefully performed animal experiments.

How do viruses cause cancer? The literature is full of interesting studies that provide insight into this question. One important study was performed by Sefton and colleagues in 1981. They found that cells transformed by Rous sarcoma virus produce a protein kinase that phosphorylates certain amino acid residues (Tyrosines) in the protein that anchors microfilaments to the plasma membrane. This protein (called *vinculin*) also plays a role in helping cells attach to the substratum. Phosphorylation of the tyrosines in vinculin may disrupt microfilament organization and cause the alterations in cell shape and adhesiveness that accompany the transformation by Rous sarcoma virus. Such studies are beginning to shed light on the cellular changes that accompany viral tranformation at the brochemical level.

Chemicals

Cancer-causing substances are called carcinogens. Many chemical carcinogens have been identified, and we shall review some here. These chemicals may cause cancer by altering genetic information, as discussed previously. Proof that a given chemical causes cancer is obtained by exposing cells to these chemicals in culture or in living animals. If a chemical causes cancer in mouse cells when administered at a carefully determined concentration, it probably causes cancer in human cells as well, because memmalian cells are quite similar from species to species. The chemical concentrations used in the living animal experiments are based on the ratio of the weight of the animal to that of a human being and the ratio of the lifespan of the animal to that of a human being. In other words, concentrations are designed to correspond, relatively speaking, to those that would build up in the human body over a long period of time.

Animal tests, however, are not the only means of determining whether a chemical can cause cancer. Many data are available from human studies examining the incidence of cancer among specific occupational groups, among other smokers, and among groupings of people. These studies have shown that cigarette tar, asbestos, benzene, nickel, vinyl chloride, 2-naphthylamine, hematite, chromium, cadmium oxide, cadimium sulphate, bis(chloromephenyl)ether, benzidine, auramine, arsenic compounds, 4-aminobiphenyl, soot, and tars can cause cancer in human. Many other substances, such as chloroform and nitrosamines, have also been implicated in human cancer. The American Cancer Society has estimated that 50% of cancer deaths could be eliminated if people stopped smoking cigarettes. This dramatic statement is a guess, but an educated one. It points to the enormous number of cancer cases that are caused by things we eat, breathe, or touch. Some cancer specialists suggest that 85% of all cancers may be induced by environmental factors!

Before leaving the topic of chemicals that cause cancer, we should point out the difficulty of testing potential carcinogens. Not all cancer-causing agents cause cancer immediately. Some cause cancer only after cells have been exposed to other substances. Some agents, for instance, work in a two-step process. One substance sets up a precancerous condition, and another substance causes the actual tumor to form. Studies have shown that long after the first agent was applied to the skin of mice, a second agent induced a tumor there. The second agent did not promote tumor development in mice to which the first agent had not been applied. These findings indicate how difficult it is to determine whether an agent is a carcinogen. Many potential carcinogens may go undetected because the right second agent is not used.

The controversy over sodium nitrite relates to its potential role in a multi-step process. Sodium nitrite, a widely used preservative in certain meat products, is not a known carcinogen. However, *nitrosamine*, a product of the reaction between sodium nitrite and amine, is a potent earcinogen. Does the human body form nitrosamine after ingesting sodium nitrite? The question is still controversial. Some studies indicate that nitrosamine is formed. It is difficult, however, to determine whether the amount formed is hazardous. Studies indicated that nitrosamine is formed when bacon containing sodium nitrite is cooked. As a result of such studies, federal regulatory agencies are now developing stringent guidelines for the use of sodium nitrite in bacon and meat products that are heated at-high temperatures. Suspected carcinogens often are not eliminated from the environment until solid data are available, and decisions to regulate hazardous substances are difficult and often based on economic considerations, as well as health-related factors. Fortunately, efforts to control carcinogens in the environment are increasing.

Tumor Types

Tumors are abnormal growths tumors grow slowly, do not spread, and do not endanger the organism. Warts and moles are examples of such tumors, which are termed benign. Benign tumors, however; can change into tumors, that can invade body tissues, often causing death. Tumors that spread are termed malignant tumors. They spread by two major means. First, the tumor cells can move to nearby tissues. Such movement is called invasion. Second, tumors can invade blood vessels; the malignant cells are then carried by the blood to distant sites in the body. Movement in the blood stream is called *metastasis*. We shall consider the mechanisms of tumor spreading in the section on cellular aspects.

Many terms are used to classify tumors. *Carcinomas* are tumors derived from epithelial tissue. *Sarcomas* are tumors derived from connective tissue. *Leukemia* is a tumor of the blood-forming tissues that produces a large number of certain white blood cells. *Melanomas* are derived from pigment cells; *teratomas* are tumors originating from germ cells. Teratomas often develop into embryo-like growths consisting of many types of differentiated tissues. Some investigators believe that teratomas are abnormally activated eggs or primordial germ cells that begin to develop like normal embryos, then go away. These investigators suggest that some of these tumors could form normal embryos if they grew in the right place in the body. In any case, teratomas are so like embryos that they offer magnificent material for studies on the relationship between embryonic cells and cancer cells, the nature of the benign and malignant states, and the nature of differentiation.

Diagnosis and Treatment

Before turning to the cellular aspects of cancer, let us say a few words about the diagnosis and treatment of cancer.

Many tumors are curable if treated before they spread. Even after they spread, many tumors can be cured with appropriate treatment. It is much more difficult to treat a tumor that has spread, however, because one must deal with numerous secondary tumors at sites often distant from the original or primary tumor. Thus, successful treatment of many cancers depends on diagnosing tumors early in their development.

Certain cancers that were major killers not too long ago are not almost always curable. One example is cervical cancer in women. Because of early diagnosis, this form of cancer is now often completely curable. The key diagnostic tool is the pap test. In this test, a smear of cervical cells is examined with a microscope. Cancer cells, precancerous cells, and normal cells can be easily distinguished. Thanks to this test, cervical cancer can often be cured with freezing or surgery before any spread has occurred—or even before a real tumor develops.

Other tumors, however—such as some lung tumors and breast tumors—are difficult to detect early. Present methods of detecting these tumors usually involve the use of diagnostic X-rays. These techniques often cannot detect a tumor smaller than about one centimeter in diameter. A one-centimeter tumor already contains about one billion cells and may have been developing for up to ten years. By the time it is detected, the tumor may have already spread. Detection of such tumors undoubtedly will improve with improving technology in the near future. Improved diagnosis will surely produce a better cure rate.

Cancer is treated by many well-established methods, as well as some new experimental ones. The well-established methods, include surgery, radiation therapy, and chemotherapy. The newer experimental methods include immunotherapy and bone marrow transplants.

Surgery is the treatment of choice if the tumor can be removed without excessive danger to the individual. Surgery can cure cancers that have not spread to other parts of the body.

Localized radiation is the treatment of choice for some tumors. For example, cancer of the voice box is often cured through radiation treatment with little speech impairment. Although very effective, surgery may drastically impair speech function. Carefully controlled, localized radiation therapy can cure cancer; however, under other conditions radiation can cause cancer.

Chemotherapy is the method of choice in treating many non-localized cancers, such as leukemia and tumors that have spread. Chemotherapy involves the use of chemicals that interfere with the processes of living cells. Some chemicals mimic metabolites needed for synthesizing DNA or RNA. When these substances are used by the cell, the cell produces defective nucleic acid, which leads to cell death. Some drugs inhibit protein synthesis; others inhibit respiration. The dosage must be carefully regulated, because these drugs are also harmful to normal cells. However, because tumor cell usually are in a state of constant growth, drugs become incorporated into these cells more easily then into normal cells. Individuals undergoing chemotherapy often display side effects (hair loss, nausea) as a result of the toxicity of these drugs. Still, many cancers, including certain leukemias, have been treated effectively with chemotherapy. A decade or two ago, leukemia was considered incurable; now many people have been free of the disease for five, ten, fifteen, or more years as a result of chemotherapy.

Immunotherapy is an experimental technique that use the body's natural defenses against tumor cells. Cancer cells often display new surface antigens called tumor-specific antigens. These antigens can be recognized by the body's immune system, just as invading bacteria are. And just as the white blood cells of the body attack invading bacteria and destroy them, they can also destroy tumor cells. For some reason, however, the immune system does not function properly in some individuals. A poorly functioning immune system can render an individual more susceptible to cancer or to persistent forms of infectious diseases. (Patients with AIDS—acquired immune deficiency syndrome—have a greatly increased chance of developing Kaposi's sarcoma, an otherwise very rare form of cancer) Immunotherapy is based on the idea that cancer can be controlled by stimulating the immune system, much as polio is through use of the Salk vaccine. Various laboratories around the world are experimenting with immunotherapy as a treatment for cancer. Immunotherapy is usually tried on patients who have little chance of recovery with conventional treatments. These patients are inoculated with a substance that acts as a generalized stimulus of the immune system, or with such a substance combined with living of dead tumor cells or with tumor cell-surface antigen material. The substance usually used as a generalized stimulus is BCG (Bacillus of Calmette and Guerin), a weakened strain of tuberculosis bacteria that has been used to treat tuberculosis. This material activates the immune system. Vaccines are also made with living or dead tumor cells or with parts of tumor cells. Immunotherapy has great potential as a cancer treatment. It is, however, very difficult to carry out. Different tumor have different tumor-specific antigens, and it is very difficult to obtain large quantities of pure human tumor cells for the vaccines. A different tumor vaccine would probably be needed for treating each specific tumor.

The last form of treatment we shall discuss is bone marrow transplants. This experimental method is used to treat certain terminal cases of leukemia. Leukemia is a disease in which certain white blood cells are produced in great number by the bone marrow. Destroying the diseased bone marrow should cure the leukemia. Unfortunately, white blood cells are essential for fighting disease. A patient without white blood cells is vulnerable to any of the thousands of diseases normally kept in check by white blood cells. The bone marrow transplant technique was developed to get around this problem. The patient's bone marrow is first destroyed with strong whole-body radiation and chemicals. Healthy bone marrow is then removed with a syringe from the bones of an identical twin or from those of a person with the identical tissue type. The healthy marrow is inoculated into the bones of the patient, colonizing them. If all goes well, the bone marrow begins to produce normal white blood cells.

A bone marrow transplant is a very drastic treatment. The patient is subjected to high doses of radiation that can have side effects. Also such transplants can be made only between identical twins or between persons who have identical tissue types, with similar antigens on the surfaces of the body cells. If the tissue types are not identical, the "graft versus host reaction may occur, in which the new white blood cells produced by the marrow transplant recognize the rest of the body as foreign and begin to destroy it. (A heart or kidney transplant may be rejected through a similar process. In that case, however, the host's white blood cells destroy the transplanted kidney or heart.)

Many cancers are curable by conventional techniques. Experimental treatments and better diagnostic methods offer new hope for the future. We shall now turn to the cellular aspects of cancer.

CELLULAR ASPECTS: CANCER CELL VERSUS EMBRYONIC CELLS

Tumor Spread and Embryonic Cell Migration

One of the major problems in dealing with malignant tumors is that cells separate from these tumors and spread to other parts of the body. In some ways this spread resembles the migration of embryonic cells. Primordial germ cells, for instance, move from other regions in the embryo to the gonad area. Neural crest cells home to many areas of the body. Optic nerve cells migrate to specific regions of the optic tectum.

The same mechanisms that control embryonic cell migration may also control tumor spread. Earlier we discussed the experiments of Barbera, Marchase, and Roth, who showed that specific adhesive recognition may control homing of optic nerves from the retina to specific regions of the optic tectum. These experiments involved rotating dorsal or ventral retina cells with dorsal or ventral chunks of optic tectum. Similarly, Nicolson and Winkelhake showed that cells from primary tumors may form secondary tumors at specific sites because the tumor cells stick best there. These experiments were done by rotating cells from tumors, such as malignant melanomas, with body tissue cells. The melanoma cells stuck better to lung cells than to other body cells. Most of the secondary tumors that developed from the line of spreading melanoma used in the experiments were in the lung. Thus adhesive recognition controls the sites of secondary tumor formation just as it controls the homing of embryonic cells to specific regions of the embryo.

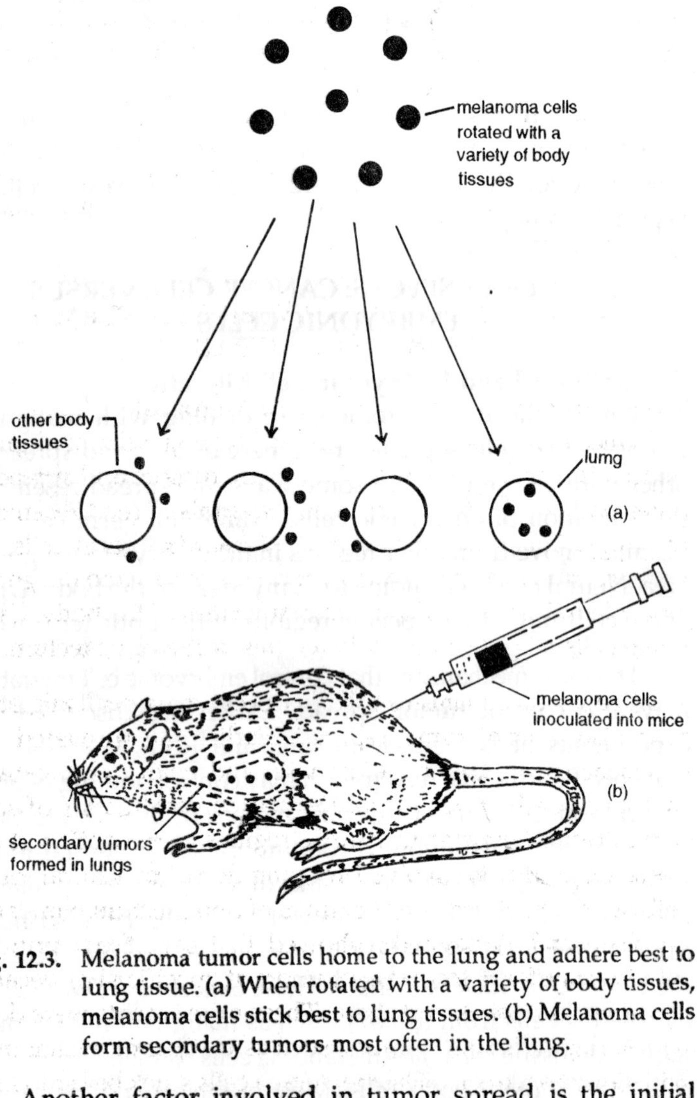

Fig. 12.3. Melanoma tumor cells home to the lung and adhere best to lung tissue. (a) When rotated with a variety of body tissues, melanoma cells stick best to lung tissues. (b) Melanoma cells form secondary tumors most often in the lung.

Another factor involved in tumor spread is the initial separation of cells from the primary tumor. Using microneedles to separate cell pairs, Coman showed that tumor cells separate more easily than do cells in the normal tissues from which the tumors are derived. Thus tumors show reduced cell adhesion (or, more correctly, tissue cohesion). Experiments by Oppenheimer's and Roseman's groups suggest that the reduced adhveness of some tumor cells may result from a decreased ability of the cells to synthesize, store, or transport L-glutamine, an amino acid that is required to form the complex carbohydrates needed for cell adhesion in the tumor lines studied (mouse ascites tumors).

Yet another important factor in the spread of tumors is the increased production of proteases, enzymes that catalyze the hydrolysis of proteins, by tumor cells. Increased quantities of protaese at the cell surface may assist the tumors in spreading and invading other regions of the body. Most tumor cells also have mobile cell-surface receptor sites. This can be demonstrated by using fluorescent tagged lectins that bind to cell-surface receptor sites that contain carbohydrate. Using the tagged molecules, one can move the receptors to one pole of the cell, just as Roberson, Neri, and Oppenheimer did with migratory embryonic cells. Thus tumor cells and certain embryonic cells both possess mobile cell surfaces may help tumor and embryonic cells to migrate.

Many normal cells contain cytoskeletons made up of organized bundles of microfilaments and microtubules. This can be observed by using fluorescent antibodies that bind to the actin of microfilaments and to the tubulin of microtubules. Most tumor cells, however, display a diffuse fluorescence when stained with these antibodies. This suggests that although mocrofilament and microtuble subunits are present in tumor cells, the subunits are not organized into distinct microfilaments and microtubules. Microfilaments and microtubules may prevent cell surface receptor-site mobility in normal cells by anchoring some of the cell-surface sites from within the cell. The lack of an organized cytoskeleton in tumor cells may allow greater cell-surface receptor-site mobility, thereby increasing the cell's ability to migrate and permitting continued growth.

We mentioned that tumor cells show less contact inhibiton of growth and movement than do normal cells. In normal tissues, cell-cell contact stops cell movement and growth. Tumors may spread partly because contact no longer inhabits cell growth and movement. Contact inhibition may be controlled by cellular cyclic AMP. Virus-transformed tumor cells often contain less cyclic AMP than do non-transformed cells. Some experiments have shown that contact inhibiton of growth and movement can be restored by treating tumor cells with derivatives of cyclic AMP. Cyclic AMP may induce contact inhibition by influencing the cytoskeleton. The relationships among the cell surface, cyclic AMP, the cytoskeleton, cell growth, and migration are of major importance to the cancer problem. One possible hypothesis regarding the nature of cell transformation is that genetic changes cause cell-surface changes, which in turn lower the levels of cellular cyclic AMP (probably by lowering the levels of the membrane-bound enzyme adenyl cyclase). Low cyclic AMP levels may then induce many of the other cellular changes associated with malignacy. Major breakthroughs in our understanding of tumor cell transformation will likely come from studies in this area.

Many theories have been proposed to explain the altered behaviour of cancer cells, including unregulated growth and spread. Only a few of these have been described in this chapter. Let us conclude by mentioning one last theory that was proposed by Holley. He suggests that a major factor in the unregulated growth of tumor cells is their increased nutrient uptake from the medium or from surrounding regions of the tumor. He suggests that if cells get a constant supply of the sugars, amino acids, vitamins, and other substances needed for growth, they will continue to grow. This hypothesis is supported by findings that many types of tumor cells take up significantly more nutrients than do their normal counterparts. The plasma membrane of tumor cells, therefore, is more permeable to nutrients. Whether this altered permeability is a cause of the transformed state or a secondary effect remains to be determined.

Embryo-Like Tumors or Tumor-Like Embryos?

We have already mentioned that teratomas are tumors derived from germ cells. These tumors are often indistinguishable from early embryos, and in fact they develop rather like normal embryos if they are maintained in the proper place in the body. Teratomas and other tumors may also carry specific embryonic antigens on the surfaces of their cells. This suggests that some of the same genes that are active in tumors are active in embryos. Normal adult cells seldom display these antigens.

Mouse teratomas can be isolated from the testes of some strains of young mice and transplanted into other mice or maintained in culture. When teratoma cells are inoculated into the body cavities of mice, they develop into free-floating clusters called "embryoid bodies." These clusters resemble normal six-day-old mouse embryos. If grown in the eye cavity of mice, they differentiate into many tissue. If maintained in the body cavity, however, embryoid bodies continue to proliferate rapidly. A variety of experiments indicate that these teratomas are derived from germ cells.

Mintz and Illmensee isolated teratoma cells and injected them into early mouse embryo blastocysts from another strain of mice. These blastocysts were then implanted into the uteri of other female mice. The embryos developed into normal, tumor-free adult mice. Some of the adult tissues were derived from the normal mouse embryo cells and others from the teratoma cells. The teratoma cells had genotypic markers; the normal blastocyst cells had different markers. The markers included strain-specific enzyme patterns, hemoglobin, immunoglobulins, and pigment. Thus normal mice developed, in part, from cancer cells! The important thing to remember about this experiment is that teratoma tumor cells can differentiate into normal adult tissues. Here we see the ultimate relationship between normal embryo cells and certain tumor cells. Both can divide and remain relatively undifferentiated under some conditions, whereas under other conditions they can differentiate into normal adult tissues.

Normal six-day-old mouse embryo cells develop into teratomas if implanted in the testes of adult mice. Older embryo cells, however, do not form tumors under these conditions. Here again we see that young embryo cells are similar to certain tumor cells. Both can grow in an unregulated way or can differentiate into normal adult tissue, depending on their environment.

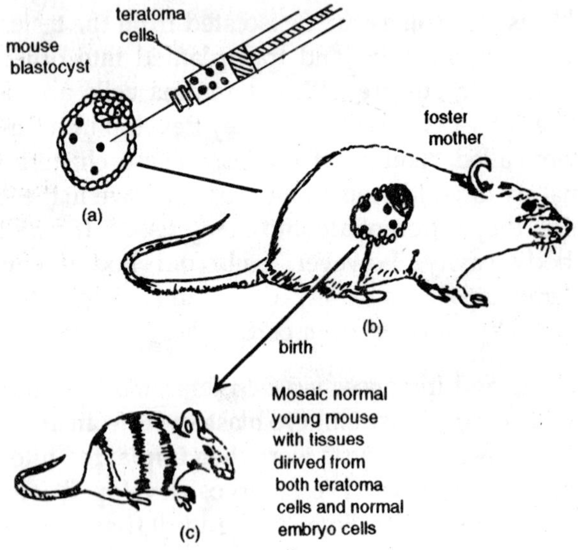

Fig. 12.4. Development of normal mouse from teratoma cells and normal mouse embryo cells, (a) Teratoma cells (black0 are inoculated into a normal mouse blastocyst (white). (b) The embryo (including the teratoma cells) is grown in the uterus of a female mouse. (c) The mouse that develops from the combination consists of normal tissues, some derived from teratoma cells (black) and some from normal embryo cells (white).

Teratomas are being used in many studies of malignant and benign states and the conditions required for cell growth and cell differentiation. Teratoma cells are excellent for these studies for two reasons. First, they can be grown in enormous numbers in the body cavities of mice or in tissue culture. Second, these cells remain malignant if maintained in the free-floating form in the body cavity, but differentiate into normal tissues if implanted into embryo blastocysts or under the skin. Thus teratomas will probably provide us with much information about the relationships between malignant and benign states and between cell growth and cell differentiation.

Recent Developments in Cancer Biology

We shall mention some developments in the area of cancer biology as a way of providing an overview of this field.

Can cancer be prevented? Cancer prevention will in all likelihood receive much more attention in the future. We already know that the number one killer cancer, lung cancer, can largely be prevented by eliminating the carcinogens taken into the lungs during cigarette smoking. Not only would the incidence of lung cancer be greatly reduced if carcinogenic cigarette smoke were eliminated, but the number of cancers of the mouth, esophagus, pancreas, larynx, bladder and kidney would also be substantially reduced. For example, the American Cancer Society estimates that among males, 97% of deaths due to lung cancer and 99% of deaths due to cancer of the larynx ares attributable to smoking.

A variety of so-called anticarcinogens that have been studied offer additional hope for cancer prevention. Let us take a moment to examine some studies.

One study, reported in 1980, indicated that two combinations of drugs (retinoic acid plus the steroid dexamethasone, and retinoic acid plus the protease inhibitor TLCK inhibited the promotion of skin tumors by specific carcinogens. Other studies showed that when a drug (2-bromoα-ergocryptine) that suppressed that secretion of the hormone prolactin was administered along with the retinoid retinyl acetate, carcinogen-induced mammary cancer was prevented in rats.

Another study showed that vitamin C and Vitamin E reduced the level of mutagens (such as the N-nitroso derivatives of intestinal lipids) in the feces of humans to 25% of control values (measured in individuals not fed vitamins A or E). The two vitamins apparently prevent the formation of carcinogens by acting as antioxidants. A study also showed that vitamin C inhibited the malignant transformation of mouse fibroblast cells by the carcinogen methylcholanthene in culture.

Vitamin A and its analogs (retinoids) apparently prevent carcinogen-induced cancers in rats. For example, in one experiment the carcinogen *N*-methyl-*N*-nitrosourea was injected into the bladders of rats in three biweekly doses. Low or high doses of the vitamin A analog 13-cis-retinoic acid were then added to the diets of some of these rats for eight months. The animals were then killed, and the surfaces of their bladders were carefully examined. The rats fed high doses of retinoic acid had many fewer and less extensive bladder cancers than did the other animals. This is just one of many studies that show that vitamin A analogs effectively prevent cancers induced by carcinogens. Other studies show that vitamin A can completely prevent lung cancers in hamsters.

The most of action of vitamin A and its analogs is not known, but some studies suggest that they prevent the carcinogen from binding to cellular DNA. In one study DNA from the tracheal epithelium of animals deficient in vitamin A bound much more of the carcinogen benzo[a]-pyrene than did DNA from animals with normal vitamin A levels. Vitamin A also stimulates the specific differentiation of several cell types. Some studies even suggest that vitamin A may enhance the carcinogenicity of some compounds. Much remains to be learned about the relationship between vitamin A and carcinogen activity.

A variety of substances other than vitamin A inhibit the action of carcinogens. Vitamic C (ascorbic acid) may inhibit the development of some cancers. Certain studies suggest that vitamin C increases the survival time of patients with some terminal cancers. Vitamin C can block the formation of carcinogenic nitrosamines (and other *N*-ritroso compounds). Nitrosamines can form in the acid conditions of the stomach through a reaction between the widely used food preservative sodium nitrite and specific amines. Vitamin C binds to nitrite and thus reduces the amount of nitrite available for forming nitrosamines. The number of cancers induced in animals fed with amines and sodium nitrite is reduced when the animals are simultaneously fed ascorbic acid.

Including high levels of riboflavin in the diet of rats fed carcinogenic azo dyes substantially reduced the induction of liver cancers. The azo dyes rapidly decrease the riboflavin content of the liver. High levels of added riboflavin presumably prevent the depletion of liver ritoflavin and protect the liver from the carcinogen.

The incidence of spontaneous mammary tumors. Leukemia, and lung cancers in mice is reduced if their caloric intake is restricted. Restricting calories also reduced the level of skin cancer induction by dibenzanthracene to less than half its value in control animals. Azo dyes induced fewer liver cancers in animals fed with saturated lipids than in animals fed with unsaturated fat. How these dietary factors reduce cancer induction is not well-understood.

Some interesting results show that the hormonal status of animals affects their susceptibility to carcinogens. When rats were treated with a variety of carcinogens, significantly fewer liver cancers formed among those from which the pituitary, thyroid, or adrenal gland had been removed. If animals are given high levels of specific hormones, liver cancer induction by such carcinogens as 2-acetylaminofluorene is greatly increased. Normal hormones probably activate the oxygenase enzymes that, in turn, activate carcinogens. If the hormone levels drop, oxygenase activity decreases and with it carcinogen activation. Hormones might also increase the rate of cell division, which could make the cells more susceptible to transformation.

In the future increased efforts will be made to develop anticarcinogens. In addition, more emphasis will probably be placed on the role of diet in cancer causation. Some forms of cancer can probably be prevented by altering what we eat and drink. For example, studies by Maugh link chlorinated organic contaminants in drinking water forms with rectal bladder, and colon cancer. Chlorine added to drinking water forms such contaminants by interacting with the organic compounds and bromine present in the water to form trace amounts of halomethanes.

At least seven carcinogens are known to be produced by fungi that contaminate foods. Aflatoxin is produced by the fungus *Aspergilius flavus* a common contaminant of groundnuts and rice stored in hot and humid areas. Aflatoxin is a liver carcinogen that can be eliminated by the proper treatment and storage of food. Nitrosamines, another group of potent carcinogens, form by the interaction of the food preservative sodium nitrite with amines, a breakdown product of protein. Some of the sodium nitrite we take in comes from natural sources, such as vegetables and our own bacteria. However, the exposure of our cells to nitrosamines would probably be reduced if sodium nitrite were substantially reduced in or eliminated from preserved food products.

Finally, studies suggest that diets high in fiber reduce the incidence of cancers of the colon and rectum. A diet high in bulk promotes frequent bowel movements, reducing the time during which carcinogens present in the large intestine contents remain in contact with the lining of the bowel, where cancers begin. Large amounts of bulk also reduce the concentration of carcinogens in the bowel. Diets high in fat are correlated with high incidences of colon-rectal cancer. Time will tell whether substantial decreases in cancer rates can be achieved by altering the diet. For example, can the incidence of colon-rectal cancer be reduced by increasing dietary fiber, decreasing dietary fat, and increasing the dietary levels of some of the anticarcinogens mentioned previously? This area of research makes sense and possibly will lead to preventive measures against some common forms of human cancer.

Tyrosine Phosphorylation and Growth Stimulation

Epidermal growth factor, insulin, platelet-derived growth factor, and some cancer-causing viruses all have something in common. All of them phosphorylate tyrosine groups in and on the cell. When the growth factors bind to cellular receptor sites, the receptors start phosphorylating tyrosines. When some cancer viruses (such as Rous Sarcoma virus) transform cells, the concentration of phosphorylated tyrosines increases many-fold. This increase is apparently due to tyrosine kinases (enzymes that phosphorylate tyrosines) that may be the transforming proteins of the viruses. We shall soon learn whether tyrosine phosphorylation is of general importance in cellular transformation and loss of growth control and, if so, how the phosphorylated product alters the cell's growth characteristics.

Expression of Oncogenes During Normal Embryonic Development

Viral oncogenes (apparently originating in the normal vertebrate genome) are required for the induction of specific virally initiated cancers, including sarcomas, carcinomas, lymphomas and leukemias. The work of Muller and colleagues shows that during normal prenatal and early postnatal development of mice, specific tissues at specific stages express normal cellular genes identical to the oncogenes of FBJ mouse osteosarcoma virus, Abelson mouse leukemia virus, and Harvey sarcoma virus. For example, rapidly proliferating placental tissue such as the trophoblast, which can invade the uterine epithelium and surrounding tissue, significantly expresses a specific oncogene base sequence. This suggests that many cancers result from the expression of cellular oncogenes that earlier played a role in such normal developmental processes as cell proliferation, trophoblast invasion and differentiation.

Human Cancer Genes

Several studies have shown that DNA fragments isolated from some spontaneous and chemically induced human, mouse, and other cancers can transform normal cells into cancer cells. In one study researchers cloned DNA fragments from a human bladder carcinoma by inserting these genes into bacterial plasmids. The cloned human carcinoma genes were then analyzed by their ability to bind other DNA sequences. The results showed that the bladder carcinoma gene has a base sequence similar to that of a specific gene carried by mouse sarcoma virus. In like manner, a human lung carcinoma gene was found to be similar to another gene carried by mouse sarcoma virus. Other studies showed that a specific phosphoprotein (called p21) encoded by a mouse sarcoma virus gene is present at high levels in cells transformed by the human bladder carcinoma gene. Thus this gene is similar in base sequence to a transforming gene carried by a mouse sarcoma virus. In addition, cells transformed by the bladder gene express an RNA that hybridizes with both the bladder gene and the mouse sarcoma virus gene. Such transformed cells contain elevated levels of the protein encoded by the mouse sarcoma virus gene.

A fragment of DNA similar to the mouse sarcoma virus transforming gene can be isolated from normal human cells. This DNA will transform cells if it is activated through linkage to a fragment of the mouse sarcoma virus gene. Cells transformed by this human gene contain within their DNA the human DNA fragment linked to the viral DNA. The transformed cells express the p21 protein. The p21 protein is apparently the transforming agent of a mouse sarcoma virus. Normal cells contain low levels of p21 and mRNA coding for this protein. These results support the notion that cancer is a disease caused by increased amounts of normal gene products.

The genetic basis of cancer is becoming clear. Continued efforts along the lines described in the final section of this chapter will probably soon pinpoint the exact mechanism(s) involved in human carcinogenesis.

Jumping Cancer Genes

We have discussed the isolated of activated human cancer genes that have base sequences similar to those of certain viral oncogenes. We noted that some of these genes code for specific proteins, such as p21, that may cause cellular transformation, investigators have shown that these transforming genes may code for a protein found in normal cells, but that the transformed cells produce it in greatly increased quantities.

In other work, segments of human bladder cancer genes were systematically exchanged with normal cellular genes that transformed cells when appropriately activated. These recombinant genes were then transferred to cultured cells. Analysis showed that a specific base—a guanine—in the normal gene was replaced by a thymine in the transforming gene. In this case a specific one-base alteration enhanced the transforming ability of the gene product.

Research has also shown that genes with oncogene-like base sequences are involved in the translocations associated with specific cancers such as Burkitt's lymphoma. One example, is the *myc* gene, which is similar to a gene carried by cancer-causing viruses. In about half of the Burkitt's lymphoma cells examined, the *myc* gene was translocated to a position close to an antibody-forming base sequence. Translocation of cancer-causing genes may activate them (or alter them) to bring about cell transformation. This may be how some cancers are induced. Decades of work have shown that many cancers are associated with specific chromosome abnormalities, including translocations. This new work may clarify the relationship between visible chromosomal changes and cancer.

New Developments in the Study of Oncogenes

The concept of oncogenes was introduced earlier in this chapter, and some of the past work in this area was described. Here we shall examine oncogenes in detail, stressing the newest developments in this area.

We noted earlier that most carcinogens cause cancer by their action on DNA. Apparently certain cellular genes, termed *proto-oncogenes,* can become activated by mutation to become oncogenes — genes that can transform a normal cell into a cancer cell. The function of proto-oncogenes in cells is not well understood, but once proto-oncogenes become active in synthesis of either altered forms of or abnormal quantities of their messages, they are termed oncogenes and can cause cellular transformation.

The work of Howard Temin decades ago suggested the concept of oncogenes. Temin proposed that cancer-causing viruses act by inserting a cancer-causing gene into the cell's genome. This notion gave rise to the concept of an oncogene. The idea was suggested that an oncogene might be present in an inactive form in normal cells, but in some way it becomes activated in cancer cells.

The actual discovery of oncogenes in human and animal cells was accomplished by applying new technologies of molecular biology, molecular genetics, and cell biology to the search for these predicted genes. This discovery, described below, was perhaps the most significant of all findings in cancer research, for it provided the beginning of an understanding of the molecular basis of cancer.

Isolation of a Human Cancer Oncogene

Weinberg and his colleagues were the first to isolate an oncogene from human cancer cells. They isolated DNA from human bladder cancer cells, fragmented it into small pieces, and added the DNA fragments to a culture of mouse cells (strain 3T3). Some of the mouse cells were transformed into cancer cells. This experiment strongly suggested that an oncogene present in the human DNA fragments was able to transform non-cancerous cells into cancer cells (Figure 12.9).

In another experiment, the DNA fragments from the human bladder cancer cells were separated into many fractions, and each was tested for the ability to transform the mouse cells. Only one DNA fragment induced transformation. This fragment contained the human bladder cancer oncogene.

We now believe that a counterpart of this bladder cancer oncogene exists in normal cells. Indeed, as noted above, a DNA fragment isolated from normal human bladder cells proved to be almost identical to the bladder cancer oncogene. The only difference is a one-base change in the coding sequence, with a guanine in the normal DNA replaced by a thymine in the oncogene. In this case the oncogene is a mutant form of a normal gene, the difference residing in only one nucleotide. It is tempting to speculate, therefore, that a carcinogen may induce bladder cancer by causing such a mutation in the appropriate gene.

Human bladder cancer is isolated

↓

DNA is isolated from cancer cells

↓

Bladder cancer DNA is fragmented

↓

Fragments are added to culture of mouse 3T3 cells

↓

Some of the mouse cells take up the DNA fragments, and some of the DNA becomes integrated into a mouse chromosome

↓

Some mouse cells are transformed

Fig. 12.5. Isolation of human bladder cancer oncogene.

Oncogenes and Cancer

The puzzle of the molecular basis of cancer is beginning to be clarified, but many questions still must be answered. One question is, how are the normal counterparts of oncogenes, the proto-oncogenes, changed into active oncogenes? One way, as shown by the bladder cancer story, is by induction of a mutation in the normal gene, causing it to become an active oncogene. However, other activation mechanisms have been discovered. One mechanism involves the translocation of the proto-oncogene to a chromosomal region that promotes gene activity, causing the newly relocated oncogene to produce excessive quantities of its messenger RNA. For example , patients with chronic myelogenous leukemina have a translocation between chromosomes 22 and 9. This causes a specific proto-oncogene to be moved away from its normal regulatory elements in chromosome 9 to the vicinity of abnormal controllers in chromosome 22. The rate of synthesis of the oncogene messenger RNA at this new location is eight times the rate when the gene is in chromosome 9. The excess amount of oncogene messenger RNA, which leads to synthesis of excess amounts of the protein encoded by the oncogene, may be a major factor in carcinogenesis.

Fragile Sites

Largely as a result of the investigations of J.J. Yunis and G.R. Sutherland, a surprising number of different specific human cancers besides Burkitt's lymphoma and chronic myelogenous leukemia are now known to be associated with one or another specific chromosome rearrangement. Among these cancers are acute lymphocytic leukemia, acute non-lymphocytic leukemia, Ewing sarcoma, and T-cell lymphoma, as well as a number of others. Virtually all individuals who suffer from a specific one of these cancer conditions have a chromosome rearrangement (inversion, translocation, or deficiency) with at least one break point in common. Even more significant is the fact that these cancer-specific breakpoints in many instances correspond to the position of a *fragile site*.

The existence of fragile sites has been known for only a relatively short-time. They are observed in dividing cells at prophase or metaphase, appearing as though a chromatid has broken. Occasionally a chromatid fragment will be completely separated from the rest of the chromosome. However, fragile sites are rarely observed in normal cell preparations; they are expressed when cells are placed in tissue culture and allowed to multiply in a medium that lacks some growth factor. *Folic acid,* which is a member of the vitamin B complex, is such a factor. Its absence from the culture medium elicits the expression of a group fragile sites scattered among the various chromosomes.

These folic acid-related fragile sites do not occur at random along the length of one chromosome or another; rather, like all other known fragile sites, each one resides at a specific location on a specific chromosome. Many of these folic acid-related fragile sites are found at the same place as some of the cancer-associated chromosome breakpoints. Of perhaps greatest interest is the fact that various oncognes are also found in the same positions as the fragile sites.

These observations give rise to the intriguing possibility that when a chromosome undergoes a chromosome rearrangement, the breakpoint occurs preferentially at a fragile site and, as a consequence, causes a mutation of an oncogene at that same location. The net result of the chromosome break is to transform a normal cell into a cancer cell. A breakpoint at another fragile site could mutate a different oncogene, producing a different cancer.

If chromosome breaks do occur preferentially at fragile sites, and if the expression of fragile sites is in fact suppressed by the folic acid normally present in the body, them supplementing the diet with folic acid may prove to be a cheap and effective way of reducing the likelihood of developing cancer. Of course, this presumes that the incidence of cancer is higher in people who have a low level of folic acid. Investigations into this question are under way in a number of laboratories.

Two Steps to Cancer

It has long been thought that many cancers occur as a result of a two-step process of *initiation* and *promotion*. This theory was suggested more than 60 years ago by Peyton Rous, two observed that cancers appeared able to remain in a dormant state until the were activated an irritating substance. The first step, initiation, is believed to involve conversion of a normal cell into a latent tumor cell. The second step, promotion, triggers uncontrolled growth of the cancer cells.

Many experiments have confirmed this concept. It was shown that when the skin of mice is painted with an agent such as methylcholan-threne or urethane, few—if any—tumors develop. If that same area is later painted with a substance such as croton oil, skin cancers develop. The first agents appear to induce latent tumor cells; they are initiators. The second agents cause the cancers to grow; they are promoters. Most initiators cause mutations in DNA : thus initiation should be a mutational event. Promotion appears to be an event in which initiated cells (latent tumor cells) are stimulated to divide. Many cancers in humans probably result from this kind of two-step process. This may be why cancers can take years or decades to develop. Precancerous conditions must come about first (that is, initiation); these can be followed years later by cancers, which develop after promotion occurs. How do oncogenes fit into this two-step model of cancer induction?

A growing body of evidence suggests that at least two different kinds of oncogenes may be responsible for the induction of cancers. Some oncogenes may bring about the initiation process (sometimes termed immortalization); whereas the products of others may cause promotion (also termed transformation). For example, normal skin cells can be transformed into cancer cells by a sequential treatment, first with *myc* oncogene DNA and then with *ras* oncogene DNA. *myc* appears to act in the initiation process; whereas *ras* acts in promotion.

Oncogene	*Source**
abl	Mouse leukemia; human leuemia cells
b-lym	Chicken and human lymphoma cells
erbA	Chicken leukemia
erbB	Chicken leukemia
ets	Chicken leukemia
fes	Cat sarcoma
fgr	Cat sarcoma
fms	Cat sarcoma
fos	Mouse sarcoma
fps	Chicken sarcoma
Ha-ras	Rat sarcoma; human and rat carcinoma cells
Ki-ras	Rat sarcoma: human carcinoma, sarcoma, and luckemia cells
mil	Chicken sarcoma
mos	Mouse sarcoma: mouse eukemia cells
myb	Chicken and human leukemia cells
myc	Chicken leukemia; human lymphoma cells
N-ras	Human leukemia and carcinoma cells
raf	Mouse sarcoma
rel	Turkey leukemia
sis	Monkey sarcoma
ros, ski, src, yes	Chicken sarcoma

*Each source is retrovirus unless a cell is indicated

Fig. 12.6. Some oncogenes and their sources.

Protein	*Oncogene*
Tyrosne kinas	*abl, fes, fgr, fps, ros, src, yes*
Other protein kinases	*etbB, fms, mil, mos, raf*
Nuclear proteins	*B-lym, fos, myo, myc, ski*
Crown factor	*sis*
GTP-pinding proteins	*Ha-ras, Ki-ras, N-ras*

Fig. 12.7. Classes of oncogene proteins.

To understand the molecular basis of cancer, we must first determine the nature of the proteins encoded by oncogenes and how these proteins are able to transform cells into cancer cells. The first part of this problem has been solved. We know the nature of many oncogene proteins. How oncogene proteins transform cells, however, remains poorly understood.

Some oncogenes code for specific enzymes that phosphorylate proteins (tyrosine kinases, other protein kinases), nuclear proteins, growth factors, or GTP-binding proteins. At this point we know little about how some oncogene proteins make cells abnormal. As mentioned above, some of the enzymes that phosphorylate proteins(protein kinases) may make cells abnormal by phosphorylating proteins, (that are involved in maintaining normal cell structure and function. For example, phosphorylation of proteins that help cells adhere to each other and maintain their internal architecture may interfere with these functions. The *src* oncogene kinase can also phosphorylate a lipid, which may lead to a whole series of changes hat eventually cause altered control of cell division. Growth factors produced by some oncogenes may stimulate continuous division and in this way lead to a cancerous state. The key to understanding the molecular basis of cancer is to discover exactly how oncogene protein products are involved in causing cellular transformation. One fact is clear, however, these proteins must interfere directly or indirectly with the regulation of cell division, because continuous, unregulated cell division is the primary characteristic of cancer cells.

The Discovery of Anticancer Genes (Antioncogenes)

A series of recent experiments has led to a most important discovery—that of the existence of anticancer genes, called antioncogenes. In this section we shall briefly review these exciting findings.

The first hint of the existence of antioncogenes came in 1969, when Henry Harris and his colleagues found that malignancy was suppressed when malignant cells and non-malignment cells were fused together. The existence of these genes was further investigated by Knudson, who analyzed the histories of 48 retinoblastoma patients. Retinoblastoma, a cancer of the eye that afflicts about 1 in 20,000 infants and young children, is curable if detected early. Children of retinoblastoma survivors, however, develop this cancer at rates as high as 50%. Knudson's studies clearly suggested that in these cases of hereditary retinoblastoma, two mutations must occur—one inherited and the other spontaneous.

In the 1970s and early-to-mid-1980s, a variety of observations suggested that Knudson's model was correct. It appeared that retinoblastoma cells have a deletion in both copies of chromosome 13. Both deletions occur in a region of the chromosome called 13q14. The same deletion was found in cells from osteosarcoma, a cancer that frequently develops in teenagers who have survived retinoblastoma. Retinoblastoma, therefore, appears to be caused by the absence of the activity of a gene or genes in the 13q14 region. Normal cells possess two copies of the so-called retinoblastoma antioncogene, one in the q14 region of each chromosome 13. Individuals who eventually develop hereditary retinoblastoma are born with one normal and one mutated chromosome 13. When the normal chromosome 13 also undergoes mutation in the 14 region in eye cells retinoblastoma develops; when the second mutation occurs in bone cells, osteosarcoma forms. In 1986, Friend and his colleagues reported success in cloning the retinoblastoma antioncogene. At that point, however, there had been no direct demonstration that this gene can actually prevent or reverse cancer.

A direct demonstration that a specific chromosome can reverse cancer was made by Weissman and colleagues in work with Wilms's tumor, a kidney cancer found in children. Wilms's tumor develops in cells that have deletions in the q13 region of both copies of chromosome 11. These workers found that when they inserted single copies of normal chromosome 11 into Wilms's tumor cells, the cells would no longer from tumors in mice. The insertion of either of two other chromosomes (including chromosome 13, which carries the retinoblastoma antioncogene) into these cells did not prevent the formation of tumors. These experiments directly implicated chromosome 11 as the carrier of the antioncogene(s) that can block the development of Wilms's tumor.

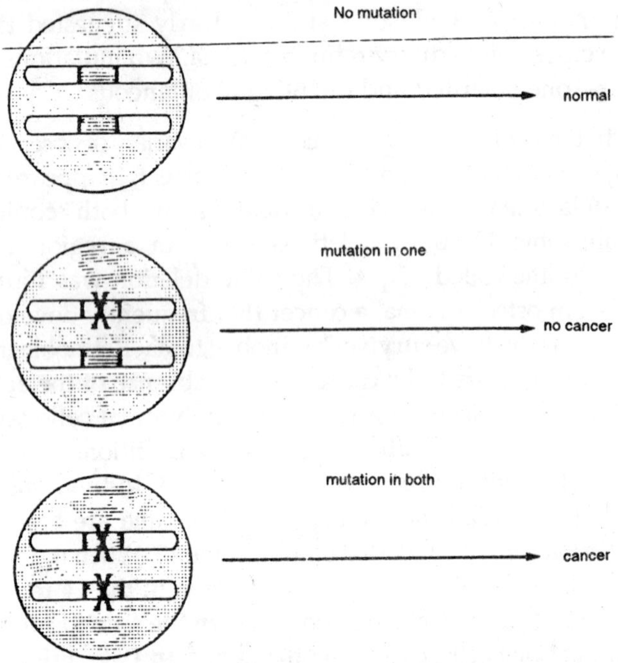

Fig. 12.8. Antioncogenes. A class of genes appears to exist that maintains the non-malignant state in cells. Mutations in the antioncogene of both homologous chromosomes are required to induce malignancy.

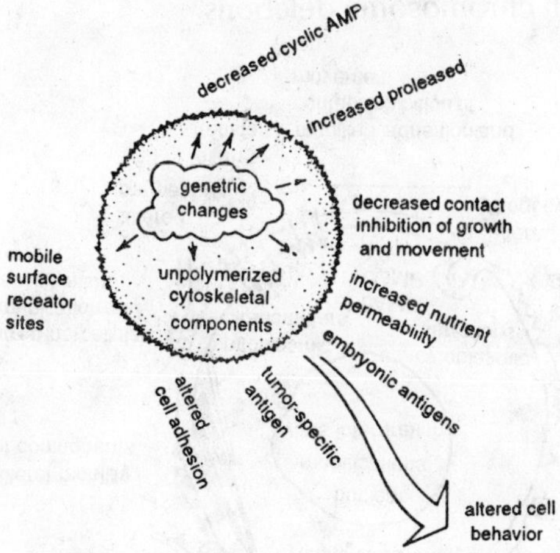

Fig.12·9. Some cellular chages associated with cancer cells.

How do these results fit into what we know about oncogenes? Oncogenes have been identified in a variety of cancers; and when inserted into some cell lines, oncogenes are able to transform these cells into cells with malignant characteristics. No direct proof exists, however, that oncogenes are the cause of human cancer. Weissman and colleagues found that there was no difference in oncogene or proto-oncogene (inactive oncaogene) expression between regular Wilms's tumor cell lines and Wilms's lines that had lost their tumorigenic potential by acquiring a normal chromosome 11 that included the Wilms's tumor antioncogene(s). No detectible expression of the oncogenes *N-myc, ki-ras, src,* or *erb* was found by Weissmar and colleagues in either of the cell lines, and *Hai-ras, N-rm* and *c-myc* were expressed at the same level in both lines.

Although there is no direct evidence at present that oncogenes cause human cancer, loss of anti-oncogenes appear to be the direct cause of retinoblastoma and Wilms's tumor, and many reports associate other human tumors, such as neuroblastoma, small-cell lung carcinoma, and accoustic meuromas, with chromosome deletions.

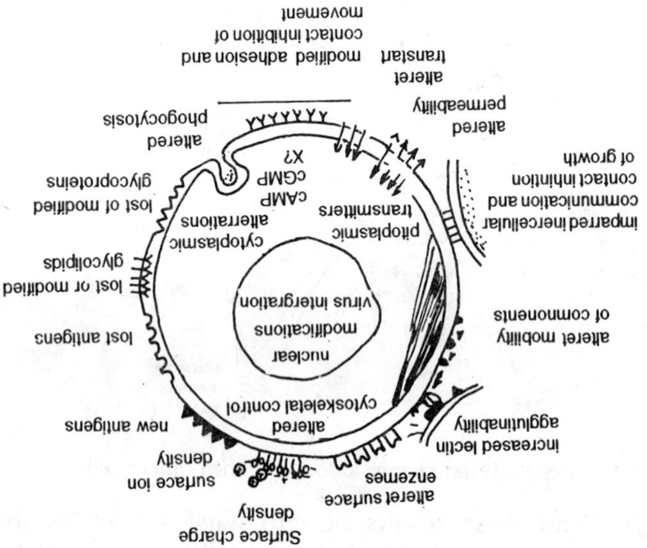

Fig. 12.10. Some changes found after a transformation to the cancerous state.

The development of human cancers appears to be a multistep process in which initiation events, which may be mutations, are followed by promotion events, which cause initiated cells to divide in an uncontrolled manner. The two-step model presented here, in which the antioncogene on each homilogous chromosome must be deleted or mutated for cancer to occur, explains the induction of cancers without the necessity of implicating dominant oncogenes. In the absence of direct evidence the oncogenes cause human cancer, the antioncogene model appears plassible. In other words, the development of many human cancers may not require activation of dominant oncogenes. Instead, it may require only the inactivation or destruction of antioncogenes present in homologous chromosomes of all normal, healthy cells.